GAINING ADVANTAGE FROM OPEN BORDERS

T0347334

Gaining Advantage from Open Borders

An active space approach to regional development

Edited by
MARINA VAN GEENHUIZEN
Delft University of Technology, The Netherlands

REMIGIO RATTI
University of Fribourg and
Università della Svizzera italiana, Lugano, Switzerland

Routledge
Taylor & Francis Group

LONDON AND NEW YORK

First published 2001 by Ashgate Publishing

Reissued 2018 by Routledge
2 Park Square, Milton Park, Abingdon, Oxon OX14 4RN
711 Third Avenue, New York, NY 10017, USA

Routledge is an imprint of the Taylor & Francis Group, an informa business

Publisher's Note
The publisher has gone to great lengths to ensure the quality of this reprint but points out that some imperfections in the original copies may be apparent.

Disclaimer
The publisher has made every effort to trace copyright holders and welcomes correspondence from those they have been unable to contact.

A Library of Congress record exists under LC control number: 2001089135

ISBN 13: 978-1-138-72851-6 (hbk)
ISBN 13: 978-1-138-72849-3 (pbk)
ISBN 13: 978-1-315-19048-8 (ebk)

Contents

Figures and Tables

xi

Contributors

Joachim Blatter, Policy Sciences, University of Konstanz, Universitätstrasse 10, 78457 Konstanz, Germany.

Enne de Boer, Faculty of Civil Engineering, Delft University of Technology, P.O. Box 5048, 2600 GA Delft, The Netherlands.

Alberto Bramanti, Department of Economics and CERTeT of Bocconi University, Via Gobbi 5, 20136 Milano, Italy.

Fabienne Corvers, Maastricht Economic Research Institute on Innovation and Technology (MERIT), Maastricht University, P.O. Box 616, 6200 MD, Maastricht, The Netherlands.

Heikki Eskelinen, Karelian Institute, University of Joensuu, P.O. Box 111, FIN-80101 Joensuu, Finland.

Henk van Houtum, Nijmegen Centre of Border Research, Department of Human Geography, University of Nijmegen, P.O. Box 9108, 6500 HK Nijmegen, The Netherlands.

Thomas Jud, Department of Economics, University of Graz, Universitätsplatz 15, A-8010 Graz, and Joanneum Research, InTeReg, Elisabethstrasse 20, A-8010 Graz, Austria.

Stefan Krätke, Faculty of Cultural Sciences, Chair of Economic and Social Geography, Europa-Universität Viadrina, P.O. Box 776, 15207 Frankfurt (Oder), Germany.

Mario A. Maggioni, ISEIS, Catholic University of Milan, Largo Gemelli 1, 20123 Milano, Italy.

Zdravko Mlinar, Faculty of Social Sciences, University of Ljubljana, P.O. Box 2547, 1001 Ljubljana, Slovenia.

Peter Nijkamp, Department of Regional Economics, Free University, De Boelelaan 1105, 1081 HV Amsterdam, The Netherlands.

Piet Rietveld, Department of Regional Economics, Free University, De Boelelaan 1105, 1081 HV Amsterdam, The Netherlands.

Arndt Siepmann, Stuttgart Region Economic Development Corporation, Privat address: Friedrichstr. 10, D-70174, Stuttgart, Germany.

Michael Steiner, Department of Economics, University of Graz, Universitätsplatz 15, A-8010 Graz.

Christian Vitta, Faculty of Economics, Università della Svizzera italiana, Via Ospedale 13, CH 6904, Lugano, Switzerland.

About the Editors

Marina van Geenhuizen is Associate Professor in the Faculty of Technology, Policy and Management of Delft University of Technology. Her research is centred in theory and practice of regional development and innovation, and in the area of new technology and transport and communication. She is co-chair of NECTAR Cluster 4 (Transport, Spatial Opportunities and Borders). Address: Faculty of Technology, Policy and Management, P.O. Box 5015, 2600 GA Delft, The Netherlands.

Remigio Ratti is Director of the Radio Televisione della Svizzera italiana in Lugano (since 1-1-2000) and Professor at the University of Fribourg and the Università della Svizzera italiana in Lugano, Switzerland. His main areas of scientific activity are regional development, transport economics, theory and policy of trans-border co-operation, and economics and institutions. He is co-chair of NECTAR Cluster 4. Address: Università della Svizzera italiana in Lugano, Via Ospedale 13, CH 6904, Lugano, Switzerland.

Preface

The present volume originates from the second study cycle on border regions and regional development within the Network of European Communications and Transport Activity Research (NECTAR). The first cycle ended in 1993 with the publication of the book Theory and Practice of Transborder Cooperation edited by Remigio Ratti and Shalom Reichman.

The editors of the present volume decided to co-operate in a second cycle on a boat trip between two Greek islands in May 1993 during the European Union Conference on Overcoming Isolation. Since then the editors have acted as "animators" for Cluster 4 of NECTAR, which is called Transport, Spatial Opportunities, and Borders. After three meetings of Cluster 4 - in Delft in 1995, and in Lugano and Zurich in 1996 - the editors decided that the new theoretical ideas and empirical research presented by the members of the cluster would provide sufficient "substance" for a new book. In particular, it was the paradigm of "active space" development that enabled us to collect a great number of relevant studies within a coherent theoretical framework.

This book differs from the first mainly in the attempt made to design a coherent paradigm of regional development, i.e. "active space" development. In this paradigm the region is seen as the outcome of a field of forces, in terms of a coherent game of initiatives and responses produced locally and towards the outside world. Openness, creative learning and concern for sustainability are central concepts of the paradigm. Like other paradigms of regional development, major emphasis is put on the institutional conditions for regional development. The novelty, however, is in the concept of "degrees of active space", meaning that the paradigm can be used in an analysis of a large spectrum of regions, ranging from transport corridor regions with low levels of sustainability to high technology regions with high levels of sustainability. The broad applicability of our paradigm also means that it can be used in more regions than border regions; but in the present book border regions are used as a "laboratory" to explore and experiment with the perspectives and frames of the paradigm.

The book comprises three parts. The first part consists of a theoretical elaboration of the "active space" approach. The second part covers an exploration of the extent to which, and ways in which, advantage is gained from openness in different border regions. The insights in this part illustrate the reason for using the "active space" approach in transport studies and regional-economic analysis. The type of cross-border interaction studied encompasses traffic, labour flow, capital flow, and inter-firm relationships. The regions involved have different locations, i.e. inner border regions and outer border regions of the European Union, and intermediate areas and peripheral areas in national spatial-economic systems. In the third part various institutional issues and policies are discussed, including their implicit or explicit contribution to "active space" development. This part is concluded with reflections on the performance of the paradigm and on emerging research paths.

Working with the "active space" paradigm in a collaboration crossing the Alps has been an interesting experience for the editors, and more so because of the excellent opportunities to enjoy "regional specialities" during the visits in The Netherlands and Switzerland. Hopefully, the readers will enjoy the results of this cross-border activity.

The editors are grateful to Christian Vitta (Università della Svizzera italiana, Lugano) and Monique van der Toorn (Delft University of Technology) for their secretarial and technical assistance, and to Miranda Aldham-Breary for language corrections. Additional support was given by the Faculty of Technology, Policy and Management of Delft University of Technology which granted additional time to the first editor to finalise the manuscript of this book.

Marina van Geenhuizen, Delft
Remigio Ratti, Lugano and Fribourg
April 2001

PART I:
INTRODUCTION

1 An Active Space Approach to Regional Development

MARINA VAN GEENHUIZEN AND REMIGIO RATTI

Introduction

Various converging trends point to an increased importance of regions in the global economy. First, the European integration and lifting of the iron curtain have opened many national political borders that were thus far partly open or almost entirely closed. As a consequence nations are loosing power in global competition, for example due to the ending of national protective measures. A second trend is the increased factor mobility and mobility of persons, goods and information. Capital and various types of information can now circulate around the globe without constraints from spatial barriers. As a result of these developments, the few remaining spatially fixed competitive forces seem the quality of the regional human capital and the quality of regionally anchored networks that support innovation (Ratti et al., 1997; van Geenhuizen and Nijkamp, 1998, 1999; Boekema et al., 2000). A core argument in this emphasis on the region is that tacit knowledge, with its crucial role in innovation, is territorially specific due to its person-embodiedness and social embeddedness; it can therefore only be transferred in close proximity of the actors involved (Morgan, 1997; Cooke and Morgan, 1998; Maskell et al., 1998). There is another, related, argument and that is the development of large parts of economic sectors towards flexible specialisation. Due to an increased complexity of production and reinforced competition in global markets, firms tend to externalise non-core activities through various types of networking and, therefore, become more reliant on the external environment. Regions thus present the base for firms that are disposed to agglomeration, externalisation and dynamic specialisation, particularly using localised pools of knowledge and labour, and close interaction with related firms. In this vein regions can be seen as a "cradle" of innovations and as a local hub that gives access to global networks, using tacit knowledge, alongside other types of knowledge.

An increased openness and mobility basically imply growing benefits from cheap factor inputs and from larger output markets. At the same time, there is a growing competition from adjacent regions or regions at a larger distance. If focusing on transport, there is also a danger from negative externalities. The "active space" approach is a relatively new approach to regional development that addresses the local-global debate. It is taken for granted in this approach that in the process of ongoing globalisation the region remains a field of forces on its own, with a strong potential influence on outcomes of global economic activity (Amin and Thrift, 1994). The perspective is thus essentially multilevel (see also Storper, 1997).

Regions as territorial systems are facing a number of structural features that influence their development over time (Nijkamp, 1990). First, regions have a distinct location in the national-economic space and the European Union, and distinct connections with other - complementary and competitive – regions, at a short distance or far away in a global context. Secondly, regions have a certain (limited) carrying capacity in terms of social and economic resources, and ecological potentials. Thirdly, regions face a certain level of multifunctionality, in terms of functional mixes of economic activity. Multifunctionality, or variety, is associated with economies of agglomeration, benefits of synergy and innovation. Carrying capacity and multifunctionality also contribute to the regeneration capacity of the economy, after shocks in development, and to the absorption capacity of new elements in society, such as migrants with different cultures and educational level. As a final point, we mention that regions encompass distinct "tissues" of institutions and, closely connected with this, formal and informal socio-economic networks and governance structures, that may support various forms of learning.

In the "active space" approach changes in openness of regions are explicitly considered in connection with the capacity of regional actors to manage these changes in terms of economic (social) benefits and sustainable growth. In this dynamic view, regions are subject to an evolutionary development with "active space" and "passive space" as two extreme outcomes. Actors in "active space" are able to take advantage of openness, by using resources and networks, while giving concern to sustainability to achieve the best balanced economic output. The key force in "active space" development is creative learning. It is learning not just in the sense of a continuous adjustment to the changing environment, it is learning in which regional actors themselves are questioning, in a critical manner, the way they function and the system functions (Morgan, 1986; Grabher, 1993; Cooke, 1998). Closely related are capacities like monitoring the environment for early-warning signals, reflexivity on own core values by organisations and the greater society, and bringing multiple skills and experiences together to solve or prevent problems. Particular regions have a strongly developed self-organising power in this respect, whereas other regions have a need for an integrator or an animator, i.e. a key person (natural leader) or key institute that connects various initiatives and networks and keep them alive (Bramanti and Senn, 1997).

The "active space" approach comes very close to other approaches to regional development that put an emphasis on the region as a site of knowledge production and use, namely that of the learning region (e.g. Morgan, 1997; Maskell et al., 1998; Boekema et al., 2000), the GREMI approach to dynamics of innovative regions (e.g. Camagni, 1991; Ratti et al., 1997) and that of regional innovation systems (e.g. Braczyk et al., 1998; see also Malecki and Oinas, 1999). The similarity of course, is the focus on learning processes and the related institutional context as major conditions for regional development.

The "active space" approach is, however, essentially different in that it is a generic paradigm, applicable to all sorts of regions. Every region can be considered in some aspects and to some degree to be an "active space" and to have an inherent capacity to gain advantage from openness. In addition, the "active space" approach is an integrative approach that includes elements such as governance and institutions, aside from innovation, firm behaviour and attitudes, etc. Finally, the present volume is specific in the selection of case studies in the empirical application, with a preference for border regions facing an increased openness and for transport dominated regions; this because of the potentially strong dynamics and impacts of an increased openness on long-term trajectories of the regions at hand. For example, an increased openness may lead to a loss of competitiveness because adjacent regions were quicker in adapting to new structures and in designing new concepts, e.g. in logistic chains. Accordingly, development trajectories of such regions might get a different orientation over time.

Empirical Application: Border Regions and Transport Dominated Regions

Border regions are facing an increased attention since the late 1980s (Ratti and Reichman, 1993; Cappellin and Batey, 1993; van Geenhuizen et al., 1996; Eskelinen et al. 1998; Anderson and O'Dowd, 1999; Boekema and Allaert, 1999). In times of an accelerated globalisation, a progressive integrating Central and Eastern Europe, and the growth of supra-state regions like the European Union (EU) and the North American Free Trade Area (NAFTA), nation borders and the associated border regions seem to derive new meanings and new opportunities. Regional economists, economic and political geographers, cultural anthropologists and sociologists, historians, etc. are increasingly giving attention to the many aspects of these changes. Various changes are centred around territoriality, i.e. the strategy to influence or control resources and people by controlling a specific area by state borders, and around an increased access into and out of this area (e.g. O'Dowd and Wilson, 1996).

We are witnessing different impacts of these changes, dependent on the type of political border in previous times, i.e. barrier- or filtering effects, contact effects or conditioning effects (Nijkamp et al., 1990; Ratti, 1993; van Geenhuizen et al., 1996; Guo, 1996; Anderson and O'Dowd, 1999; Paassi, 1999; Scott, 1999). For example, barrier- and filtering effects of political borders may disappear meaning that interaction, that was strongly penalised or partially impossible in previous time, is now free or increasingly free to occur. In addition, contact effects based upon opportunities of contact between different institutional and/or socio-economic subsystems in previous times may become even stronger and lead to co-operation in various fields. However, the conditioning effects of political borders, in terms of institutional divergences, technological development, language and culture, mental maps of regional actors, etc., may be so strong that, long after the lifting of political borders, there are still barriers to interaction. Taking advantage of open borders also means that regional actors have to cope with policy influences that originate from higher levels or from other sectors. In other words: cross-border interaction, particularly cross-border co-operation, takes place in a multilevel policy arena. Thus, various remaining barriers may prevent actors in border regions to gain advantage from openness; these barriers are summarised in *Table 1.1.*

In Europe, the type of barriers or conversely of opportunities that prevail are to some extent connected to the location of the border regions at hand, i.e. in the national economic space and in the EU, particularly in EU inner border areas or outer border areas.

Table 1.1 A classification of barriers in cross-border (economic) co-operation

Barrier Type		Illustration
Political borders	-	Limited access for persons and goods
		Time-consuming cross-border formalities
Economic interaction	-	No opportunities (e.g. no complementarity)
		High level of interaction costs
Institutions	-	Different fiscal regimes
		Different property regimes
Power	-	Obstruction from higher policy levels or different sectors
Culture	-	Different language and vocabulary
		Different management style
Identity	-	Distorted mental maps
		Negative sentiments based on (recent) history
Transport and communication	-	Missing (deficient) links in rail and road systems
		Deficient service levels
Technology	-	Different standards and specification
		Different development levels

Accordingly, the following types of border regions can be identified, of which a selection is presented in this volume:

- In national economic cores and central in the EU; an example is the Dutch region of West-Brabant facing the region of Antwerp in Belgium.
- In the periphery of national spatial-economic systems and central in the EU; examples are the Swiss region of Ticino and adjacent Insubria in Italy (see *Ratti*, Chapter 2, and *Maggioni and Bramanti*, Chapter 8).
- Close to the national economic core and peripheral in the EU; examples can be found in Scandinavia (Southeast of Finland) and Greece, both facing outer borders of the EU.

- In the periphery of both systems; examples are the northern Finnish-Russian border regions (see *Eskelinen*, Chapter 10) and, perhaps less pronounced, the German-Polish Oder-Neisse regions (see *Krätke*, Chapter 11).

As a final point, it needs to be emphasised that, in a global world, taking advantage of new openness by no means implies cross-border interaction on a regional scale only. Interaction may equally mean the overlooking of border areas at the other side and starting co-operation with non-border regions in the adjacent country, or in other countries than the adjacent ones. Cross-border interaction is thus a multilevel phenomenon. At the same time, following from the inherent character of space, encompassing different economic and social sectors, cross-border interaction is also a multisector affair.

Aim and Outline of the Book

The present book is the second book on border regions in the context of NECTAR research. The first - "Theory and Practice of Transborder Cooperation" - by Ratti and Reichman (1993) can be seen as an exploration of borders and border-related effects, among others aimed to establish various typologies. The current book now moves from merely descriptive analysis of empirical facts to a more comprehensive and, from a theoretical point of view, more satisfactory approach. In the book a paradigm is presented - the "active space" approach - and various border regions are used as a "laboratory" to "experiment" with frames of analysis and perspectives of the paradigm.

The book falls into three parts. The first part consists of a theoretical elaboration and underpinning of the "active space" approach. The second part covers an exploration of the extent and ways in which advantage is being gained from openness in different types of interaction. Accordingly, this part answers the question whether there are grounds for an "active space" policy (*Table 1.2*). The third part discusses institutional issues and policies at different policy levels, and their implicit or explicit contribution to "active space" development.

Table 1.2 Contributions in Part II and Part III

Author (s) and chapter	Focus	Type of regions/ scale level	(Cross-border) flow/activity
Rietveld (5)	Use of openness	Dutch border regions National scale	Traffic flow (part. business)
De Boer (6)	Use of openness	Dutch-German Ems/Dollart region	Traffic flow (part. public)
Van Houtum (7)	Use of openness	Dutch–Belgium regions of Zeeland and Genk/Eeklo	Co-operation between firms
Maggioni and Bramanti (8)	Trajectory of openness and integration	Swiss-Italian regions of Ticino and Insubria	Labour flow
Van Geenhuizen (9)	Use of openness Conditions for integration	National scale and regions in Central and Eastern Europe	Capital flow Co-operation between firms
Eskelinen (10)	Use of openness Conditions for integration	Finnish-Russian border regions (Karelia)	Various flow (trade, traffic)
Krätke (11)	Conditions for integration	German-Polish Oder-Neisse region	Co-operation between firms
Mlinar (12)	Conditions for integration (history, institutions, policy)	Italian-Croatian-Slovenian region of Istria	Various types of interaction
Blatter (13)	Conditions for integration (multilevel policy arenas)	Examples from the United States-Canada border and German borders	Various types of interaction
Van Geenhuizen and Nijkamp (14)	New conditions in policy making for regional development	Transport dominated region of Rotterdam	Seaport activity and – related activity
Ratti and Vitta (15)	Local governments co-operation	Swiss border region of Locarno	Tourism and – related activity
Steiner and Jud (16)	Policy to foster small high-tech firms	Austrian border region of Styria	High-tech industry
Siepmann (17)	Location decision analysis for a free enterprise zone	German border regions of Lower Saxony	Not applicable
Corvers (18)	EU border region and innovation policy Policy networks	No focus	Not applicable

Part I

In the second chapter, *Ratti* elaborates the regional "active space" approach by conceiving the region as the outcome of a "field of forces", i.e. a coherent game of initiatives and responses that regional actors are together able to produce locally and towards the outside world. The conceptual novelty is the dynamic approach, in which the degree of "active space" stands for a transitional state in a development trajectory defined on the dimensions of openness and concern for sustainability. In an empirical exploration, the author takes the development of the Chiasso region in the Swiss-Italian border area, between 1950 and 1990, as an example. Ratti summarises various advantages of the approach, such as the general applicability to all regions and the integrated perspective, including diverse disciplinary contributions. At the same time, the author realises that a formalised approach to the non-economic variables concerning the way openness is managed, such as learning processes and networking, requires a great effort in finding useful indicators.

Steiner (Chapter 3) first discusses the traditional neglect of firm behaviour in traditional (micro)economic theory and then moves to the viewpoint of evolutionary economics on regional development. Accordingly, he puts an emphasis on the importance of variety and of the small firm segment in the regional economy. Small firms, or small decentralised units, generate variety as opposed to the often inert behaviour of large firms. In addition, the author places the "active space" approach in the tradition of institutional economics, particularly based on the recognition that economic behaviour takes place within a cultural milieu while being part of this milieu.

Bramanti (Chapter 4) elaborates governance as an important factor in "active space" development. After a brief discussion of the historical example of the Hanseatic league, the author addresses governance in terms of the conditions under which it contributes to a better performance of the regional economic system. Governance is seen by the author as the capacity of political institutions, jointly with civic society, to articulate, through public policy and representation, conflicting trends within the political process. With regard to the global competition of regional production systems, governance structures have to advance the production and re-production of tacit knowledge and learning processes, both being inter-organisational and territorially embedded. The active role of a co-ordinator, or animator, is regarded decisive in transcending narrow-mindedness and egocentrism of powerful local actors. A learning-by-monitoring attitude is also seen as essential to preserve a positive development trajectory of the regional economy. Finally, the author addresses the risk of institutionalisation of governance structures, with regard to the need for flexibility and loose mechanisms of coupling, and related with this, the important role of social capital in building appropriate governance. The latter issues form part of a research agenda proposed by the author.

Part II

Part II consists of various empirically-based contributions focusing on the use of openness, with a specific eye on impacts of remaining border barriers (*Table 1.2*). In this vein, *Rietveld* (Chapter 5) investigates cross-border traffic flows in the case of the Netherlands at different spatial levels and for different modes. First, the author discusses an analytic framework that includes both distance effects and border effects in terms of homogenisation and flow (discontinuity), and symmetric and a-symmetric development. In the following sections, the author measures the border effect empirically. For medium- and long distance connections, air and rail, reduction factors amount to some 30 to 40 per cent for cross-border links. In addition, this analysis indicates smaller border effects for business and freight transport compared with traffic based on other motives. In public transport, supply frequencies crossing borders are lower compared to those in other regions, leading to an additional negative effect on cross-border interaction. Thus, border effects remain substantial, indicating that, despite the economic integration in the European Union, not all border-related obstacles have been removed. Rietveld raises also the point of lack of knowledge about cross-border interaction for coastal boarder regions.

The contribution of *De Boer* (Chapter 6) has an almost exclusive focus on traffic flow using public modes of transport in the Dutch-German border regions, particularly the Ems-Dollart Region in the North. First, the author discusses the regulatory changes in public transport at either side of the border. He uncovers a difference in pace of change between the Netherlands and Germany, with more rapid changes in the former. Further, the author concludes that the supply of services in the Ems-Dollart Region is the poorest among all Dutch-German border regions; a situation that can partly be explained by the sparse population in this region. In general, there is a lack of motives for cross-border interaction, due to language differences and institutional differences such as school systems. However, one motive

manifests itself strongly and positively in the most southern and densley populated Dutch-German border region, namely commuting based on higher wage
levels in Germany and relatively cheap housing in the Netherlands, and on
opportunities to benefit from these cross-border differences. Compared with
this situation, the Ems-Dollart Region is thus clearly a region where
disappearing political borders have not led to substantial cross-border
interaction, due to missing opportunities in the regional economy, such as in
the housing market and in labour market.

Van Houtum (Chapter 7) also investigates the extent in which regional
actors can take advantage of the increased openness of political borders. His
case study deals with cross-border co-operation between firms located in the
regions of Zeeland Flanders and Central and North Zeeland (The
Netherlands) and that of Gent/Eeklo (Belgium). In this area, with a
relatively longstanding openness, there is no language difference, and
economic and financial policies are to some extent connected. However, the
image of a political border seems surprisingly hard to overcome, in that it is
rooted in the minds of people. Only one third of the firms has a co-operation
relationship in the neighbouring country. It seems, however, that a physical
barrier, the Westerschelde estuary, separating Zeeland Flanders from the
remaining Netherlands, causes locally some more cross-border interaction
with Belgium than elsewhere. In addition, the use of a channel, connecting
Gent (Belgium) with Terneuzen (The Netherlands) and the North Sea seems
to neutralise the mental barrier effect of the border. The study thus finds
grounds to plea to extent theory of cross-border interaction barriers with a
mental factor. Mental distance impacts of (former) political borders seem to
influence the potential of gaining advantage from open borders

The next contribution focuses on the movement of people and the
labour market in a long-term development view. *Maggioni and Bramanti*
(Chapter 8) analyse labour market dynamics in the Canton of Ticino in
Switzerland, including cross-border flows from Italy. In a theoretical
analysis the authors adopt a population-ecology approach to sustainable
development and competitiveness. In an empirical part they connect this
with "active space" development of Ticino. In the development trajectory of
the Ticino labour market, the authors identify an initial stage starting with a
highly fragmented market and strong entry barriers, followed by a future
intermediate stage in which human capital is upgraded through training and
education, including a more free movement of labour, and ending in a final
situation of integration of labour markets on the European level.

Van Geenhuizen (Chapter 9) investigates cross-border capital flows in a study of foreign direct investment (FDI) in Central and Eastern Europe. This study indicates that both the stage of transition of the economy of these countries and their proximity to Western Europe matter in attracting FDI. Accordingly, one may conclude that economic and institutional barriers, and distance, including cultural distance, have an influence on cross-border FDI in Central and Eastern Europe. Inflow has been largest in Hungary in the years after 1989 but Poland is running up most recently. Although the level of analysis of FDI patterns in this chapter is mainly national, the question is posed how investments can contribute to integration and "active space" development in the different regions of Central and Eastern European countries. There seems to be a long way to "active space" development due to the fact that the major motives for investment are concerned with market share increase and cheap labour, not with knowledge. Moreover, particular shortages in the regional support space, stemming from the communist past, still exist today, particularly missing learning networks. Good opportunities however emerge where FDI connects with local available knowledge that is competitive in a global economy.

Part II also includes two studies with a focus on border regions at the edge of the European Union. *Eskelinen* (Chapter 10) investigates the prerequisites and first experiences of cross-border co-operation between Russian Karelia and regions in Eastern Finland. These border regions are facing a clear divide due to past political developments, with relatively poor prospects of future integration on the short term. First, the author considers the major institutional changes influencing regional policy for border areas in Finland, such as the transfer of responsibility for regional development to regional councils, as bottom-up organisations based on municipalities. In addition, there is the policy for near region co-operation, focusing on Russian Karelia, and also on the Murmansk and Leningrad regions, and since 1996, the EU Interreg programmes. The author proceeds with an analysis of transport and economic relations across the border. There has been a clear increase in interaction, but the weak performance of the Russian Karelian economy, due to factors such as an export based on a few staple commodities and the absence of foreign investors in the paper and pulp industry, inhibits the rise of strong relationships. Given such serious constraints, the author addresses various scenarios in the adaptation to openness. One of them is based on the legacy of peripherality, i.e. a "passive" corridor region, whereas another one is based on the absence of any competitive economic activity, even transport. Because traditional

Finnish industries that would appreciate low cost labour have almost entirely disappeared, there is almost no ground for cross-border co-operation today. There is one exception, namely forest industry firms on both sides of the border that are expected to exploit forests on an ecologically sustainable basis. All in all, there has been an increase in openness and improvement of conditions for cross-border co-operation but it remains to be seen whether the region, without the necessary economic complementarity, can evolve towards a functional region.

For a partly comparable case, the German-Polish mid Oder-Neisse border regions, *Krätke* (Chapter 11) addresses the economic impact of the transition and investigates various barriers to regional-economic integration. One important result is that on both sides of the border there is a trend to divert economic development from the border areas to Berlin and to Poznan. In general, there is also a trend for establishing supra-regional co-operation and trans-national co-operation, overlooking one or both sides of the border regions at hand. In fact, cross-border interaction between regional firms in the border area turns out to be small. In addition, as a specific traffic corridor on a European scale, the border regions suffer strongly from negative externalities. With regard to policy instruments to improve the situation, Krätke expresses serious doubts on positive impacts of free enterprise zones on the regional economy, because of the lack of regional embeddedness of the firms that tend to locate here. He proposes an alternative option which compares with the "active space" development, namely improving endogenous skills in existing (mostly) traditional industries and a shift away from mass production to advanced, high quality products.

Part III

Part III encompasses various studies with a focus on conditions and policies in "active space" development. In a historically and sociologically-oriented analysis of border regions *Mlinar* (Chapter 12) addresses that border region development needs to be considered today from a network linkages perspective and multilevel power-sharing. In an illustration using the case of Istria, with Croatian, Slovenian and Italian parts, he stresses the role of distrust and need for confidence building in gaining benefits from new opportunities. Non-parity roles related with differences in size and power of countries and actors on both sides of the border, is also a potentially influencing factor. The author proposes a number of strategies for opening up of border regions and incorporating them in the European integration processes.

In a similar vein, *Blatter* (Chapter 13) draws attention to the policy forces under which border regions develop. In this context he proposes a spatial multi-actor multilevel framework of political arenas that influence political cross-border interaction and institution building. A few examples of these arenas are the sub-national cross-border arena, the international/continental arena, the vertical intergovernmental arena, and the intersectoral arena. In order to reduce the concomitant structural complexity in "active space" development the author arrives at various policy advice. First, regarding cross-border policies he recommends to differentiate in the design of co-operation institutions according to the policy character of the issues involved, e.g. conflicting problems or consensus issues. Secondly, on the operational level he advises to refrain from comprehensive big programs and to focus on small projects that are easy to implement.

Part III proceeds with a contribution that falls outside the present scope of border regions but clearly addresses the management of openness in a transport dominated region. *Van Geenhuizen and Nijkamp* (Chapter 14) investigate the region of Rotterdam, with the largest seaport in Europe. In an empirical part, they observe that competition between ports, particularly in container throughput is growing and that hinterlands become increasingly difficult to demarcate. The region accordingly faces the challenge to preserve a high level of competitiveness, and at the same time, to find ways to improve the use of its learning capability to move towards more innovative economic activity. In this context, the authors discuss various conditions of learning-based economic development and policy making that may support the transition to new policies in the region and a turn in its development trajectory. This work contains some attempts to operationalise policy-making aspects of "active space" development, in terms of policy content and policy process. For example, the policy making process preferably incorporates methods to deal with uncertainty about future development, such as innovative scenario analysis and experiments. The chapter concludes with the observation that where innovations in policy making are introduced, institutional barriers may arise that cause a fall back to "comfortable", traditional solutions and approaches.

Ratti and Vitta (Chapter 15) introduce a new concept and operational elaboration of co-operation between local governments, i.e. a holding company for public shareholding (HCP). This solution matches with "active space" development because of its network strategy that increases flexibility and because of its new governance structure that responds to the region's changes in internal and external contexts. The study underpinning the concept took place in the Swiss region of Locarno, dominated by tourism. The focus is on the need to facilitate co-operation among municipalities and other public bodies aiming to develop and manage with private parties initiatives that go beyond the municipal level of public interest, this mainly in connection with the strategic activity of tourism. The holding company is controlled by public entities and is a structure that respects democratic principles: the decision to acquire HCP capital and the designation of members of the board of directors are the competence of municipal governments. The authors conclude that the HCP may trigger new dynamics in the Locarno region that improves its competitive position. The possibilities of the HCP to open up to nearby areas, would also allow collaboration among regions facing similar problems and finding themselves in a collaborative-competitive position.

Further, *Steiner and Jud* (Chapter 16) investigate the development of a specific border region in Austria, i.e. Styria. They draw attention to opportunities and threats for new technology-based firms, because of the major role of these firms in renewal of the regional economy. The authors analyse the specific problems these firms are facing and the response given to this situation in the "technology policy concept for Styria". In the latter context, the authors discuss three policy strategies to improve the situation. Existing support schemes need to be directed more closely to the problems of the firms at hand and better co-ordinated to fit the needs of firms during the process of establishment and first growth. Moreover, there is a need for additional services to improve the relation with customers, collaborative behaviour and access to private funding.

Siepmann (Chapter 17) focuses more explicitly on a specific regional development tool, i.e. the free enterprise zone, and the best location of such a zone in Saxony in Germany. A free zone is seen as a transport related instrument, aimed at functioning as a logistic platform between Saxony and the adjacent Central European regions. The author develops an evaluation framework in order to support the decision between various alternative locations. He analyses various transport-related indicators to identify the location with the highest level of openness and sustainability possible from two alternatives, i.e. the traditional market place of Leipzig and Dresden in the Elbe/Labe Euregion. This study is one of the firsts to gain experience in the operationalisation of some key concepts of "active space" development, such as the concept of openness. The author arrives at the conclusion that

Dresden is the best location, because this city has proved over the past years to perform better in a socio-economic sense, due to its supportive learning capability and institutional structures.

Finally, in the contribution of *Corvers* (Chapter 18) the question is addressed how the "active space" approach compares with various regional policies of the European Union. The emergence, aims and instruments of the Interreg initiative, in border regions, and of the RITTS/RIS scheme, regional innovation policy, are discussed. In fact, the recent approach of Interreg complies with the "active space" approach in bringing all relevant actors together in order to improve self-organising and synergy. Particularly, the RITTS/RIS experiment places innovation in the core and mobilises all relevant regional actors in various stages of innovation policy. The RITTS/RIS scheme also places the region in a global framework, by putting an emphasis on the consequences of global trends for the region and on necessary action based upon this. The approach particularly compares with the "active space" approach in assigning a crucial role to learning and problem-solving behaviour. Experience to date indicates a great importance of the scope of manoeuvre of the project promoter, determined by resources made available and his/her personal qualities. In both Interreg and RITTS/RIS schemes a policy approach is adopted that puts an emphasis on the role of policy networks in achieving particular objectives in the region. In the context of a network approach to regional policy, Corvers gives some rules of thumb in order to make policy networks move to advance "active space" development, while recognising that in border regions national government may remain a crucial factor.

In a concluding chapter (Chapter 19) *van Geenhuizen and Ratti* review the main issues in gaining advantage from open borders and implications for "active space" development. In addition, they propose a number of new research paths.

References

Amin, A. and Thrift, N. (eds) (1994), *Globalization, Institutions, and Regional Development in Europe*, Oxford University Press, Oxford.

Anderson, J. and O'Dowd, L. (1999), 'Contested Borders: Globalization and Ethno-national Conflict in Ireland', *Regional Studies*, vol. 33(7), pp. 681-696.

Anderson, M. and Bort, E. (eds) (1998), *The Frontiers of Europe*, Frances Pinter, London.

Boekema, F. and Allaert, G. (eds) (1999), *Cross-border activities on the move* (in Dutch), Van Gorcum and Comp., Assen.

Boekema, F., Morgan, F., Bakkers.S. and Rutten, R. (eds) (2000), *Knowledge, Innovation and Economic Growth*, Edward Elgar, Cheltenham.

Braczyk, H-J., Cooke, P. and Heidenreich, M. (eds) (1998), *Regional Innovation Systems. The role of governances in a globalized world*, UCL Press, London.

Bramanti, A. and Senn, L. (1997), 'Understanding Structural Changes and Laws of Motion of Milieux: A Study on North-Western Lombardy', in R. Ratti, A. Bramanti, and R. Gordon (eds), *The Dynamics of Innovative Regions. The GREMI Approach.* Ashgate, Aldershot, pp. 47-73.

Camagni, R. (1991), 'Local 'milieu', uncertainty and innovation networks' in R. Camagni (ed), *Innovation Networks. Spatial Perspectives*, Belhaven Press, London, pp. 121-144.

Cappellin, R. and Batey, R.W.J. (eds) (1993), *Regional Networks, Border Regions and European Integration*, Pion, London.

Cooke, P. (1998), 'Introduction: origins of the concept', in H-J. Braczyk, P. Cooke and M. Heidenreich (eds) (1998), *Regional Innovation Systems. The role of governances in a globalized world*, UCL Press, London, pp. 2-25.

Cooke, P. and Morgan, K. (1998), *The associational economy: firms, regions and innovation*, Oxford University Press, Oxford.

Eskelinen, H., Liikanen,I. and Oksa, J. (1998), *Curtains of Iron and Gold:Reconstructing Borders and Scales of Interaction*, Ashgate, Aldershot.

Geenhuizen, M. van, and Nijkamp, P. (1998), 'Improving the knowledge capability in cities: the case of Mainport Rotterdam, *International Journal of Technology Management*, vol.15 (6/7), pp. 691-709.

Geenhuizen, M. van, and Nijkamp, P. (1999), 'Regional policy beyond 2000: learning as device', *European Spatial Research and Policy*, vol. 6 (2), pp. 5-20

Geenhuizen, M. van, Knaap, B. van der, and Nijkamp, P. (1996), 'Trans-border European networks: shifts in corporate strategy?', *European Planning Studies*, vol. 4(6), pp. 671-682.

Geenhuizen, M. van, and Ratti, R. (1998), 'Managing Openness in Transport and Regional Development: An Active Space Approach', in K. Button, P. Nijkamp, and H. Priemus (eds), *Transport Networks in Europe. Concepts, Analysis and Policies.* Edward Elgar, Cheltenham, UK, pp. 84-102.

Grabher, G. (1993), 'Rediscovering the social in the economics of interfirm relationships', in G. Grabher (ed), *The Embedded Firm. On the Socioeconomics of Industrial Networks*, Routledge, London, pp. 1-31.

Guo, R. (1996), *Border-Regional Economics.* Springer, Heidelberg.

Malecki, E.J. and Oinas, P. (eds) (1999), *Making connections: technological learning and regional economic change*, Ashgate, Aldershot.

Maskell, P., Eskelinen, H., Hannibalsson, I., Malmberg, A. and Vatne, E. (eds) (1998), *Competitiveness, Localised Learning and Regional Development. Specialisation and Prosperity in Small Open Economies*, Routledge, London.

Mlinar, Z. (1992), 'European Integration and Socio-Spatial Restructuring: Actual Changes and Theoretical Response', *International Journal of Sociology and Social Policy*, vol. 8.

Morgan, G. (1986), *Images of Organization*, Beverly Hills CA, Sage.

Morgan, K. (1997), 'The Learning Region: Institutions, Innovation and Regional Renewal', *Regional Studies*, 31 (5), pp. 491-503.

Nijkamp, P. (1990), *Sustainability of Urban Systems*, Avebury, Aldershot.

Nijkamp, P., Rietveld. P. and Salomon, I. (1990), 'Barriers in spatial interactions and communications. A conceptual exploration', *The Annals of Regional Science*, 24, pp. 237-252.

O'Dowd, L. and Wilson, T.M. (eds) (1996), *Borders, Nations and States*, Aldershot, Avebury.

Paasi, A. (1999), 'Boundaries as Social Practice and Discourse: The Finnish-Russian Border', *Regional Studies*, vol. 33 (7), pp. 669-680.

Ratti, R. (1993), 'Spatial and Economic Effects of Frontiers', in R. Ratti and S. Reichman (eds), *Theory and Practice of Transborder Cooperation*, Helbing & Lichtenhahn, Basel, pp. 23-49.

Ratti, R., and Reichman, S. (eds) (1993), *Theory and Practice of Transborder Cooperation*, Helbing & Lichtenhahn, Basel.

Ratti, R., Bramanti, A, and Gordon, R. (eds) (1997), *The Dynamics of Innovative Regions. The GREMI Approach*, Ashgate, Aldershot.

Scott, J.W. (1999), 'European and North American Contexts for Cross-border Regionalism', *Regional Studies*, vol. 33 (7), pp. 605-617.

Storper, M. (1997), *The regional world*, Guildford Press, New York.

2 Regional Active Space: A Regional Scientist's Paradigmatic Answer to the Local-Global Debate

REMIGIO RATTI

In the process of globalisation of the economy and society there is a risk that the regional debate will move into the perspective of a "passive space" approach. The analysis of spatial phenomena shows, however, that a region is the result of a "field of forces," and can even be considered an "active space." The author proposes the concept of "active space" as the regional scientists' paradigm and an answer to the local-global debate. He defines the notions and the logic of the regional "active space" and suggests an outline of a model for theoretical analysis. The regional "active space" and its level of activity are presented as being the result of different mutually interrelated regulation processes, i.e. the functional spaces of firms in terms of market, production, and support space, as well as the rules and governance of the territoriality.

Introduction

The concept of "World Economy," according to Wallerstein (1974) and Braudel (1985), obviously does not carry today's date; however, if the debate on reciprocal relationships between "local" and "global" has taken centre stage today, this is not just because of the nature and broadness of the globalisation process acting on the economy and society, it is above all because the preponderant direction of relations seems to be inverse (Reich, 1993): if in the past the process was seen as "local" moving towards "global", today the predominant orientation is rather "global" towards "local". The latter aspect is particularly important for regional sciences, considering that the predominance of the "global" tends to obscure the "territorial"; so the territorial aspect may be seen and experienced even more as a "simple support space" of phenomena, the explanation of which lies elsewhere. Territorial space will be characterised, then, only by the product of functional logic of economic activity.

Surely, following Vidal de la Blanche, the geographers have taught the value of the concept of living space, and the scientific debate on the relationship between territory and innovation has allowed, based particularly on the work of economists and regionalists on industrial areas, innovative milieu and the flexible economy, but also on the work of sociologists (Granovetter, 1985; Grabher, 1993) and institutional economists (Williamson, 1985), the discovery of a "territorial space as actor of development" (Rallet, 1988). In this light, space, far from being a simple passive support, becomes an active element in the explanation of territorial development processes. We thus speak of "regional active space", attributing to this concept the role of a true paradigm.

In this chapter we want first to discuss the fundamental problem areas of "regional active space", interpreted as a useful instrument for the understanding of possible solutions to the debate on the dichotomous relations between "global" and "local". We will subsequently present a simplified model that describes the logic of an "active space" in a process of territorial economic development.

The Regional Active Space: a Proposal for a Paradigmatic Approach

The nation-state has represented, as long as the great majority of exchanges of goods and services took place in the classic framework of international trade among companies belonging to different domestic production systems, the leading territorial level in the mediation between local and global. Today, it is clear how this is increasingly less possible; to the contrary, new emerging spatial configurations are actively influencing the productive organisation and the strategies of individual actors and companies: this is the emergence of the regional level and territorial logic, which take on the role of complements to macroeconomic approaches. Thus, among the minority of economists who integrate the spatial aspect in their reasoning, most support (with Perrin, 1983; 1992) a formalisation of the meso-analytical spatial level in the general theory. The concept of "regional active space" can represent a valid instrument for meeting this aim.

The concepts of "regional active space" as a "force field space"

As mentioned, economists, geographers and sociologists, belonging to different schools of thought, but all of them aiming to understand what takes place in the spatial reality, by now consider that, contrary to what takes place according to the concept of a "passive space", the development trajectory of a region interacts with the technological development trajectories. In particular, the region becomes integrated in a process of creation/destruction, diffusion, or concentration of technological innovation. This is the view of "active space" (Ratti, 1992) that takes on the role of a paradigm, i.e. of a research orientation starting with a series of hypotheses that can not yet be consolidated.

We prefer to talk explicitly about "active space" rather than use other notions such as "space as actor" and "factor of development", because, irrefutably, this active role is not taken on by the region intended as a pole space characterised by the existence of a sole and sovereign decision-making centre, but rather this active role of the region is the fruit of a "force field" (Ratti, 1980), i.e. a coherent game of initiatives and responses that the actors of a given territorial space are together able to produce within the space and towards the outside. Each region is potentially an "active space", but what counts in reality is the *degree* to which it is. Precisely here one can find the main conceptual novelty with respect to those who have up to now considered the cases of industrial districts or innovative milieus as exceptions in a context where space remains a simple support for the sectoral functional logic. As already outlined in research with van Geenhuizen (1998), "active" regional space is situated in a force field characterised by three aspects: openness, concern for sustainability, and the creative ability of a specific territorial system (*Figure 2.1*).

Openness is intended as the entering into communication between the different components of a system (regional or city) and between the components and the system centres. It implies the ability to take advantage of economies of scale and scope, in particular to overcome physical, socio-economic, cultural or political trade barriers and impediments.

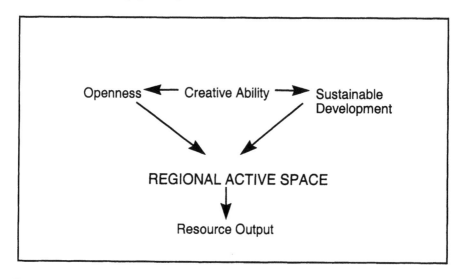

Figure 2.1 Regional active space as a force field model

Source: van Geenhuizen and Ratti, 1998.

Sustainability here is considered in the territorial system's capacity to sustain environmental development, and in its socio-cultural content. If in reality the openness potentially enriches the local culture, an overly strong external influence could be perceived as a danger to society due to the loss of cultural cohesion (integration) and social stability (Mlinar, 1992). Sustainability thus represents a limit to this danger. Finally, creative ability or "territoriality" represents the aspect determining the local system's ability to respond to internal and external challenges with objectives, norms and strategic behaviour. Creative ability is based on knowledge capital and on regional collective learning processes that also find an equilibrium between cohesion and openness, and between internal and external networks (Storper and Scott, 1993).

All together, regional "active space" is the result of a force field, the degree of output of which depends on its ability to generate an adequate and complex response to system dynamics; this response is expressed by strategic behaviour capable of innovation related to openness and internal cohesion in order to gain economic benefits from the potential for sustainable development of a territorial space. The degree or value of the active component of space will therefore depend on the combination of factors worth studying and modelling. This is in conformity with the conclusions of various regional approaches according to which numerous, very differentiated, models coexist in a single global area (Benko, 1995).

The axes on which the "active regional space" lies

The vertical axis, between global and local, is the axis of the force field that can be described in terms of "territoriality" (for example, Raffestin, 1986; Pecqueur, 1993) (*Figure 2.2*). In human sciences, territoriality is a paradigm that expresses a complex and dynamic relation between a human group and its environment. It is characterised by a body of principles, rules and strategic behaviour aimed, in general, at searching for dynamic coherence which can be used to give a society and a specific territory the capacity to define its sustainable development trajectory. The territoriality of a country or a region therefore represents a construction, a socio-cultural, economic and political product, and also a complex process through which a society develops its ability to respond, internally and externally, to changes.

The horizontal axis is the axis of companies' functional spaces which we consider a composite of classic production and market spaces, and a third category of relations, namely, support space. Support space (Ratti, 1995) is defined as a whole body of relations outside the market or preceding it. More precisely, it comprises three types of qualified relations: those connected to factors of production such as labour, capital, technology, and in particular researchers and educators (trainers); those determined by alliances and informal agreements with suppliers and customers; and those deriving from relationships that firms maintain with institutions and government, and thus correlated with "territoriality". A part of this space can be territorialised; the other, instead, has network-like characteristics and concerns discontinuous space. The support space plays the role of a connection, above and beyond the price system, between the production space and the market space; and it characterises the development trajectory of a specific territorial reality (region, city).

In conclusion, "regional active space" is, as a result of the entrepreneurial and territorial strategic behaviour of a region's actors, the mirror of a specific society's culture of development.

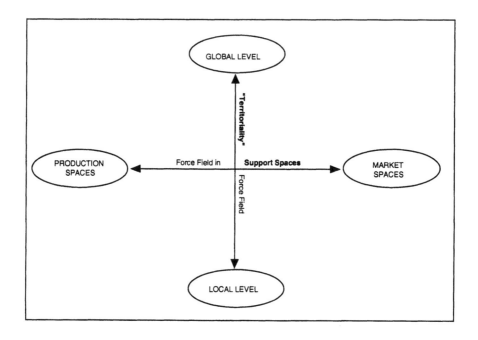

Figure 2.2 Regional active space as a meso-economic force field

Source: Ratti, 1997.

Factors and instruments for determining and managing "regional active space"

"Regional active space", in terms of territoriality and support spaces, can be analysed by adopting the instruments of the methodological fields of:

- the new institutional economy
- decision-making and governance models
- the logic of network co-operation/competition, and
- the economy of strategic resources.

Each of these approaches contributes to the paradigm of "regional active space" (RAS) as displayed in *Figure 2.3*. The first approach is concerned with the interrelation between institutions and organisations and is at the centre of "active space" interests. Although the new institutional economics is not yet able to provide a well-established methodological corpus (Parri, 1995), the interrelation between "institutions" and "actions" is at the centre of interest today. The processes of economic liberalisation or deregulation tend to modify the rules of the game and the strategic behaviour of a region's individual and collective actors. In particular, it is necessary to pay attention to complementarity, as well as to substitution games, between the role of institutions and the role of organisations.

Institutions correspond to the formal rules of the game, i.e. constitutions, legislation, regulations, or informal rules such as implicit social pacts, procedures/applications, leading to a sort of unwritten law (Williamson, 1991). Organisations refer instead to actors such as political parties, representatives of the state, groups of businesses, unions and entrepreneurs. These will be treated further on, under the heading "governance". Institutions can represent both limitations and opportunities for "regional active space" development (North, 1990), and their function, by setting the rules of the game, is to reduce uncertainty. Examples of this can be found in the definition of property rights and contracts, antitrust legislation or temporary receivership. The analytical instruments applicable to this area of interrelations between institutions and actors are those of risk and uncertainty reduction, as well as those of the distribution and negotiation of duties between different institutional levels through game theory. Thus, in a phase of deregulation and privatisation and of debate over the competencies among the various political-institutional hierarchies (Anderson, 1992), the influence of institutional factors no longer seems to be contested. Similarly, the optimal quality of institutions does not seem to be guaranteed by rational choice processes and not even by spontaneous evolutionary or historical processes such as those described by Hayek or North (Parri, 1995). In our view, the institutional factors in the inventory model of the force field make a regional space "active".

The second point is concerned with the importance of governance in the "active space" *(Figure 2.3)*. An emerging approach is undoubtedly that of governance and the analysis of its structure (Amin and Thrift, 1993; see also, Bramanti in this volume). First, we identify, if we accept the debate on the social construction of the market (Bagnasco, 1988) and the postulates of Coase (1937) on organisational forms, co-operation as the third mode along with the firm and the market. According to Williamson (1985) and Lipietz (1992) we find ourselves in the realm of transaction costs, due to imperfect information, entrance barriers, opportunism and control costs.

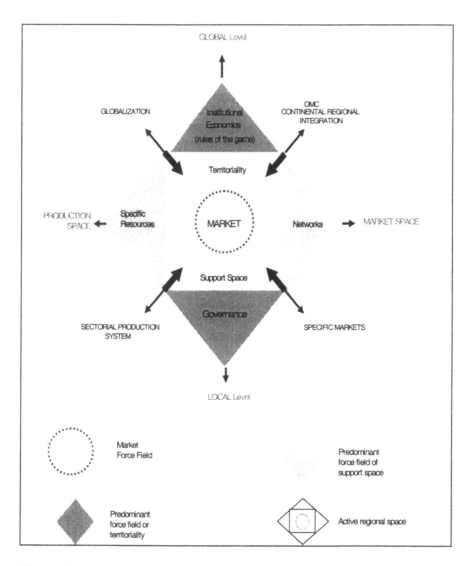

Figure 2.3 Active regional space and its essential elements

Source: Ratti, 1997.

The "regional active space" can be significantly influenced by these factors and by the ability (or lack of it) to translate governance relations, by nature part of the field of companies' functional spaces and not necessarily in line with the needs of a territorial logic, into a coherent management system.

According to Storper and Harrison (1992), the structures of governance involve different typologies, such as presence/absence of a leader, central-peripheral relations, co-operative or non-co-operative relations, which again are key factors in establishing the degree and intensity of logic that make a territorial space "active". In particular, governance structures seem to be able to orient the movement law of innovative regions, with above all a "core ring with co-ordinating firms" type of structure (Bramanti and Senn, 1997). The aim of this first element of governance is thus represented by the structure of companies' relations; it plays a considerable role in defining the dynamics of the territorialised productive system. The second element is represented by organisations. The typology of relations between social partners must also be analysed from a political and economic viewpoint. In connection with the institutional rules of the game, it determines the fundamental strategic behaviour at the actor level (sectoral corporatism, pluralism). An excellent historical example that is still timely is given by consensual practices and democratic corporatism, which constituted a significant part of Switzerland's development (Ratti, 1995). The object of this second field of governance concerning organisations will be based, for example, on the framework of labour contract conditions and wage negotiation rules and will be manifested in the participatory process leading to the formation of a socio-political consensus. This point can thus be concluded with Lipietz (1992: 53), who affirms that governance comprises '...all forms of regulation that are neither commercial nor state'. Governance '...is civil society, minus the market, plus the local political society, the prominent, the municipalities'.

The third point deals with functional relations within "active space". At the level of economic actors, a first category of "off-market" relations concerns the search for openness by constructing networks of qualified functional relations, upstream (suppliers) or downstream (customers). The network is in fact (Maillat et al., 1993) '...a whole, composed of selected and explicit ties, with preferential partners, included in a perspective of a company's market relations and its search for complementary resources, having as a main objective the reduction of uncertainty'. It is necessary to underline how the creation of a support space, through a similar strategy of creating a network, concerns the search for complementary resources that accompanies the gain in economic benefits from specific resources. The network thus becomes a channel with the following aims (Camagni, 1991):

- To reach sufficient economies of scale, through links in R and D and in the structures of marketing or production systems.
- To control certain key components of the market, so that the ability to adapt/react quickly is ensured.
- To verify the development trajectory of crucial components, so that a continuous process of adaptive innovation is ensured.

The contribution of support spaces thus created, consists of the structuring of a network economy, whose value lies in its flexible and interconnected architecture, allowing overall management of individual relations.

The final point to be discussed here is gaining economic benefits from specific resources in the "active space". Another area of relations in which the definition of support spaces acts is that of gaining economic benefits from a region's specific resources. These are characterised by a considerably more strategic feature than generic resources, such as low or unskilled labour reserves, availability of buildings and land, fiscal benefits, etc. Among the specific resources for which it seems particularly interesting to develop "off-market" strategic relations and that possess a strong stimulating impact on innovative behaviour (Perriard, 1994), we cite the following:

- The competence of labour which obviously represents the basic value of specific resources.
- The presence of universities and research centres, above all as far as their indirect effects on human capital and regional culture are concerned, as well as the support given to technological transfers.
- Access to efficient transport and communication networks.
- The possibility to turn to consultants and information services and the existence of a diversified urban base.
- Access to financing in view of capital and risk.
- Regional amenities, such as the natural and constructed environment, and cultural and educational offerings.

Finally, another category of relations that contribute to the definition of a support space, those that come into play along with collective, public or private actors of the territorial environment, brings us to a particular aspect of governance approaches, and in particular of institutional economics (Granovetter, 1985). The instruments for analysing strategic behaviour of actors towards institutions also depend on important methodological approaches that go from the classical analysis in terms of imperfect competition, to game theory, up to theories of non-profit organisations (Bluemle and Schauer, 1995).

Definitively, in the context of an "active space", the new deployment of sectoral support spaces allows a regional system to gain economic benefits from the region's specific resources as a content space, e.g. for labour, capital, technology, and to add complementary resources, which represent the fruit of the region as a relations space. The "regional active space", and, in particular, its degree of activity, will therefore be the result of a force field; that is, of numerous regulatory processes that interact, such as the market, company support spaces, the rules of the game and of the governance of territoriality. We point out, finally, that each region is potentially active, although to different degrees.

The Logic and Dynamics of Regional Active Space: a Draft Model of Theoretical Analysis

In the first part of this chapter, we presented the foundation of the concept of "active space" as an element of the paradigm of analysis and political action for regional development in the current context of the globalisation of the economy and society. Markets, companies' support space and the ability to develop regional "territoriality" together constitute the regulatory factors of a complex system of relations between society's territorialised spaces and the companies' functional spaces. We then saw how some new approaches, in particular multidisciplinary ones such as networks, governance, and new institutional economy, support the ongoing development of the "regional active space" (RAS), from an empirical as well as theoretical model point of view. As far as its empirical value is concerned, the application of the new concepts has turned out to be particularly fruitful in a number of specific occasions: the application of support space to the study of innovative milieus (Ratti and D'Ambrogio, 1992; Ratti and Alberton, 1997), that of territoriality to the Swiss reality (Ratti, 1995) as well as the more recent and complex application in the study of transportation (van Geenhuizen and Ratti, 1998). In this second part, we would like to stay at this theoretical level, well aware of the extreme broadness of the field of study and of the possibilities to model; that is, in conformity with the paradigmatic nature of the concept of "active space".

The trajectory of "regional active space" development

We define the variation in degree of RAS using the following three aspects: openness, concern for sustainability, and creative ability. We start with a simplified model based on the relation between openness and sustainability (*Figure 2.4*).

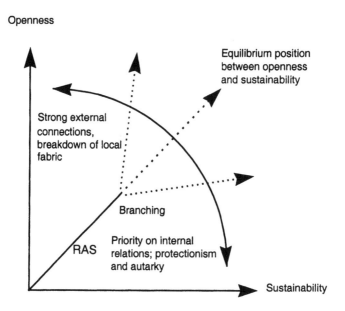

Figure 2.4 The trajectory of regional active space development

Source: Ratti, 1997.

We can assume a non-linear relation between openness and sustainability. Indeed, an overly high level of openness could turn out to be unsustainable for a region's development, while at overly low levels its development could be blocked by autarky. In general, however, openness represents the possibility to turn to complementary resources and to gain greater economic benefits from specific resources; but the question that arises spontaneously is the following: with which logic will a change in the degree of regional openness translate into a greater or lesser degree of RAS?

The logical process of the "active space"

Figure 2.5 shows how our reasoning starts with a simplified situation, recognising the fact that a typology of cases is possibly starting with the different curves shown in the quadrants of the figure. Box 1 is the market box. The market economy is strongly dependent on its degree of openness. Transaction costs represent an instrument for measuring market access. The form and slope of the curve, which here represent the relation between the degree of openness of a region and transaction costs, normally present decreasing transaction costs. To simplify matters, we will not consider other cases here. Let us suppose that there is a variation in the degree of openness, which goes from t0 to t1. The consequences of this openness can above all be interpreted in terms of transaction costs. Normally, a higher degree of openness generates lower transaction costs.

A decrease or increase in transaction costs means, according to the concepts of the new industrial economy, a benefit (or loss) measured in terms of a different form of regulation of the market/hierarchy/alliance triad, and, more importantly, in terms of a different degree of integration (Box 2). In this quadrant, the ability of companies to adapt their support space to the new situation can be seen in the curve. In the case shown in the figure, this ability to adapt, and hence the sectoral development trajectory, are positive.

The variation in transaction (market) costs and the degree of a company's integration, the success of which is linked to the specificity of support spaces, have repercussions on regional development in terms of RAS grade reached, as a function of "territoriality"; in other words, the ability of governance and also of institutions to face and take advantage of change. Box 3 indicates, depending on the curve, the trajectory of territorial development. Again, the case in the figure is an example of a positive trajectory, i.e. an innovative milieu. In many cases, however, the curve could be very different, with a greater slope and therefore less sensitive to external changes, or with an inverse slope, in the case of declining regions.

Box 4 indicates the result, i.e. the arrival point in the development trajectory in terms of the degree of RAS. In the specific case shown in *Figure 2.5*, an increase in the region's openness is translated into an increase in regional sustainability and as a repercussion, an increase in the degree of "active space".

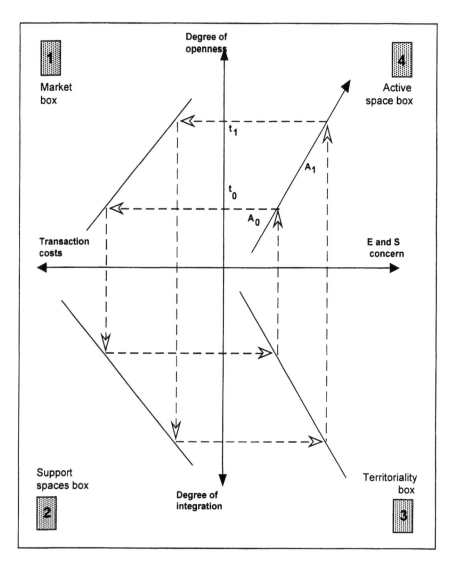

Figure 2.5 The logic of the active space (simplified case)

Source: Ratti, 1997.

Empirical application

The framework analysed in the previous section is suitable for an empirical application provided that the configuration in the relations indicated in each box can be constructed, or at least be drafted and described. A first application of the framework took into consideration research conducted over several decades on the development of a border region (Ratti, 1971, 1982, 1995) and was formalised in a recent study on new European logistics (van Geenhuizen and Ratti, 1998). The application concerned the empirical case of the Chiasso area, a very important border point about fifty kilometres north of Milan between Switzerland and Italy, a region situated along the European rail route and the north-south motor way across St. Gotthard. This case is described by distinguishing three periods (*Figure 2.6*).

The starting point (t0) corresponds to the 1950s, the eve of the establishment of the European Common Market. The border between Switzerland and Italy is still a barrier-line, a point where all goods must halt for customs procedures, often entailing a shipping interruption, with a change of the shipment and/or reshipment of goods. In this situation a region's degree of openness is therefore very limited, and due to the border, transaction costs are high (box 1) because market integration is weak (box 2); however, at the regional level this situation constitutes an advantage for agents in specific markets such as customs shippers and international service activities, often concentrated in one point. A network of specific relations is organised and exists in Chiasso, i.e. the bridge function indicated in curve 1 in box 3, with entrance costs at a relatively high level. The degree of support and coherence of regional economic activities with respect to external dynamics ("active space") is good.

The progressive realisation and success obtained in the first phase of the integration of the European industrial economy (t1) has as a consequence a significant increase in trading within the European Community and within the continent; however, the transportation infrastructures, customs procedures and international trade, including financial transfers connected to import/export, have not yet changed radically. Even if the procedures are softened, the position of Chiasso remains strong in Italy, and local operators gain further economic benefits from their know-how, putting themselves in an almost monopolistic position or in an innovative milieu (Ratti, 1995). In this period, around the 1970s, the situation of the Chiasso region is that of a region giving an active response to changes in the external scenario: openness causes transaction costs to decline (box 1); economic integration processes have begun (box 2), followed by import/export service operators and transit traffic across the Alps. The Chiasso region therefore, develops and affirms its own ability to manage these flows through strategic alliances and, as a transit area, brings out the value of its economic culture, particularly open to the needs of international trade. The result is also a certain enrichment of the region that reinforces, with the creation and the development of banking and asset management activities, its position as a gateway region (box 3). This outcome can be seen in the improvement of the situation in the scale of appreciation of "active space" (box 4).

The third period of analysis (t2) relates to the 1980s, in which economic processes - openness - are increasingly influenced by the creation of the single European space. Transaction costs fall, thanks to new communication technology, to computerisation, to EU directives (box 1), along with the ability of sector actors, above all Italian, to operate outside of domestic conditions and without feeling the need for intermediaries. What is changing is definitely the nature of integration processes that shift from commercial trading processes between domestic economies to a globalisation process of logistic activities. These processes can lead to breakdowns in relation to traditional territorial structures. For this reason there is a branching in box 2. In the new scenario of integrated logistics a location along a border point is in principle reduced to the role of a passage point. Within Europe, true logistic regions are coming into existence, which are in turn linked to some bridgeheads. In the context of our empirical example, the logistic region is that of northern Lombardy with the Milan actors fundamentally taking on the role of leader. Transit across the border is increasingly devalued and takes on the characteristics of a transit corridor (box 3).

The resulting view on the specific Chiasso region's ability to respond in terms of "active space" logic can be seen in box 4. This box shows a substantial loss in the region's ability to manage in relation to the external context, with a break in the milieu effect. The region progressively loses the collective capacity to create specific factors and complementary resources connected to the network strategies. What results, therefore, is a smaller degree of appreciation of the "regional active space" (box 4). It needs to be emphasised that a pro-active policy for evaluating changes in trends might have allowed the region to remain on a positive trajectory and to maintain its gateway functions, in particular through an innovative strategy aiming at rendering functional services to the new needs of integrated logistics. The goal of this operation would have been that of finding other niche products and of substituting specific factors connected with the border, for example, modifying and developing the potentials connected with the capacity to manage the communication, transport and distribution networks.

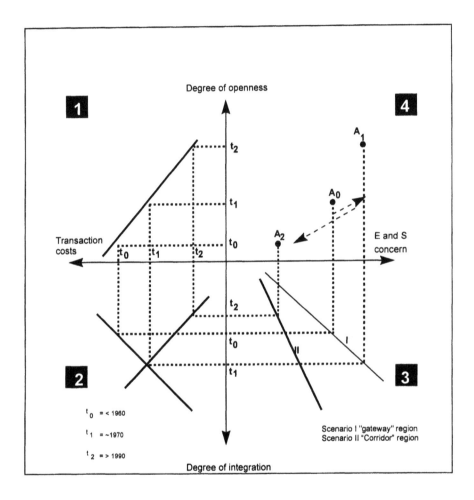

Figure 2.6   Example of an active space trajectory: the case of the Swiss-Italian border region

Source: van Geenhuizen and Ratti, 1998.

General Conclusions

Fundamentally, the concept of "regional active space", created above all as a reaction to the concept of support space, proves to involve a degree of pertinence and generality decidedly greater than what was imagined up to now. Indeed, the contribution of the "active space" concept appears, under various conditions, to be scientifically rich and promising due to the following features:

- The ability to explain an open and dynamic regional reality; in particular thanks to the consideration of a "force field" space.
- The applicability to each region, and not only to textbook cases of industrial districts or innovative milieus, in the sense that each region has at its disposal, actually or potentially, a degree of "active space", be it small or large.
- An integrating role, i.e. it includes the gaining of economic benefits and it includes diverse disciplinary contributions in a common orientation, which makes it a research paradigm.
- Its theoretical and empirical possibilities; on a theoretical level the "active space" tends to put the market, the socio-economic actors and the agencies of regional territoriality in a logical circuit; on a practical level it presents itself as an effective connecting thread for regional development analysis and policies.

Efforts are already underway to measure the degree of openness or to reach representative indicators of the quality of (eco-)systems and their sustainability in environmental and socio-economic terms (e.g. Camagni et al., 1995; and various attempts in this volume). As a next step, the analytical construction of a model and of trajectory curves regarding the measure of the degree of integration of firms (support space box) or of the degree of sustainability (territoriality box) calls for further research. We have already proposed a theoretical framework which appeared to be very useful in empirical studies within the study of the border-barrier in some initial attempts (Ratti and Reichman, 1993). In our recent modelling attempt we pose the problem concerning preference for support strategies, such as alliances and agreements, in terms of transaction costs, control costs and creativity potentially leading to technological innovation and preference for integration.

Finally, it is precisely the formalisation of territoriality, with numerous non-economic variables, that poses the greatest problems in view of the construction of a curve indicating a territory's development trajectory; however, despite the degrees of freedom in the construction of these curves, it is easy to demonstrate the interest in the logical formalisation applied to empirical cases (van Geenhuizen and Ratti, 1998). To conclude, the RAS seems to offer a new and promising approach in the analysis and design of policies for regional dynamics. In addition, it integrates numerous partial concepts, such of those of functional support space and territoriality, and presents itself as a possible answer, from regional scientists, to the local-global debate.

References

Amin, A. and Thrift, N. (1993), 'Globalization, Institutional Thickness and Local Prospects', *Revue d'Economie Régionale et Urbaine*, 3, pp. 405-427.

Anderson, J. (1992), *The Territorial Imperative - Pluralism, Corporativism and Economic Crisis*, Cambridge University Press, Cambridge, pp. 34-35.

Bagnasco, A. (1988), *La costruzione sociale del mercato*, Bologna, Il Mulino.

Benko, G. (1995), 'Les théories du développement local', *Sciences Humaines, Numéro spécial Régions et mondialisation*, Paris, Février/Mars.

Benko, G. and Lipietz , A. (eds) (1992), *Les régions qui gagnent*, PUF, Paris.

Blümle, E. and Schauer, R. (1995), *Non profit-Organisationen - Dritte Kraft zwischen Markt und Staat*, Universitätsverlag, Linz.

Bramanti, A. and Senn, L. (1997), 'Understanding Structural Changes and Laws of Motion of Milieux: A Study on North-Western Lombardy', in R. Ratti, A. Bramanti and R. Gordon (eds), *The Dynamics of Innovative Regions. The GREMI Approach*, Avebury, Aldershot, pp. 47-73.

Braudel (1985*), La dynamique du capitalisme*, Arthaud, Paris.

Camagni, R.(ed) (1991), *Innovation Networks. Spatial Perspectives*. Belhaven Press, London.

Camagni, R., Capello, R., and Nijkamp, P. (1995*), 'Sustainable City Policy: Economic, Environmental, Technological'*. Paper presented at the International Urban Habitat Conference, Delft, 15-17 February 1995.

Coase, R.H. (1937), The nature of firm, *Economica*, 4, pp. 386-405.

Gaudard, G. (1987), 'Regional Economic Development and the Future of Environment', in G. Pillet and T. Murota (eds), *Environmental Economics*, Genève.

Geenhuizen, M. van, and Ratti, R. (1998), 'Managing Openness in Transport and Regional Development: An Active Space Approach', in K. Button, P. Nijkamp and H. Priemus (eds), *Transport Networks in Europe. Concepts, Analysis and Policies*, Edward Elgar, Cheltenham, pp. 84-102.

Grabher, G. (1993), *The Embedded Firm. On the socio-economics of industrial networks*, Routledge, London.

Granovetter, M. (1985), 'Economic action and social structure: the problem of embeddedness', *American Journal of Sociology*, vol. 91 (3), pp. 481-510.

Lipietz, A. (1992), 'Le local et le global: personnalité régionale ou interrégionalité?' in C. Courlet (ed), *Industrie et territoire. Les systemes productifs localisés*, IREPD Actes du Colloque Industrie et Territoire, Grenoble.

Maillat, D., Quévit, M. and Senn, L. (1993), *Réseaux d'innovation et milieux innovateurs: un pari pour le développement régional*, GREMI/EDES, Neuchâtel.

Mlinar, Z. (1992), *Globalisation and Territorial Identities*, Aldershot, Avebury, pp. 15-35.

North, D.C. (1990), *Institutions, Institutional Change and Economic Performance*, Cambridge UK.

Parri, L. (1995), *'Le istituzioni nelle scienze economico-sociali: una Torre di Babele concettuale'*, Working Paper DSS Papers SOL 3-95, Brescia.

Pecqueur, B. (1993), 'Territoire, Territorialité et Développement', in *IREPD Actes du Colloque Industrie et Territoire*, Grenoble.

Perriard, M. (1994), *Infomotion et dynamique spatialisée de l'économie*, Editions Universitaires Fribourg, Suisse, Fribourg.

Perriard, M. (1995), *Toward a Measure of Globalisation*, UCSA/CRESUF, Working paper, Santa Cruz/Fribourg.

Perrin, J.P. (1992), 'Pour une révision de la science régionale. L'approche par les milieux', *Revue Canadienne des sciences régionales*, vol. XV (2), pp. 155-199.

Raffestin, C. (1986), 'Territorialité: concept ou paradigme de la géographie sociale?', *Geographica Helvetica*, 2.1, pp. 91-96.

Rallet, A. (1988), 'La région et l'analyse économique contemporaine', *Revue d'économie régionale et urbaine*, 3, pp. 365-380.

Ratti, R. (1980), *Investimento pubblico ed effetti economico-spaziali*, Editions Universitaires, Fribourg.

Ratti, R. (1991), 'Small and medium-size enterprises, local synergies and spatial cycles of innovation', in R. Camagni (ed), *Innovation Networks, Spatial Perspectives*, Belhaven Press, London, pp.71-88.

Ratti, R. (1992), *Innovation technologique et développement régional*, IRE/Méta Editions, Bellinzona.

Ratti, R. (1995), *Leggere la Svizzera. Saggio politico economico sul divenire del modello elvetico*, ISPI/Casagrande Editore, Milano/Lugano.

Ratti, R. (1997), 'L'espace régionale actif: une réponse paradigmatique des régionalistes au débat local-global', *Revue d'économie régionale et urbaine*, 4, pp. 525-544.

Ratti, R. and D'Ambrogio, F. (1992), 'Processus d'innovation et intégration locale dans une zone périphérique', in D.Maillat and J.C. Perrin (eds), *Entreprises innovatrices et développement territorial*, GREMI/EDES, Neuchâtel.

Ratti, R. and S. Reichman (1993), *Theory and Practice of Transborder Cooperation*. Helbing & Lichtenhahn, Basel.

Ratti, R. and Alberton, S. (1997), 'Structural Trajectories of InnovativeMilieux: The Case of the Electronic Sector in Ticino', in R. Ratti, A. Bramanti and R. Gordon (eds), *The Dynamics of Innovative Regions, The GREMI Approach*, Ashgate, Aldershot, pp. 75-109.

Ratti, R., Bramanti, A. and Gordon, R. (eds) (1997), *The Dynamics of Innovative Regions, The GREMI Approach*, Ashgate, Aldershot.

Reich, R. (1993), *L'économie mondialisée*, Paris, Dunod.

Storper, M. and Harrison, B. (1992), 'Flexibilité, hiérarchie et développment régional: les changements de structure des systèmes productifs industriels et leurs nouveaux modes de gouvernance dans les années 1990', in G. Benko and A. Lipietz (eds) (1992), *Les régions qui gagnent*, PUF, Paris.

Storper, M. and Scott, A.D. (1993), *The Wealth of Regions: Market Forces and Policy Imperatives in Local and Global Context*, Lewis Center For Regional Policy Studies, Working Paper N° 7, UCLA, Los Angeles.

Storper, M. and Walker, R. (1989), *The Capitalist Imperative. Territory, Technology and Industrial Growth*, Blackwell, Oxford.

Wallerstein, I. (1974), *The Modern World-System. Capitalist Agriculture and the Origin of the European World-Economy in the Sixteenth Century*, Academic Press, New York.

Williamson, O. E. (1985), *The Economic Institutions of Capitalism,* Free Press, New York.

Williamson, O. E. (1991), 'Economic Institutions: Spontaneous and Intentional Governance', *Journal of Law, Economics and Organisation*, 7 (2), pp. 159-187.

3 The Importance of Being Active. A Comment on the Active Space Approach from an Austrian Perspective

MICHAEL STEINER

Introduction

The nineteen-eighties and nineteen-nineties marked, as an almost paradoxical result of an enforced internationalisation of the (European) economy, the re-emergence of regions as economic entities in their own right and as promoting agents of economic development. For a long while regions were, economically and politically, considered to be irrelevant entities, just better or worse off parts of eminent national economies. Regions were regarded as "passive" subdivisions of the respective national economies. This, theoretical, negligence of regions as economies sui generis was to an important degree due to the dominance of macroeconomics in the nineteen-sixties and nineteen-seventies. With the re-emergence of a microeconomic perspective many of the structural problems of regions found a new interpretation.

There were other reasons supporting this new perspective: profound changes in the national and, as a consequence of new dimensions of global communication and networks, in the international economy, new technological developments, and new needs for political, economic and cultural identity on different levels. New identities bring about new challenges and responsibilities: regions can no longer rely exclusively on national policies, economic competition becomes increasingly a competition between regions; and this competition takes up new forms and new dimensions.

Regional differences in competitiveness, their causes and forms then become a predominant concern. Why nations and regions differ in their development, and wealth, is of course a standard question in economics and regional science. One of the standard answers is to point to the different structures and resources of the respective economies. In recent times these answers have become more and more differentiated; emphasis has been laid upon special features of regional resources: locational quality, knowledge and/or use of technology, and concomitant innovative activities, co-operation within and by means of networks and clusters. Pointing to these features the interpretation of differences in regional potentials became more "active" and, at the same time, more "complex": there are no one-dimensional factors that influence regional development, the "active space" has to be interpreted as a "force field" (van Geenhuizen and Ratti, 1998; Ratti, in this volume). This force field is to be regarded as '...a coherent game of initiatives and responses that the actors of a given territorial space are together able to produce within the space and towards the outside' (Ratti, in this volume, p. 23). It is characterised by three aspects: openness, concern for sustainability and the creative ability of a specific territorial system.

The output of this force field depends on the regional actors' ability to generate a mix of cohesion, innovation and strategic behaviour in a systemic-evolutionary context. The possible combinations of these factors account for the fact that numerous local models can coexist in a single global area. The "degree of active space" depends on managerial ability and regional creativity.

In this comment managerial and creative ability and its importance for "territoriality", interpreted as the local system's ability to respond '... internally and externally, to changes' (Ratti, this volume p. 25), will be underlined from an evolutionary "Austrian" perspective and special reference will be made to the necessity of variety and firm size as two of the potential indicators and driving forces of the logic of "active regional space". With this comment I want to make suggestions for some of the open questions of an "active space" approach pointed at by Ratti, namely the necessity of model construction and research on indicators adequate for the multiple observations and information that the topic of "active space" calls for.

Extended Concept of the Market

One of the theoretical roots of an "active space approach" can be found in the emphasis of the Austrian school of thought on the creative function of the market: the market is an instrument that transfers incentives for economic changes. Hayek, especially, underlined the explorative potential of the market, its capabilities do not rest so much in the allocation but in its flexibility to adjust to new situations. The market is an optimal process of exploration, it is a development process. In the same sense, Schumpeter always regarded the market as a process and not as a state: economic competition is not price-taking behaviour under conditions of perfect competition, but a process with winners and losers. The procedural character of the market brings about, by its creative destruction, innovations. These innovations are not only of a technical kind, but also have an organisational character, thus opening new markets. Compared to these dynamic aspects the pure state and allocative advantages of the market are loosing importance. In this vein Streißler (1980: 8) noted: 'The market is predominantly an optimal process of exploitation, beyond this a process of development, a process of change of data'.

The creative function of the entrepreneur who has to create markets, and who does not exclusively react to a given set of data, and who has to master production under conditions of uncertainty, is at the centre of this approach (Casson, 1982). This personal element is a dominant feature of the market economy: the entrepreneur is the one who explores hitherto unknown opportunities and who realises them, who creates innovations connected with entrepreneurial risk and hazard. He or she is the one who explores future needs, the one whose informative function is of importance especially in the case of "thin" markets, where the number of participants in the market is too small to warrant a fairly stable intertemporal development of prices. The emphasis of the allocative function obstructs the understanding of the importance of the entrepreneur and of questions concerning the person of the entrepreneur, the variety in his or her behaviour, and the size of firms.

Extended to questions of regional development and of locational behaviour of entrepreneurs, this creative dimension leads to forms of behaviour surpassing an allocative determinism. Entrepreneurs are "indecisive" with regard to their location, they are "market-creating", they can themselves induce change and are not restrained by locational costs and static allocative efficiency. Hence, approaches relying on static principles of regional distribution of economic activity have to be replaced by evaluations of regional potentials in the framework of the creative function of the market (Steiner, 1990a, 1990b). This process-oriented perspective gives special importance to the exploratory and experimental role of small and young firms within regions; differential behaviour and variety are essential elements of the dynamic efficiency of markets and for the development of regions. Or in the language of an "active space approach", a simple mainly passive support space has to be replaced by territorial space as actor of development (Ratti, this volume).

Traditional Neglect of Firm Behaviour in Economic Theory

The typical firm in traditional (micro)economic theory is without dimensions and exhibits "representative" behaviour, a behaviour which is determined by the conditions firms are facing. What is going on inside the firm, how decisions are made, how they correspond to the outside is rarely considered, or only in a very restricted sense; in economic theory therefore, the firm is usually a "black box", furthermore, a box whose size is hardly paid attention to. Recent developments have qualified this picture of the firm in traditional economic theory as follows:

-   In industrial economics the dominating structure-conduct-performance paradigm with a strong deterministic influence of structure on conduct has been replaced by an approach that stresses the indeterminacy of the elements of structure and postulates the opposite causality: structure follows strategy (for surveys on recent developments, see Schmalensee and Willig, 1989; Fisher, 1991; Peltzman, 1991; Braulke, 1992; Schwalbach, 1994). This also changed the interpretation of (industrial) structural influences on the growth of regions (Steiner, 1990a).
-   The theory of the firm, since Coase's seminal contribution in 1937 a long neglected subject, has been extended by game theoretical approaches, by principal-agent aspects, by transaction cost theory and capital market and financial considerations (Kay, 1991). In a regional context this has led to the emphasis of transactional approaches of regional policy (Cappellin, 1992).
-   A broad literature on strategic management (Rumelt et al., 1991), of competitive advantages of firms within a regional and national context (Porter, 1980, 1990), of business strategies (Chandler, 1966, 1990) emphasised the importance and diversity of firm strategies.

- Since Birch (1979), new attention has been paid to the importance of firm size and to the role played by small and medium sized firms for employment and economic growth. This aspect was soon taken up by regional economics pointing to the role of small and medium sized enterprises for regional development (Keeble and Wever, 1986).

Game, not Players

There are two reasons for this past neglect of economic theory. The first resides in the specific interest: in economics the firm is a player in a game with many actors; the interest is in the game and in its outcome, not so much in the specific play or in the result of the play for single actors (Nelson, 1991). It is not Fiat, but the car industry, not Olivetti, but the computer industry that is of interest to the economist. Or to put it in the interpretative framework of A. Marshall: economists are interested in the wood, not in the single tree. The second reason is the basic theoretical orientation. Economists start with the question: what is the essence of economic activity? The basic answer: the optimal allocation of resources with given preferences and given technology, the allocation of given resources to preferences. This implies a specific decision theory of the firm (Nelson and Winter, 1982): firms are faced with a given and known set of possibilities and therefore have no difficulties in setting those actions that are best for them. Thus their activities, within a given context, are to a large degree predetermined. This does not mean that these activities may not be different: what a firm does is determined by specific constellations, e.g. market structures and possibly also by a few unique characteristics, such as a privileged access to new technology and locational circumstances. However, as soon as these structural determinants are altered, the behaviour is also altered. There are differences in behaviour, but they do not have an essential autonomous quality (Nelson, 1991). Or to slightly change Marshall: tree follows wood.

It is from this principal theoretical orientation that the dominating understanding of a market economy results: the economy as a set of markets. It is marked by the "law of demand and supply", by the idea that the reactions of supply and demand on many simultaneously existing markets imply a process leading to equilibrium, that these reactions lead to market clearing. The focus is on the resulting allocative efficiency; the dominating question is that of whether the market mechanism is sufficient to secure this efficiency. Economic theory has, both on a macro and on a micro level, pursued this problem with a high level of subtlety and with high degree of mathematical elegance. In this context, the variety of firms is a determinant of the (market)structure. The size of firms is an element of market structure, impeding allocative efficiency via oligopoly and monopoly, and an element of technical relations leading to possible economies of scale.

Static versus Dynamic Efficiency

For certain questions and problems, this perspective of a market economy is useful and adequate; but it cannot explain sufficiently economic and structural change, it only accounts for a part of economic activity. We need a different interpretative framework than the one offered by traditional economic theory to be able to evaluate the phenomenon of (regional) change, of technological progress and the role which firms play in that process. A market economy is more than a set of markets working towards a balance of supply and demand. More important is the idea of the market as a process that pushes change. The important aspect of a market economy is its ability to initiate change from within. This innovative ability is the essential element creating growth and welfare. A market economy is about dynamic rather than static efficiency; it is the ability of a market economy endogenously to create change.

These ideas based on Schumpeter and Hayek have led to an "evolutionary" perspective of economic change (see Nelson and Winter, 1982; Witt, 1993; Nelson, 1995). The argument for our context basically runs as follows (Metcalfe, 1989): economic change is driven by the variety of economic results between competing, alternative possibilities to fulfil needs. On the other hand, variety of economic results depends on the variety of technical and organisational forms. Innovations introduce new varieties; yet imitation and competition consume variety so that economic progress and economic change depend on the balance of these two factors. Variety and diversity are therefore the main forces of economic progress in the context of a competition oriented market economy. This evolutionary perspective focuses, as an essential explanatory variable, on the behaviour of economic agents. The behaviour of firms, and that of consumers and people working within firms, is not only different, it is also subject to change. It is subject to the phenomenon of "learning by doing" and, as such, changes in the course of being exercised. Diversity here, in contrast to the above mentioned perspective, is possible and prevalent under identical circumstances: different reactions of firms happen in the same context.

Variety versus Representativity

Traditional theory points to "representative" behaviour; from an evolutionary perspective however differences are the object of interest. The principal question is not what firms have in common but why they behave differently. Differences in the sense of discretionary chosen behaviour imply that there are obviously no fully binding limits to behavioural options, both in the short and in the long term, so that firms with different conducts or with differences in some decisive points are able to survive within the same economic environment. These differences are the result of different strategies influencing the decisions on different levels of the firm. The question of whether representative behaviour, or instead differences in behaviour, are the focus of analysis, signifies an essential shift with far reaching consequences. As long as firms are essentially regarded as similar and identifiable by a few characteristics, all variation around the ideal is seen as accidental "outliers".

From an evolutionary perspective, these "outliers" are the typical elements. It is the variety in the system which is the driving force of the economic process. Only because of variety is selection possible, selection is the mechanism through which some of the firms are better able to resist the pressure of the economic environment and the pressure of competition. In the context of technological competition this means that the economic environment is the basis for selecting competing technologies; this environment creates criteria which enable the direct evaluation of rival products and processes and thus indirectly an evaluation of producing firms.

Trial and Error

From an evolutionary perspective, entrepreneurial behaviour is marked by search processes under uncertainty. Economic behaviour therefore becomes essentially experimental behaviour: the activities of economic agents are a process of trial and error. The essential element of experimental behaviour is to behave differently. There is a great variety of possible modes of conduct. It would therefore be wrong to misunderstand this evolutionary perspective as a social Darwinism, as a "survival of the fittest"; it is rather a "non-survival of the unfit". This allows for differences, also in a regional interpretation: there is no unique form of being "fittest", both firms and regions may develop different strategies to survive. This perspective also does not correspond to the biological concept of evolution: in the field of biology an experiment represents blind variation, in the field of economics this would not be sufficient. Here variation is (mostly) intended and guided; it is an intended and guided process of attempts to anticipate the future.

Variety and Size

The above processes are connected to the aspect of size as a further essential element. Imitations destroy diversity, they consume variety. Therefore a mechanism is needed which is able to generate variety. Here small firms have a big advantage, they are an instrument against the inertia of the big ones, against too stable forms of behaviour. Small firms also have a much more rapid feedback mechanism than big ones. For these reasons small firms play an essential role in the generation of innovation within an economy.

The experimental ability of an economy depends on how decentralised and competitively organised it is. Here lies the importance of small and decentralised units. The experimental ability is the essential advantage of a market economy that is its biggest strength. The more differentiated the forms of behaviour, the smaller the units of behaviour, the more it is possible to pursue experiments with small volume and therefore with low costs and little resources. Successful experiments bring high rewards. This is the principal incentive to undertake experiments. At the same time this reward is a signal for others to do the same, either to imitate or to innovate themselves; and with a dominance of small firms, inadequate experiments, and these are more numerous than successful ones, can be stopped with limited costs and a limited loss of resources.

Active Behaviour and the Culture of Regions

There are no final conclusions to be drawn from these commentary remarks. Just one final remark: an "active space approach" may lead to a better understanding of the "global-local" debate. It points to the fact that economic activity does not only happen in a virtual process of globalisation, it also happens in real regions and in real space. This space is more than a passive allocative factor, it also possesses a creative function. As a force field, it has a further reaching influence on the decisions of economic agents: it conditions the form of their behaviour. An "active space" as a proposal for a paradigmatic approach then stands in the tradition of institutional economics which has always emphasised that economic action is neither independent of time nor of space. 'For the institutional economist behaviour is cultural behaviour. Economic behaviour takes place within a cultural milieu and is, in fact, a part of that cultural milieu. In the work of the institutionalist this is of extremely great importance; for he not only explains individual behaviour in terms of culture, but he is concerned with economic behaviour in general as an aspect of culture. In other words, economic behaviour patterns are those aspects of cultural behaviour that are concerned with earning a living.' (Hamilton, 1953: 81).

In an "active space approach" regions stand for different environments which shape the behaviour of economic agents. They also stand for a "cultural milieu".

References

Birch, D. (1979), *The Job Generation Process*, MIT, Cambridge.

Braulke, M. (1992), *Quo Vadis Industrieökonomik? Stand und Entwicklungsperspektiven der Industrieökonomik*, Ifo-Studien - Zeitschrift für empirische Wirtschaftsforschung 38, pp. 255-270.

Cappellin, R. (1992), *Patterns and Policies of Regional Economic Development and the Cohesion among the Regions of the European Community*, Lisbon University, Finisterre.

Casson, M. (1982), *The Entrepreneur. An Economic Theory*, Robertson, Oxford.

Chandler, A. (1966), *Strategy and Structure*, Anchor Books Edition, New York.

Chandler, A. (1990), *Scale and Scope: The Dynamics of Industrial Capitalism*, Harvard Business School Press, Cambridge MA.

Fisher, F. M. (1991), *Organizing Industrial Organization: Reflections on the Handbook of Industrial Organization*, Brookings Papers: Microeconomics, pp. 201-225.

Geenhuizen, M. van, and Ratti, R. (1998), 'Managing Openness in Transport and Regional Development: An Active Space Approach', in K. Button, P. Nijkamp and H. Priemus (eds), *Transport Networks in Europe. Concepts, Analysis and Policies*, Edward Elgar, Cheltenham UK, pp. 84-102.

Hamilton, D. (1953), *Newtonian Classicism and Darwin Institutionalism - A Study of Change in Economic Theory*. University of New Mexico Press, Albuquerque.

Kay, J. (1991), 'Economics and Business', *Economic Journal*, vol. 101, pp. 57-63.

Keeble, D. and Wever, E. (eds) (1986), *New Firms and Regional Development in Europe*, Croom Helm, London.

Metcalfe, S. (1989), 'Evolution and Economic Change', in: A. Silberston (ed.), *Technology and Economic Progress*, Macmillan Press, London, pp. 54-85.

Nelson, R. (1991), 'Why do Firms Differ, and how Does it Matter?', *Strategic Management Journal*, vol. 12, pp. 61-74.

Nelson, R. (1995), 'Recent Evolutionary Theorizing about Economic Change', *Journal of Economic Literature*, vol. 33, pp. 48-90.

Nelson, R. and Winter, S. (1982), *An Evolutionary Theory of Economic Change*, Harvard University Press, Cambridge MA.

Peltzman, S. (1991), 'The Handbook of Industrial Organization: A Review Article', *Journal of Political Economy*, vol. 99, pp. 201-17.

Porter, M. E. (1980), *Competitive Strategy: Techniques for Analyzing Industries and Competitors*, The Free Press, New York.

Porter, M. E. (1990), *The Competitive Advantage of Nations*, The Free Press, New York.

Ratti, R. (1996), *The Active Space: A Regional Scientist's Paradigmatic Answer to the Local-global Debate*. Paper presented at the 36th European Congress of the Regional Science Association, Zurich, August 26-30.

Rumelt, R., Schendel, D. and Teece, D. (1991), 'Strategic Management and Economics', *Strategic Management Journal*, vol. 12, pp. 5-29.

Schmalensee, R. and Willig, R. (eds) (1989), *Handbook of Industrial Organization*, 2 volumes, North-Holland, Amsterdam.

Schwalbach, J. (1994), 'Stand und Entwicklung der Industrieökonomik', in M. Neumann (ed), *Schriften des Vereins für Socialpolitik*, Bd. 233, Duncker & Humblot, Berlin, pp. 93-109.

Steiner, M. (1990a), *Regionale Ungleichheit*, Böhlau, Wien/Köln.

Steiner, M. (1990b), 'How Different are Regions? An Evolutionary Approach to Regional Inequality', in K. Peschel (ed), *Infrastructure and the Space Economy. Essays in Honour of Rolf Funck*, Springer, Berlin, pp. 294-316.

Streißler, E. (1980), 'Kritik des neoklassischen Gleichgewichtsansatzes als Rechtfertigung marktwirtschaftlicher Ordnungen', in E. Streißler and C. Watrin (eds), *Zur Theorie marktwirtschaftlicher Ordnungen*, Mohr/Siebeck, Tübingen, pp. 89-112.

Witt, U. (ed) (1993), *Evolutionary Economics*, Edward Elgar, Aldershot.

4 Governance for Territorial Development: What is It About?

ALBERTO BRAMANTI

This chapter contains an exploration of a particular dimension of territorial competitiveness and firms' performance, i.e. the governance structure of local production systems. Governance is addressed on a threefold basis. Under which circumstances does governance structure make the territorial system perform better? What is the role of government within a territorial-specific governance structure? To what extent is the "institutionalisation" of loose mechanisms of interaction appropriate? The answers offered follow a four step articulation: (i) a distinction is made between local government and governance, the first being an "ingredient" of the latter, in order to reinforce and motivate the role of governance within the "active space" approach; (ii) we reflect briefly on the multidimensional space of application of the concept of governance; (iii) the role of "local" governance in "global" competition is highlighted; and, finally (iv) a research agenda is presented on the theme considered as a fundamental building block of territorial competitiveness and sustainable development of spatial aggregates of actors.

Introduction

It is widely accepted that Fordism is no longer the dominant paradigm of socio-economic co-ordination. The "back-to-the-past" visions are becoming increasingly isolated even if the "old regime" has far from completely disappeared (Amin and Thrift, 1994; Scott, 1998) and the new one is, up to now, simply an emerging paradigm (Storper, 1997; Storper and Salais, 1997).

Instead of the clear-cut "one best way" of production and organisation of production systems, a multitude of new dynamic elements has come to light in the context of territorial systems of production and innovation (Ratti et al., 1997), giving origin to more loosely defined clusters of inter-firm and inter-actor relationships (Eber, 1997). These webs of links are frequently "heterarchical"; where heterarchy (Cooke and Morgan, 1998) is the condition in which network relationships pertain, based on trust, reputation, custom, reciprocity, reliability, openness to learning and sustainability (Brunetta and Tronti, 1995; Fukuyama, 1995; Mutti, 1998).

Concepts that invoke some form of collaboration are now acceptable in economics; indeed they have become highly fashionable. Economic development cannot be successfully achieved or sustained without co-operation between actors. Hence, the dimension of networking may be interpreted as a relational system of shared knowledge in which interdependence is "a weak point turning into a virtue".

Networks and networking, as main representations of the relational approach, imply complex and lasting relations based on trust and reciprocal communication and exchange structures. This prompted us to investigate a particular dimension of territorial competitiveness[1] and performance: the *governance structure* of territorial systems, the mechanism articulation of which directs the modelling and evolving of the actors' interrelations on a territorial basis.

Governance, throughout the chapter, is not regarded as a universally applicable blueprint for territorial success, but it certainly, in addressing the "relational" dimension of development, plays a central role in contemporary theory. Within a territorial perspective the relational approach has important impacts on organisational models. The actors' embeddedness and their behaviour develop and create a social environment, a sort of "cultural software" that is continuously fed and modified in a continuous process of interaction; but a second consequence seems to be even more radical: the diffusion of networks reshapes traditional territorial identities[2] based on

---

1 Competitiveness here will refer to the ability of an economy to hold stable or increasing markets shares — while sustaining stable or increasing standard of living for those whom participate in it — *jointly with* maintaining or expanding employment in a qualitatively satisfactory way.

2 Territorial identity is defined by two criteria: differentiation and continuity. Whatever differentiates a territory from others and leaves it unchanged across time creates identity; but territory is also a significant source of symbolic and resource power (Mlinar, 1992).

contiguity, homogeneity and clearly identifiable borders. The transformation under way contributes to the metaphorical shift from identity as an island towards an *identity as a cross-road* (Mlinar, 1992), with the discovery and development of trans-national historical, social, and cultural identities.

Notwithstanding the dramatic increase of "globalism" within the economy, which basically means that there is nothing inherent in transactions that makes geographical proximity necessary, such a trend is not correlated with a corresponding decline of "localism" (Veltz, 1996; Taylor and Conti, 1997). We are escaping from the unproductive "either/or" controversy, which denies the commonality and economic interdependence of the over-mentioned dyad "local/global". The two concepts blur and overlap as part of a single continuous spectrum but provide useful benchmarks for noting that, many times, for both individuals and firms, small-scale territorial community is the most important level of rootedness and the departure point for innovative processes and competitiveness.

So, the territorial milieu is not just a locus for delivery of services, it develops conventions and relations, associated with a specific production system, which, in turn, may affect the long-term evolution of technologies and organisations in that region. At the same time, it provides the level of government closest to the citizen and has a role in representing the concerns and views of the locality.

We experience, in fact, a double shift from the nation state to supernational institutional arrangements on the one side, and from the nation state to re-emerging sub-national, and some times even international, regional economies on the other side. Many questions remain unanswered but these two opposite movements suggest a much more complex adjustment mechanism than the simple market exchanges (Benco and Strohmayer, 1997; Braczyk et al., 1998).

Governance: a Far-away Coming Story

Governance, as a philosophical concept, and governance structures, as empirically and historically correlated objects, are not easily definable. We simply opt for looking at the real world to see that a panoply of collective actors, i.e. research and higher education institutes, private R and D laboratories, technology transfer agencies, chambers of commerce, business associations, vocational training organisations, relevant government agencies and appropriate government departments, constitute the basis for an integrative governance arrangement (Danson, 1997). Associative governance, typified by clubs, forum, working parties, consortia and partnership models, involves something like a shift from a state regulation of economic affairs to a degree of self-regulation by the economically and socially qualified group (Cooke and Morgan, 1998).

Governance, as a way of co-ordinated self-interest and micro-economic choices, of resolving disputes, of assigning property rights, of distributing power among community's members, of enforcing entitlements, of assuring social embeddedness, of pursuing shared visions of development, of producing relational public goods, of strengthening trust, confidence and participation, is the object of the present chapter.

In the following sections we aim at differentiating government, and particularly local government, from governance, the first being an "ingredient" of the latter, in order to stress and motivate the role of governance within the "active space" approach; at briefly reflecting on the multidimensional space of application of the concept of governance; at highlighting the role of "local" governance in "global" competition; and, finally, at pointing out a research agenda on the theme as we consider it a fundamental "building block" of territorial competitiveness and sustainable development of spatial aggregates of actors. Before placing governance within the "active space approach" we have chosen to open a small historical view, i.e. on the Hanseatic league, as we consider that experience to be a valuable witness of what governance is, of what are the strong points as well as the problematic weaknesses.

The discrete glamour of the history

An original socio-economic formation was founded and consolidated during the XII to XVII centuries: the Hanseatic league. It was a sea trade society, developed around the Baltic sea, and functioned in an original manner by combining a mix of opportunism, self-interest and reciprocal trust in a way that guaranteed the fulfilment of important economic objectives (Dollinger, 1964; Pichierri, 1997).

The Hanseatic league was a peculiar actor functioning with the economic purpose of strengthening and defending the interests of sea merchants belonging to it. Notwithstanding some quasi-structural features, the Hanseatic league undersigned treaties as well as it declared war. The Hanseatic league maintained loose characteristics: non-contiguity among members, boundlessness, and loose membership with a fluctuating number of associates[3]. The Hanseatic league developed a common culture and language, "Nederdeutsch", which constituted a prerequisite and a consequence of its collective identity. Moreover, the league acted outward as a collective actor, while it behaved internally as a network where the single nodes showed relationships differentiated by nature, intensity and duration. In this economic model, transport and communication are no longer a simple service, they represent an essential component of the production way of the Hanseatic league. Thus logistic excellence is a fundamental departure point for an outstanding economic performance (Pichierri, 1998). If these are the main distinctive features of this original model, it is stimulating to confront the experience of the historical Hanseatic league with the present reflection of governance in contemporary territorial systems. What are the teaching and the legacy of that experience?

The most striking aspect of the Hanseatic league is that it is properly an "interests community", or, as we use to define today, a network, which developed an organisational device implying common decisions, up to the point where the "interests community" became a collective actor. Furthermore, the single nodes belonging to the net remained autonomous and developed a "co-operation among egoists" where heterarchical linkages, the exit option and learning by monitoring, allowed control of transactions giving rise to *ad hoc* arrangements instead of permanent linkages.

The most interesting conclusions that can be drawn from the successful historical co-operation of the Hanseatic league seem to be the following:

-   The degree of embeddedness was high in the league but it did not evolve towards the institutionalisation of routines, norms, codes and rules. A relatively low degree of institutionalisation preserved the system against lock-in mechanisms and "petrification" of the networking relations.
-   The organisation of internal relationships was strongly "outward oriented": the openness of the league is not open to discussion, as witnessed by the joint presence of Protestant and Catholic towns within the league. Thus, there was a strong preference towards bargaining, compromise and co-operation instead of simply, but possibly costly, "fighting against adversaries".

---

3 During the time of the Hanseatic league, 500 years, some 200 towns belonged to the league but no more than 70 at any one time.

The composition and re-composition of regional aggregates

What can be learnt from the Hanseatic league is that appropriate governance structures allow the formation of "progressive coalitions" within territorial systems, and enable the strengthening of the "active space" towards a sustainable development. We are quite distant today from such a form of flexible and variable "functional organisation", even though, we are attempting to form in Europe a number of loosely coupled organisations aimed at co-ordinating their efforts to reach more complex and interdependent objectives. A great example of such co-ordination can be observed in the bilateral and multilateral associations of regions sharing a common border and belonging to different nation-states. Generally speaking, such a co-operation is motivated by the identification of common problems and/or interests jointly with the conviction that collaborative action is required to address these problems and to solve them.

Trans-national networks of regions seem, thus, to be a tool for those steering functions which regions are asked to implement and manage. This is a typical example in which the institutional agreement seems to be rather poor, due to the considerable differences in the level and structure of competencies between regions, where less institutionalised co-ordination among actors, i.e. shared governance structures[4] may reach suitable results and makes appropriate pressure on the respective national government.

The extent to which it is appropriate to "institutionalise" governance structures, i.e. endow them with procedural routines, standardised mechanisms of solving controversies, and structured organisations with defined operative tasks, is an open question. Best practices around Europe suggest that this is not the appropriate way to respond to rapid changes and flexibility requirements (Wannop, 1995; Whitley and Kristensen, 1997). One visible implication of this experimental regionalism is that advances in regional planning and the most outstanding strategic planning decisions commonly occur outside established systems of government. They often take place at the interstices of these systems and frequently through *ad hoc* initiatives, consultants' reports and often through other transitory sources of advice and policies.

---

4 The Upper Rhine Valley seems to be a positive model for cross-border co-operation. It is characterised by a tightly-knit network bringing together political, social and economic actors able to manage a large number of co-operation projects.

A new set of post-industrial beliefs and values reinforces other pressures to develop new organisational forms. A system of local control and co-ordination made up of any combination of formal governmental agencies, civil associations and organisations, and private-public partnership, in short an appropriate governance structure is then invaluable to provide a many-sided boost to processes of regional economic development and growth (Bonomi and De Rita, 1998). This is the main reason why the next section is devoted to develop the logical shift from govern-*"ment"* to govern-*"nance"*.

From Government to Governance: the Active Space Approach

A preliminary consideration, before deepening the core argument of the section, is devoted to a definition of the term region, which has an important implication for this debate. It is worth taking time to distinguish between regionalisation and *regionalism*.

> 'Regionalization refers to a process whereby national governments or the EU define regional policies for, or impose them on, regions. Something is done for the regions but it is often top-down, centralized and technocratic. Regionalism is an "ism": it refers to an ideology and to political movements, which demand greater control over the affairs of the regional territory by the people residing in that territory. It is essentially bottom-up, decentralized, and political.' (Keating and Loughlin, 1997: 5).

Regionalism has been promoted in Europe as a response to reassertion of cultural and historical identities. We are facing a reconfiguration of political space in which territory continues to feature largely, without as yet a new territorial hierarchy replacing the old. So the regional question is studied largely as a problem of policy formulation and implementation (Humes IV, 1991; Saxenian, 1994; Barnes and Ledebur, 1998; Perulli, 1998). In addition, we have passed from the national embeddedness of economic institutions to their nestedness within a multilevel system (Mlinar, 1992). The most competitive firms or regions are not mirroring the market, they are struggling to manufacture consensus, trust, collective forms of governance, and long term vision.

The nation-state level remains arguably "sandwiched" between regional and supra-national levels, and there is a dramatic need for the emergence of an intermediate level of government between the centre and the basic municipal or communal level (Batley and Stoker, 1991; King and Stoker, 1996; Jeffery, 1997). As a matter of fact most existing forms of local government are not particularly well equipped to deal with the complexities of regional economic governance. New and urgent demands for shareable planning spring from local society and the decision-making mandates. The administrative arrangements are simply not designed to respond effectively to these demands. Thus, local government has to move forward looking instead for a limited concern for better administration and efficient services which remains certainly a must for territorial competitiveness. Greater attention and resources have to be given to how to nurture the local economy to meet the challenge of global competitiveness.

Since the domain of externalities lies, by definition, beyond the discretionary control of any private agent, relevant forms of social co-ordination are essential if the best possible performance of the regional economy is to be secured (Braczyk et al., 1998). Formulating a discourse around govern*ance* rather than govern*ment*, therefore, involves a double leap away from the previous debate:

- from a purely structural concern to include the *process*, and
- from purely governmental focus to a *broader set* of civic, political, voluntary, and related capacity.

Governance is understood as the capacity of political institutions, jointly with civic society, to articulate, through public policies and democratic representation, conflicting trends within the political process. Whereas government refers to sovereignty and political autonomy, governance refers to social mediation and institutional interdependence. Governance engages in the task of recognising needs, in a complex intergovernmental policy-making system, in aggregating and co-ordinating many affected public and private interests. Governance applies to the "meso", from the Greek *mesos* for middle, and describes in abstract terms a "decision space" rather than a level of government.

The emergence of regionalism suggests new ways of governance to tackle sectoral, temporal or micro-macro contradictions (Humphries, 1996).

Thus, the role of government becomes that of catalysing[5] new solutions through society, by integrating the complementary resources of the different actors. The majority of regions have now a range of public and quasi-public authorities and agencies engaged in a wide variety of economic and social development functions.

Regional institutions certainly remain important in defining the issues, in mobilising resources, in providing differential access, in mediating interests and in implementing policy; however, civil society has the ability to promote consensus on development and distribution, and to regulate competition and co-operation. Social cohesion is necessary to bridge individual and collective forms of rationality (Becattini, 1998; Heller et al., 1998).

Governance and "active space": the ebb and flow

Within this cultural background, i.e. the shift from regional*isation* to regional*ism*, it is not difficult to link the passage from government to governance using the 'active space' approach. The "active space" represents a field of stimuli and answers, a blending of social and territorial cohesion, a tense force-field between innovation and strategic behaviour (Ratti, 1997).

Cohesion, innovation and strategic behaviour are three different ingredients of a unique phenomenon: the *collective* process of change (OECD, 1996; Cooke and Morgan, 1998). Change, in fact, is not the outcome of actions undertaken by isolated individuals; different social conditions lead to different reactions to innovation and to the development of what can be defined as an "innovation prone" (or averse) society (Rodríguez-Pose, 1996).

There is a widespread and outstanding literature focusing on the role of social norms, inextricably linked to multilevel institutions, in shaping individual and collective behaviours. Economic behaviour is characterised by competition, and competition is fostered by innovation. Thus, successful innovation is the key to maintain or gain competitive advantages. Innovation is not carried out, however, by individual actors, such as technicians, entrepreneurs, or even firms; it is first and foremost a collective response

---

5 In the literature we find a full panoply of definitions of this special role: facilitator, stimulator, experimental-provider, monitor, co-ordinator, ombudsman, contributor, translator, imitator, an so on.

where technical competencies and problem solving attitudes match dynamically with evolving needs. All that goes hand in hand with a view of technology as an interplay between hardware, software, "organisation-ware", "human-ware" and other types of invisible assets. These invisible assets are rooted in organisational routines, concerning co-ordination and problem-solving, which are frequently a locus of conflict, governance, and a way of codifying microeconomic incentives and constraints.

Therefore, the key for a truly sustainable development requires coherent behaviours, appropriate governance structures, the emerging of progressive coalitions, in one word a strong functioning regional "active space" characterised by its three fundamental dimensions: openness, sustainability and creative capability (van Geenhuizen and Ratti, 1998). One possible weak point may exist even in high-trust, intensely networked territorial systems. As a matter of fact, long-term strategies of collaboration require common behavioural expectations and understanding among actors.

> 'These institutional features are embedded in forms such as institutional memory, insider-outsider practices, and obligations. Organizations such as firms, artisans' associations, trade unions, trade associations, and local chamber of commerce then tend to privilege consensus and denigrate dissonance. The absent of dissent delays "creative destruction" processes and creates a barrier to innovation.' (Cooke and Morgan, 1998: 75).

Consequently, in a number of cases the single "exit" option, i.e. the selection process produced by competitive forces on the basis of the ability to imitate and adopt innovative behaviour, may be transformed into a collective "voice" option, where a strong group adopts a sanction-applying behaviour which obstructs positive evolution: this is the interplay of regressive coalition in innovation averse societies. Conversely, positive, collective behaviour may be strengthened by a specific institutional context. An appropriate regional "active space" may provide adjustment mechanisms and social conditions acting as a social filter, which determines the rhythm at which any society adopts innovation and transforms it into economic activity.

Theoretical and empirical literature on the matter shows that innovative development seems to be easier when the interacting group is large and open. This is only apparently counter-intuitive. Instead, it shows full coherence with Olson's (1982) view on "exclusive" and "inclusive" groups: the more members help in achieving a good the better, *if* the good is a positive-sum good. Thus, it is the nature of the collective good produced by the group that is crucial. Inclusive groups are the best reply in the presence of relational goods, club goods, positional goods, or whatever the label is for those ingredients within innovative localised processes, which are not completely excludable and generate spillovers and externalities.

Governance structures arise to assure the provisions, i.e. collective goods offered by a group to its members, and the entitlements, i.e. the rules governing the right to enter the system, and to use "voice" to change the system (Bianchi and Miller, 1994). In any "unique" social filter, however, innovative and conservative components are combined: weak entrepreneurship, inadequate age structure, lack of appropriate skills in the labour force and social instability and conflict tend to be the causes behind the formation of rigid social filters which prevent the assimilation of innovation.

A Kaleidoscopic Articulation of Governance

It would be beyond the scope of this chapter to try comprehensively to discuss the results and achievements of a surging wave of publications addressing the multilevel articulation of governance (Hooghe, 1996; Bullman, 1997). Anyway, in this section, two different and complementary interpretations of governance arrangement are depicted, with the aim to widen the object of the analysis that needs to be comparative in nature.

We have already addressed the theme of the "European-national-regional-local" puzzle, which seems to be more and more complex as: *a)* individual units at any level in the hierarchy are subject to multiple and overlapping allegiances in their relations to units at the next higher level, and *b)* short-circuits between non-adjacent tiers are growing numerous.

We firstly pay attention to the world of enterprises. "Going it alone" in economic development is not an option likely to achieve success except in small-scale or very specific activities. Given the multifaceted character of global competition, success increasingly depends on the extent to which all actors can be mobilised and their resources and know-how made available. Thus it is decisive to explore the ways in which firms govern those specific and general purposes in inter-organisational networks to which they belong. Secondly, we consider the labour markets' governance, all the more important as human resources are at the very heart of knowledge production, innovation, competitiveness and, consequently, social and economic well-being of citizens.

Inter-organisational network and learning processes

Firms will forge networking relations to govern their access to those resources and capabilities, which will reduce their dependence or, otherwise, improve their competitive position. Inter-organisational networks represent one institutional form of co-ordinating or governing economic exchange relations among actors (Eber, 1997). When, why and how firms engage in inter-organisational networking has been researched by a variety of disciplines, such as economics, sociology, information systems, organisation studies, marketing, business strategy, but the focus pursued hereafter, coherently with the emphasis of the "active space" approach, is on *institutional factors*. These factors involve incentives as well as restrictions, with a strong impact on the networking relationships emerging in a specific institutional context and on actors' behaviour.

The nature of the micro-level ties has been widely studied according to the threefold approach linked to resource flows, mutual expectations, and informational flows among firms. Complementary to this first side of the coin, our interest is devoted to the institutional-level ties, through which actors co-ordinate their relations. On this side we have to distinguish the: *a)* aspect of governance structure, i.e. the distribution of property rights over resources; and *b)* the co-ordination mechanisms employed by actors when allocating these resources.

Property right structure is often contractually fixed, establishing a specific distribution of rights over resources, outcomes, and incentives. The co-ordination mechanism relates to the way in which the existing interdependencies are managed: they govern behaviour by establishing rules of conduct and by providing and structuring information that then guides behaviour (Grabher, 1993; Mutti, 1998). The consequence is that relationships between companies are much "thicker" than is depicted by any market model.

All this has been duly internalised even in large and multinational enterprises; no matter how globalised an enterprise is, and what the parent company's nationality is, multinational corporations are showing a greater sensitivity towards the differences in specific features of individual territorial units. They respond to the world market with a closer attention to the local milieu in the different areas where their units operate. They have learned the lesson: "when in Rome, do as the Romans do".

Firms can preserve and improve their competitiveness by creating, exchanging and using knowledge, both codified and tacit. As codification makes knowledge more easily transferable over space, it gradually becomes available globally, and this process erodes some firm's advantages (Maskell et al., 1998). What is not eroded, however, is the non-tradable/non-codified result of knowledge creation: the embedded tacit knowledge. As a significant part of tacit knowledge is territorially specific[6], and the easiest way to exchange some form of tacit knowledge is to build trust-based relations between firms, a step-wise building of trust-like relations between actors in industrial territorial systems is an important means to sustain knowledge-based competitive advantage. Thus, a business environment that enhances trust will always make an economic difference. Within this context, an appropriate co-ordination mechanism may generate and foster collective learning processes, which are valuable according to the degree of uniqueness and strongly decisive for territorial competitiveness as well as firms' performance.

Work systems governance: a decentralised approach

Work systems are mainly the predictable outcome of both particular kinds of interest groups and of certain institutions (Whitley and Kristensen, 1997). They are constituted by interconnected aspects of task organisation and control, workplace relations between social groups, and employment practices and policies. It is widely accepted that societal institutions affect how work systems are organised and this evidence has helped to generate the concept of "national systems of governance".

As far as workplace relations are complex and systemic phenomena substantially varying among different institutional contexts, we find place even for their regional regulation, as a widespread literature witnesses (Regalia, 1997). However, as skilfully shown by Sabel (1996) in this field we are in the area of experimentation, and we hope, also, of increasing returns, where firms, institutions and associations with which they are allied

---

6 That means territorially embedded, i.e. strongly rooted in the specific social and institutional setting of a place.

can no longer take the efficacy of any feature of their internal organisation and their relations to others for granted.

Even if there is a large variety of specific goals pursued by local authorities, industrial relations and work systems regulation represent a specific field of encounter of political process, administrative and legislative, and the autonomy of interests organisations. In this process the participants strengthen themselves, even if not always in a symmetrical way, and try actively to reach a wide social consensus. From this point of view the outcome of the regional work systems governance may be red as repertoires of instruments and accumulation of "social capital" and relational networks useful in searching for solutions to complex and dynamic problems of "governability".

Social groups involved in production can discover how to play different games from those connected with opportunistic behaviour and free-riderism by learning from experience and continually monitoring previous results. The profusion of regional round tables, customer-supplier conferences, local summit meetings between unions, industrial associations and local government support and "foster the fitting" and suitability of the system to the changing environment, all are offering resources and incentives for the development of interaction models among social parts on a more participatory basis.

In particular, in the work field system, a central practice has recently become widespread: "social dialogue" or "concertation"[7], which under favourable conditions gives rise to a "social covenant"; however, the possibility to reach co-ordinated agreement is not the outcome of accidental or fortuitous circumstances; on the contrary, it is linked with learning which, in turn, is involved in chains of interactions that "endogenously" produce rules that actors agree to honour in the face of future uncertainties.

Axelrod (1984) shows that the best strategy in interaction is "tit-for-tat", whether or not there are underlying gains in spite of possible distributional conflict. Even in this case the main teaching of a multiplicity of national/regional experimental governance work systems seems to be the necessity to maintain a low level of formalisation and dirigism without pretending to institutionalise the relations and strongly diminish the

---

7 'Concertation results from making current commitments involving sacrifices and risks with the prospect of realising future gains jointly with others, but about which distributional conflicts may well arise.' (Storper and Salais, 1997: 294).

autonomy of the single component of the societal institutions and organisations.

Local Governance for Global Competition: the Role of Business Interests

Governance structures interact with natural and human resources, with regional culture, and with formal institutions, in constituting a specific set of regional and local capabilities that contribute to the competitiveness of firms.

Regional cultures facilitate the socialisation of individuals in various ways, and are the bearers of significant informal knowledge effects, which enable them to operate as important repositories of tacit know-how. In this direction collective action is frequently essential to sustain the cultural-cum-economic virtues of specific territories and to safeguard their production and reputation from the negative influence of free-riding behaviours (Immerfall, 1998; Perulli, 1998).

One major role of governance is to anchor regional development policy and planning to public-private partnership, rather than to public authority. The aim of this partnership approach is to create a flexible mechanism of growth, of those public-private partnerships, in order to establish a strategic agenda for change.

The local differentiation of the regulation models can be explained on the basis of the endogenous logic of inter-organisational networks and institutions functioning. Through the interplay of disembedding and re-embedding, a given location changes and its business and professional association, unions and politicians concerned with regional development policies are faced with the task of creating an innovative local environment. In addition, a territorial-specific governance structure has to be built jointly with the right setting of administrative boundaries (Bennet, 1989). As a matter of fact, government carries on a decisive role in maintaining the institutional environment, stabilising the economy, filling the gaps, and leading catalysing co-operation.

Regions can neither utilise an over-night improvised model of governance, nor simply copy some "best practice" experienced around the world. The limits to a voluntary, patent-remedy type approach to institution building are demonstrated by the failure of most Silicon Valley imitations: only few territories, if indeed any at all, are able to integrate innovative forms of organisation into old collective orders. It may be easier to meet the challenge of institutional re-engineering via many small-scale changes than to attempt to implement a long list of major changes.

In the new competitive scenario, governance structures have to trigger, support and foster the production and re-production of tacit knowledge and learning processes which are both inter-organisational and territorially embedded (Scott, 1998). Within governance structures local government still plays a relevant role, aside from the business networks and their representations; but institutional endowment is undoubtedly a constructed part of an area's capabilities. Thus, local actors play a relevant role in building their own future. This is the reason why we now dedicate special attention to a specific declination of governance, i.e. *business interests governance*.

Among enterprises and other "private actors" some specific conventions, i.e. implicit rules of action and co-ordination, generated by cultures, habits, tradition and path-dependency, have emerged with the aim to mobilise economic resources, to organise production systems and factor markets, to enable the rise of specific patterns of decision making processes (Chisholm, 1998).

Between the level of individual action and governmental actions there is an intermediate level of "collective action", businesses acting together, motivated by the willingness to manage problems which individually they can not achieve. The specific forms, which such collective action can take, vary under different problem solving activities and variable institutional/organisational setting. As a matter of fact, partnership for local development can be generated primarily as a local synthesis, although regional and supra-regional level agencies provide both the context of regulatory frameworks, and are often key animators of activities. An important feature of these forms of governance is that, to a large extent, they are self-regulating. Members, both single actors as well as organised business representations, are responsible for developing a vision, mission, and goals and for initiating and managing work activities. At the same time, partnership does not necessarily imply that all actors will play equal roles, nor that absolute and complete consensus[8] is required to realise given projects or activities.

---

8 The very important thing is that no actors could exercise a true right of veto.

Among the most recent initiatives widespread around Europe[9] which are formally and substantially a good example of effective territorial governance, we must mention the *territorial employment pacts*[10] designed to tackle one of the major concerns faced by the large majority of communities of territorial systems: unemployment problems.

The initiative to develop pacts was designed to highlight the centrality of locally based partnership to combat unemployment, and to mobilise all available resources in favour of an integrated strategy for the creation of new employment opportunities. The creation of such "social covenants", at different territorial levels, has shown to be indispensable in achieving effective, useful contacts and collaboration between the different components of the territorial system, and the cornerstone of the covenant must be an awareness of the role assigned to each actor[11].

One of the most decisive gains of the European experiment is that there is no single model of partnership. The very challenge and chance is to ensure the commitment of the players to the common process of local development and job creation. Ultimately, success depends on the existing mutual benefit and added value for each partner with the strong beliefs that acting individually the expected outcome could not otherwise have been achieved.

Conclusion: towards a Research Agenda

In the present chapter we have examined the concept of governance within regional economics addressing the issue to preserve dynamic competitiveness of territorial systems. Governance goes hand in hand with inter-organisational and territorial networks.

A major distinction has been made between regionalisation and regionalism, only the latter being a true course towards the "inflated" but

9 Following a less "centralised" approach we can find extremely interesting "best practices" in the US. One of the successful experiences which has been carefully studied is "The New Baldwin Corridor Coalition" at Steelton, Pennsylvania (Chisholm, 1998).

10 In response to the Commission's proposal, the Florence and Dublin European Council agreed to implement some 60 pacts on a trial basis, with the aim of developing new approaches to deal with the unemployment concern throughout Europe (EC, 1997).

11 Here the public counterpart, political and administrative governments, can no longer continue to dictate rules without confronting with the required increasing flexibility at every level in the economy.

still politically interesting concept of a "Europe of regions". To some extent a parallelism has emerged between this first alternative, i.e. regional-*isation* vs. regional-*ism*, and a second one which is absolutely crucial within the followed "active space" approach, i.e. the distinction between govern-*ment* and govern-*ance*.

Governance has been defined as a way of co-ordinating self-interest and micro-economic choices, of resolving disputes, of assigning property rights, of distributing power among community's members, of enforcing entitlements, of assuring social embeddedness, of pursuing shared visions of development, of producing relational public goods, of strengthening trust, confidence and participation. In this respect, government is a fundamental ingredient of the larger decision-making space of governance. The merit of the governance approach to local development and territorial competitiveness rises from its capacity to shift the focus of the analysis onto "actors". It is an agent oriented approach more than a factors exploring one, with an emphasis put on complex, dynamic, non-linear, externalities producing, relations which are at the very heart of contemporary economics (Maskell et al., 1998).

Even if we have made important progress in the understanding of the links web, feed-backs and path-dependent loops arising in a territory among the different agents, organisations and institutions, we are still a long way from having a full comprehension of mechanisms governing the re-production of the features assuring dynamic competitiveness and sustainable development. In the present context of "non-congruent federalism"[12] (Hooghe, 1996; Danson, 1997; Jeffery, 1997), how can we arrange governance and policy so that good decisions can be accountably made about the future of regional systems?

In the chapter we have addressed three main questions and even if the answers offered are neither entirely original, nor fully exhaustive, we have entered the world of conventions and relations showing their fundamental proximity-inducing effects which give rise to "asset specificity" which, in turn, fosters territorial competitiveness. In addition, the same questions should be regarded as an appropriate index for a new research agenda on a manifold basis.

---

12 Obviously, the variance among Europe is very high, but in a number of cases we have to confront dramatic difficulties in passing from a soft regionalisation process toward a strong regionalism course.

The *first* issue may be summarised as follows: under what circumstances does governance structure make the territorial system perform better? We have stressed that competition and market are not free-floating social phenomena. They are rather shaped and made possible by an underlying framework of institutions and conventions. These are pre-conditions for learning in terms of common regional culturally-based rules of behaviour, language of engagement and collaboration, accepted but tacit codes of conduct between firms, enabling the development of trust, which is in itself essential for innovative collaboration.

A precondition is that of a "truly-boundedness" of territory (Bennet, 1989). The ideal situation is that of an industrial and productive networking which is precisely matched by administrative boundaries. In addition we need a good "division of labour" between government and associative governance, resting on a horizontal rather then a hierarchical organising principle. Furthermore, the emerging of a co-ordinator is decisive, i.e. insightful, charismatic leaders, managed to transcend the frequently narrow-mindedness and selfishness of local oligarchs. Finally, a learning-by-monitoring attitude[13] is very welcome. As outcomes of policies, programs, and planned actions are non-deterministic, due to the complex, interlinked character of territorial development problems, the capacity to monitor outcomes constantly is strongly required; this, in turn, permits rapid changes in activities in order to move the system in a positive direction.

The *second* issue is the type of role of governments within a territorial-specific governance structure. Should it lead the play, or should the private actors maintain the leadership of the process? Here, the focus is on *government* as a fundamental ingredient of governance, playing a role of "primus inter pares". Its own challenge is to establish and maintain two different economic dynamics: the first one is technological trajectory, i.e. enabling an endless learning process, and the second one is the trajectory of conventions. This second point expresses the content of that set of "relational goods", or "soft public goods", which are all the more important to foster learning processes within the system. It is precisely the existence of an element of "publicness" in the creation of territorial assets which enables the local government to be active in territorial development without

---

13 Following Sabel (1996) we call learning-by-monitoring a system of co-ordination because of the way it links evaluation of performance to reassessment of goals.

presuming to know more than the economic agents about how to compete in the global economy.

The third issue addressed inquires the extent to which the "institutionalisation" of loose mechanisms of interaction is appropriate. Are there any risks in this institutionalisation, to preclude future adjustments when required by eroded localised competencies? The answer offered derives from a first generation of trials and errors, where the recurrent regularity is the maximum degree of looseness. Without this perceived flexibility excessive embeddedness may promote a petrification of the supportive tissue and, hence, may pervert networks into cohesive coalition against more radical innovation with the resulting inability to face further change. This is not to deny the complexity of the phenomenon nor to belittle the importance of differences that remain unexplained, but we have incontrovertible examples of strongly institutionalised regions which are facing the challenge of complementing, or perhaps even replacing, their institutionally stabilised and trust-based co-operative relations with new co-operative networks[14].

Finally, once more, we have to stress the role of *social capital* (Putnam, 1993; Brunetta and Tronti, 1995; Mutti, 1998) as a very important ingredient in building appropriate governance: local civic values provide a common set of social rules and social control to regulate behaviour in civil society. Where these civil values are well diffused it is easier to engage in collective action for the achievement of the collective good. Where these values do not exist, the pursuit of the collective good is much more difficult to organise, and individual ends predominate over collective ones.

Social capital is not a simple over-night construction, as is mockingly suggested by the following passage[15]:

> 'How come you got so gorgeous a lawn?' 'Well, the quality of the soil is, I dare say, of the utmost importance.' 'No problem.' 'Furthermore, one does need the finest quality of seed and fertilisers.' 'Big deal.' 'Of course, daily watering and weekly mowing are jolly important.' 'No sweat, just leave it to

14 The most cited case in literature is the development of the Ruhr area, where: 'a high degree of personal cohesion let to limited perception of innovation opportunities and left no room for "bridging relations". In this sense, the strongly embedded regional networks insidiously turned from ties that bind into ties that blind.' (Grabher, 1993: 24).

15 The quotation has been taken from Maskell *et al.* (1998: 69) where the point is the recognition that any localised capability will always be a function of history. And the same is all the truest for social capital.

me!' 'That's it."' 'No kidding?!' 'Oh, absolutely, there is nothing to it old boy; just keep it up for five centuries.' (Dierickt and Cool, 1989: 1507).

We have to be patient and, at the same time, strongly determined, obstinate and resolute, tenacious towards the final purpose. After all, territorial competitiveness and sustainable development are not child's play. Indeed, we see ahead a promising research agenda.

Acknowledgement

The present chapter has been produced within the frame of a "Basic Research Project" — carried out by the author at Bocconi University, Milan — devoted to the study of governance structures and territorial systems of production and innovation. I would like to thank the editors for a number of valuable suggestions and improving criticisms on a previous version while remaining the sole responsible person for the present version.

References

Amin, A. and Thrift, N. (eds) (1994), *Globalization, Institutions, and Regional Development in Europe*, Oxford University Press, Oxford.

Axelrod, R. (1984), *The Evolution of Cooperation*, Basic Books, New York.

Barnes, W.R. and Ledebur, L.C. (1998), *The New Regional Economies.* Cities & Planning, Sage Publications, Thousand Oaks.

Batley, R. and Stoker, G. (eds) (1991), *Local Government in Europe. Trend and Development*, Macmillan, London.

Becattini, G. (1998), *Distretti industriali e Made in Italy. Le basi socioculturali del nostro sviluppo economico*, Bollati Boringhieri, Torino.

Benco, G. and Strohmayer, U. (eds) (1997), *Space and Social Theory. Interpreting Modernity and Postmodernity*, Blackwell Publishers, Oxford.

Bennet, R. (1989) (ed), *Territory and Administration in Europe*, Pinter Publishers, London.

Bennet, R. and Kreb, G. (1991), *Local Economic Development*, Belhaven Press, London.

Bianchi, P. and Miller, L. (1994), 'Innovation, Collective Action and Endpgenous Growth: An Essay on Institutions and Structural Change', *Dynamis*, Quaderno 2/94, IDSE-CNR, Milano.

Bonomi, A. and De Rita, G. (1998), *Manifesto per lo sviluppo locale. Dall"azione di comunità ai Patti territoriali*, Bollati Boringhieri, Torino.

Braczyk, H-J., Cooke, P. and Heidenreich, M. (eds) (1998), *Regional Innovation Systems*, University College London Press, London.

Brunetta, R. and Tronti, L. (eds) (1995), *Beni relazionali e crescita endogena*, Società per l'imprenditorialità giovanile SpA—Fondazione G. Brodolini, Roma.

Bullman, U. (1997), 'The Politics of the Third Level', in C. Jeffery (ed), *The Regional Dimension of the European Union. Towards a Third Level in Europe?*, Frank Cass, London.

Chisholm, R.F. (1998), *Developing Network Organizations: Learning from Practice and Theory*, Addison Wesley, Reading.

Cooke, P. and Morgan. K. (1998), *The Associational Economy. Firms, Regions, and Innovation*, Oxford University Press, Oxford.

Danson, M. (ed) (1997), *Regional Governance and Economic Development*, European Research in Regional Science, no. 7, Pion, London.

Dierickt and Cool (1989), 'Asset Stock Accumulation and Sustainability of Competitive Advantage', *Management Science*, pp. 1507.

Dolliger, P. (1964), *La Hanse. XII-XVIIe siècles*, Aubier, Paris.

Eber, M. (ed) (1997), *The Formation of Inter-Organizational Networks*, Oxford University Press, Oxford.

European Commission (1997), *Territorial Employment Pacts. Examples of Good Practice*, EC Structural Funds, Bruxelles.

Fukuyama, F. (1995), *Trust. The Social Virtues and the Creation of Prosperity*, Hamish Hamilton, London.

Geenhuizen, M. van, and Ratti, R. (1998), 'Managing Openness in Transport and Regional Development: An Active Space Approach', in K. Button, P. Nijkamp and H. Priemus (eds), *Transport Networks in Europe. Concepts, Analysis and Policies*, Edward Elgar, Cheltenham, pp. 84-102.

Grabher, G. (ed) (1993), *The Embedded Firm. On the Socioeconomics of Industrial Networks*, Routledge, London.

Jeffery, C. (ed) (1997), *The Regional Dimension of the European Union. Towards a Third Level in Europe?*, Frank Cass, London.

Keating, M. and Loughlin, J. (eds) (1997), *The Political Economy of Regionalism*, Frank Cass, London.

King, D. and Stoker, G. (eds) (1996), *Rethinking Local Democracy*, Macmillan, Houndmills, Basingstoke.

Heller, F. *et al.* (1998), *Organizational Participation. Myth and Reality*, Oxford University Press, Oxford.

Hooghe, L. (ed) (1996), *Cohesion Policy and European Integration. Building Multi-Level Governance*, Clarendon Press, Oxford.

Humes IV, S. (1991), *Local Governance and National Power*, Harvester Wheatsheaf, London.

Humphries, C. (1996), 'The Territorialisation of Public Policies: The Role of Public Governance and Funding', in OECD *Networks of Enterprises and Local Development*, LEED, OECD, Paris.

Immerfall, S. (ed) (1998), *Territoriality in the Globalizing Society. One Place or None?* Springer, Berlin.

Maskell, P., Eskelinen, H., Hannibalsson, I., Malmberg, A. and Vatne, E. (eds) (1998), *Competitiveness, Localised Learning and Regional Development*, Routledge, London.

Mlinar, Z. (ed) (1992), *Globalization and Territorial Identities*, Avebury, Aldershot.

Mutti, A. (1998), *Capitale sociale e sviluppo. La fiducia come risorsa*, Il Mulino, Bologna.

OECD (1996), *Networks of Enterprises and Local Development*, LEED, OECD, Paris.

Olson, M. (1982), *The Rise and Decline of Nations*, Yale University Press, New Haven.

Perulli, P. (ed) (1998), *Neoregionalismo. L'economia-arcipelago*, Bollati Boringhieri, Torino.

Pichierri, A. (1997), *Città Stato. Economia e politica del modello anseatico*, Marsilio Editori, Venezia.

Pichierri, A. (1998), 'Stato e identità economiche locali', in P. Perulli (ed), *Neoregionalismo. L'economia-arcipelago*, Bollati Boringhieri, Torino.

Putnam, R.D. (1993), *Making Democracy Work*, Princeton University Press, Princeton.

Ratti, R. (1997), 'Lo spazio attivo: una risposta paradigmatica al dibattito locale-globale', in A. Bramanti and M. Maggioni (eds), *La dinamica dei sistemi produttivi territoriali: teorie, tecniche, politiche*, Franco Angeli, Milano.

Ratti, R., Bramanti, A. and Gordon, R. (eds) (1997), *The Dynamics of Innovative Regions. The GREMI Approach*, Ashgate, Aldershot.

Regalia, I. (1997), 'Prospettive di regolazione regionale delle relazioni di lavoro', Paper presented at *La molteplicità dei modelli di sviluppo nell'Italia del Nord*, Parma, 6-7 novembre.

Rodríguez-Pose, A. (1996), 'Innovation Prone and Innovation Averse Societies: The Passage from Innovation to Economic Performance in Local Economies', Paper presented at the *VII International Conference on Economics and Policies of Innovation*. Cremona, June, 11-13.

Sabel, C. (1996), 'Learning-by-Monitoring: The Dilemmas of Regional Economic Policy', in OECD *Networks of Enterprises and Local Development*, LEED, OECD, Paris.

Saxenian, A. (1994), *Regional Advantage*, Harvard University Press, Cambridge MA.

Scott, A.J. (1998), *Regions and the World Economy*, Oxford University Press, Oxford.

Storper, M. (1997), *The Regional World*, The Guilford Press, New York.

Storper, M. and Salais, R. (1997), *Worlds of Production. The Action Frameworks of the Economy*, Harvard University Press, Cambridge MA.

Taylor, M. and Conti, S. (eds) (1997), *Interdependent and Uneven Development. Global-Local Perspectives*, Ashgate, Aldershot.

Veltz, P. (1996), *Mondialisation villes et territoires. L'économie d'archipel*, Presses Universitaires de France, Paris.

Wannop, U.A. (1995), *The Regional Imperative. Regional Planning and Governance in Britain, Europe and the United States*, Regional Policy and Development no. 9, Jessica Kingsley, London.

Whitley, R. and Kristensen, P.H. (eds) (1997), *Governance at Work. The Social Regulation of Economic Relations*, Oxford University Press, Oxford.

PART II:
REASONS FOR AN
ACTIVE SPACE APPROACH

5 Obstacles to Openness of Border Regions in Europe

PIET RIETVELD

Border related obstacles appear to play a large role in international transport flows. A typology of possible backgrounds to such obstacles is given in this chapter. Empirical results are shown for various transport modes: car, bus (public transport), train, and plane. Special attention is given to business and freight transport because these most accurately reflect the effects of borders on the organisation of economic activity on different sides of borders. We find some evidence that border effects are smaller for business and freight transport than they are for other travel motives. This seems to be an indication that the development of openness in regional development has proceeded further than is sometimes thought. The gap between domestic commercial interaction and cross-border commercial interaction, however, remains significant.

Introduction

The ongoing process of economic integration in Europe has led to a reduction in the importance of borders as a factor discouraging spatial interaction in terms of flows of goods, persons and information. Borders do, however, have various implications for spatial interaction (cf. Ratti and Reichman, 1993). They correlate with fiscal and institutional differences, and also with cultural and language differences. Fiscal differences can be changed relatively easily by changing fiscal laws; but other types of differences are much more difficult to change. Therefore it is interesting to investigate more carefully the impact of borders on spatial interaction, and more specifically the impact of changes in the nature of borders as they are taking place in Europe today. This leads to the question of what types of border effects remain after the process of European integration which has emphasised the harmonisation of fiscal and legal dimensions.

Borders tend to function as obstacles in spatial interaction since interactions with foreign neighbouring regions tend to be weak. It would be too easy to conclude that this will always have a negative economic impact on a border region. Borders can lead to a certain degree of isolation, interpreted as high transport costs, which discourages interregional trade. Firms producing in a border region are relatively isolated and thus do not face competition from firms at the other side of the border. Hence border regions may experience a certain level of protection because of their location. Indeed, borders have often been constructed to provide protection against alien forces at the other side, the Great Wall of China being an excellent example.

As outlined in Rietveld (1994) the higher degree of competition implied by reducing the obstacle effect of a border has various effects. Some sectors will be hurt by the increase in competition and may decline. Other sectors may find new opportunities to export. In terms of employment effects the balance is not clear. For consumers a similar result obtains: some products will become cheaper because of increased competition, but other products may become more expensive when the firms located in the region find that it is more profitable to export their products to other countries. For the aggregate of consumers and producers in all regions the reduction of obstacles to trade is beneficial, in terms of consumer surplus and profits, but the distribution of the two may be uneven across regions.

The above reasoning is based on a simple interregional trade model. This does not exhaust the relevant economic perspectives on borders, however. Other relevant economic aspects of border regions concern the markets for inputs, labour and knowledge. For example, borders will induce difficulties in cross-border commuting. Borders may also discourage cross-border co-operation between firms. The latter would imply that borders also function as obstacles to the diffusion of knowledge that may hamper the vitality of firms located in border regions.

The term "border region" actually covers a wide variety of regions according to sectoral composition, income levels and infrastructure endowment. At the European level there is certainly no reason to equate border regions with problem regions. If one considers the problem regions in the various EU countries, one observes that many of them are not border regions (cf. Armstrong and Taylor, 1993). Concerning the position of border regions in transport networks and trade flows we observe that particular places near borders which are suitably located have the potential to become export nodes. These locations may benefit from borders because they function as gateways to the neighbouring country so that a concentration of economic activity is induced; however, such a gateway position is only possible for a limited number of places. Most places in border regions will not have such an opportunity and will have to search for other strong points to exploit.

Our conclusion is that the impact of borders on economic activity in a region can attain many forms varying from positive to negative. In studies of the impact of borders on regions one should take into account differences among sub-regions. Reduction of the obstacle effects of borders as taking place in the EU provides a challenge to the firms involved to exploit the opportunities offered. An appropriate term seems to be the managing of openness. In this respect it is useful to point to a certain bias in the discussions on border regions. If borders are interpreted as semi-permeable lines in space which discourage interaction with neighbouring regions, seacoasts can be given a similar meaning. Coastal regions have an even bigger disadvantage compared with regions located at national borders, since the communication partners at the other side of the line are virtually absent, whilst with border regions they are not absent, but only more difficult to reach. Of course, a location on a seacoast gives certain advantages to a region: coastal regions are well located to exploit sea related resources, such as oil, fish, and tourism. In addition they may host seaports; but, especially given the declining economic importance of sea transport relative to other transport modes, coastal border regions may be in a less advantaged position compared with land border regions (cf. Rietveld and Boonstra, 1995).

The aim of the present chapter is to investigate the nature of the obstacles implied by borders. In addition we will do some empirical investigation of the level of the border effect: to what extent do borders really discourage spatial interaction between regions? This chapter is organised as follows. In the next section a typology of border related obstacles is discussed. This is followed by two sections in which we present results on obstacle effects of borders for the European aviation and railway sector. Then attention is devoted to the impact of border related obstacles on business traffic by car and to the role of borders on border crossing car and public transport trips.

A Typology of Border Related Obstacles

A framework to analyse the impact of borders has been developed by Cattan and Grasland (1992). In this framework (see *Figure 5.1*), two factors are distinguished which have an impact on places in space: distance and borders. The impacts of distance and borders is specified for two types of variables, i.e. *state variables*, relating to the situation in a certain place, and *flow variables* relating to the interaction between different places. Two possible effects of borders are considered. They lead to:

- non-homogeneities among places at different sides of the border, and
- discontinuities in flows between places at different sides of the border.

Distance has a similar impact on places and interaction, but its effect is much more gradual.

In research attention is usually focused on the upper part of *Figure 5.1*. For example, the impact of distance, or travel costs, on transport flows F_{ij} has been widely studied in the context of spatial interaction models. Spatial autocorrelation analysis has been used as a similar tool in the investigation of similarities between places. In this case the dependent variable is a similarity index S_{ij}. The role of borders has usually been neglected in this context. For similarity indices it would mean that similarity depends on distance and also on whether or not two places are at the same side of a border. For flows, borders would also have a potential impact in addition to distance.

The two aspects: similarity of places and flows between places are clearly related. For example, places may be different because one place may have adopted an innovation the other did not adopt. An improvement of communication will usually stimulate equal patterns of innovation adoption. Thus a reduction of the obstacle effects of borders may lead to an increase in similarities between places at different sides of the border. Such a parallel development is not guaranteed, however. For example, a reduction in trade barriers will usually stimulate specialisation in production processes, leading to a decrease in similarity of economic structure between places or regions.

In the present chapter we focus on the impacts of borders on flows, i.e. on obstacle effects of borders. The impacts on similarities will not be treated here (for an example in the field of demographic fertility indices refer to Decroly and Grasland, 1992).

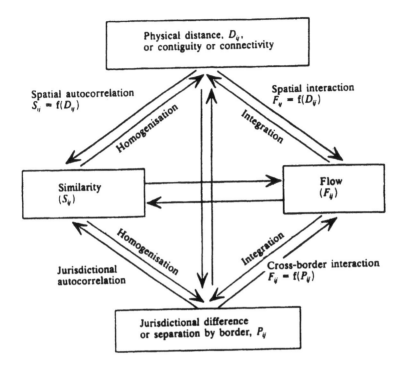

Figure 5.1   A methodological framework for the analysis of obstacles and discontinuities

Source: Adapted from Cattan and Grasland (1992).

Border related obstacles can be defined to exist when the intensity of interaction in space suddenly drops at places where a border is crossed. Various reasons for the existence of obstacle effects can be distinguished (see also van Geenhuizen et al., 1996). *Table 5.1* contains some of them.

Table 5.1 Reasons for existence of obstacle effects of borders

-   Weak or expensive infrastructure services in transport for international links
-   Preferences of consumers for domestic rather than foreign products and destinations
-   Government interventions of various types
-   Lack of information on foreign countries

The first type of border related obstacle effect concerns the *supply of transport and communication services*. This effect is expressed in the form of various types of costs. If one computed generalised costs, one would observe a discontinuity in these costs when a border is crossed. The generalised costs consist of two main components: monetary expenditures and time related costs. An example where there is an extra *monetary* burden related to international transport compared to domestic transport is in the airline sector. The reason is that international regulatory agreements often limit the supply of international services so that tariffs are higher. In international road transport cabotage and quota systems may lead to inefficiencies and hence to high tariffs. In international rail transport the lack of co-operation between national railway companies leads to relatively high international tariffs. In telecommunication a similar tendency can be observed: international tariffs are often much higher than long distance domestic tariffs, even though the distance between the communication partners may be very much the same.

Most cases of supply related transport costs concern the *time* component. Take for example the road network: international links are underdeveloped in the road system, as can be seen in the Alpine region. This leads to detour factors which may be somewhat higher in international transport compared to domestic transport. In railway infrastructure one observes that countries start investments in high speed rail for domestic links, such as in France, Germany, and Spain. Only at a later stage are international links added. This means that the speed of services between major links in the same country is faster than between comparable links in different countries. Another example can be found in the field of telecommunication. There is a lack of supply of telecommunication infrastructure in the former USSR. This leads to high failure rates when one wants to establish contacts with another country and this means that one loses much more time compared with calls to other destinations. Another example from the field of telecommunication is found in certain developing countries where international calls are not automated which also leads to time losses.

The above examples concern time-related obstacle effects due to the absence of a sufficient infrastructure. A somewhat different obstacle effect is due to the way infrastructure is used. For example, train services at international links usually have lower frequencies than at comparable national links. This means that the international traveller faces higher inter-arrival times which lead to higher waiting times or a less efficient use of time abroad. A similar case holds true for international airline services. Rail transport provides other examples of obstacle effects. Technical incompatibility in railway systems due to differences in gauge, for example between Spain and France, or voltage, for example between Germany and The Netherlands, leads to time losses when passing the border because one has to change carriages and/or locomotives.

The second group of obstacle effects concerns the *preference of consumers and producers* for domestic interactions compared to international interactions. Such a preference may be based on taste: for example in food consumption one can observe clear differences in national habits, leading to a disincentive for the international trade in certain food products. Language, ethnical and cultural differences can lead to a strong preference for trade or communication partners from the own country compared to other countries. This holds true for consumers, and for firms. As indicated by Hofstede (1980), there are substantial cultural differences between certain groups of countries which makes co-operation between firms in different countries difficult. Another example is found in governments in their role of final consumer which may give priority to producers from the own country in the procurement of equipment, weapons, business services, etc.

The third group of border related obstacle effects concerns *regulations or interventions of national governments*. These interventions can have both a monetary and a time effect. Examples of monetary effects are the costs of getting a visa or special taxes levied on people crossing the border. An interesting example of the latter is found in Indonesia where every Indonesian citizen leaving the country has to pay an amount of some US$ 100. This tax was imposed in order to discourage cross-border shopping in Singapore. Another well-known example of a monetary effect occurs with fiscal obstacles where import duties lead to a disincentive to import products from abroad. A similar effect occurs when excise duties on particular products are different. Another example of a regulation leading to higher costs when trade takes place internationally is related to currencies. The (possible) introduction of the European ECU aims at removing this cost, but as long as this has not yet been realised, banks will continue to charge customers for the change services they provide. In addition hedging costs for firms operating in international markets may be substantial. Firms have to follow certification procedures to introduce particular new products in a country. If each country has its own procedure this will lead to additional costs and the possibility of delays. A related problem is that countries often differ in the specification of the requirements certain products must satisfy. This leads to the need to adapt products to particular national standards which obviously has a cost increasing effect. A well-known example is the difference between the UK and other European countries in the choice of which side of the road is used leading to differences in automobile design.

Time related obstacles of an institutional nature concern the waste of time due to getting visa, waiting at customs offices, waiting at borders etc. Avoidance of border delays is very important for firms working with a just-in-time concept. It may induce the selection of domestic rather than international suppliers. To these time losses must be added the time needed for extra paper work in the case of international trade.

The fourth reason for the existence of obstacles relates to *lack of information on foreign destinations*. Lack of information always plays a role in the intensity of spatial interaction, but in border-crossing interactions it is more severe. For example, many newspapers, data banks and information systems have a clear national orientation. Acquiring additional information is possible, but it gives rise to costs in terms of money and time. Personal information networks also often have a domestic bias. The information people have is strongly influenced by interaction patterns in the past. Thus information related obstacles to international interactions depend on the other types of obstacles mentioned above; they can be said to reinforce them. Since the stock of information is built gradually, the historical component of obstacle effects may be expected to be substantial.

In the above list of factors leading to obstacle effects of borders we find both symmetric and a-symmetric effects. Symmetry occurs when spatial interaction is reduced in both directions to the same extent. There are also examples where the effect is a-symmetric. The reduction takes place in both directions, but not to the same extent. Still another possibility is that borders lead to a decrease in interaction in one direction and an increase in the other direction. In this case one might speak of an adverse border effect. Cross-border shopping is an example. Another example can be found in tourism, where certain tourists prefer foreign locations above otherwise identical domestic locations because they are more interesting. In the context of *Figure 5.1* this means that spatial heterogeneity stimulates international flows.

Border related obstacles are not the only obstacles which may exist in space, however. For example, migration flows in a country with several ethnic or language groups, each having their own home region, will be biased towards the own region (see Cattan and Grasland, 1992 for a case study in former Czechoslovakia). Telecommunication flows may also be biased within countries towards regions with the own language, as found for example by Klaassen et al. (1972) for Belgium, and Rossera (1990) and

Donze (1993) for Switzerland (see also, Capinieri and Rietveld, 1997). In the present chapter we will, however, focus on border related obstacles.

Obstacle Effects of Borders in the Airline Network

Consider two airports at a certain distance apart, frequencies of flights between these airports will tend to be higher when they are located in the same country compared with the situation where the two airports are in different countries. There are two main reasons why this is true. The first one relates to the *demand* side. Demand for international air traffic over a certain distance is smaller than demand for domestic air traffic over the same distance. This is a consequence of the various obstacle components of borders discussed in the previous section. The second reason relates to the *supply* side. Regulation in the airline system tends to reduce the number of flights in international linkages.

In this section we discuss a numerical estimate of the extent to which these effects occur. The method used is the quasi-experimental approach. In this approach one compares a pair of objects, in this case airports (A, B), with another pair (A, C). The airports B and C have been chosen in such a way that they are identical in all relevant economic characteristics. In addition, the distance between A and B is equal to that between A and C. The only difference is that A and B are located in the same country, which is not the case for A and C. By comparing the frequency of flights between A and B with that between A and C one can isolate the impact of borders. One of the factors which has to be controlled in the approach is the availability of all alternative transport modes. For example, the number of flights between Brussels and London is much larger than between Brussels and Paris. The reason is that no rail or road connection existed, at the time of the measurement, between Brussels and London so that the share of air traffic on this link is very high. Thus, an obstacle in a certain mode (road) appears to function as an incentive to use another mode (air). Another factor which might interfere is the different position of airports in hub and spoke networks.

The advantage of the quasi-experimental approach is that one does not need to formulate and estimate a model to isolate the border effect. An obvious disadvantage of this approach is that one will never find airports which are entirely identical according to all relevant features. One is forced therefore to use airports which are only approximately identical which produce noise in the outcomes. We applied the approach outlined above for some 20 pairs of airports in Europe (cf. Rietveld, 1993). The reduction factor for international flights is in all cases smaller than 1: international flights are consistently less frequent than domestic flights. The average value of the reduction factor is about .30. This means that against ten flights a day on a certain domestic connection there are only about three international flights to a similar destination at a similar distance. This is a clear indication that border effects play a role in aviation networks.

Obstacle Effects of Borders in Rail Transport

For rail transport we have followed an approach similar to the one discussed in the previous section. In *Table 5.2* we present the results of frequencies for a number of comparable city pairs in Europe. Among a set of 11 comparisons there are 9 where the international frequency is clearly lower than the domestic frequency; for two pairs we happen to find equal frequencies: Hamburg-Essen with Hamburg-Arhus, and Nurnberg-Heidelberg with Nurnberg-Linz. The reason for the high score of the international link in these cases may be that it is part of an important international corridor. For example, Nurnberg-Linz is part of the corridor Frankfurt-Vienna. Based on a larger set of data, Boonstra (1992) found that the average reduction factor is equal to 0.44. This means that against ten trains a day on a certain domestic route there are four or five international trains to a similar destination at a similar travel time away.

In most cases crossing a border in Europe means that one also crosses a linguistic border, but there are exceptions. A further analysis of the data reveals that the reduction factor is indeed different for countries where the same language is spoken (average value 0.57) and countries where different languages are spoken (average value 0.38) (cf. Boonstra, 1992).

Cross-Border Obstacles to Business Trips

Business trips are an interesting case from the perspective of border effects on economies of regions. Data on business trips are fragmented; there is no common data base on cross-border business trips in Europe. This is a barrier for conducting a European wide analysis. In the present chapter we present some results on cross-border business trips and obstacle effects of borders for trips originating from The Netherlands.

Table 5.2   Frequency of railway connections between equivalent pairs of European cities

Railway stations (pair)	Country frequency per day		Reduction factor
Amsterdam-Groningen	NL-NL	20	
Amsterdam-Oberhausen	NL-DE	14	0.70
Hamburg-Essen	DE-DE	14	
Hamburg-Aarhus	DE-DK	14	1.00
Essen-Hannover	DE-DE	19	
Essen-Amsterdam	DE-NL	12	0.63
Innsbruck-Salzburg	AU-AU	25	
Innsbruck-Augsburg	AU-DE	6	0.24
Saarbrücken-Koln	DE-DE	17	
Saarbrücken-Paris	DE-FR	6	0.35
Köln-Mannheim	DE-DE	33	
Köln-Utrecht	DE-NL	12	0.36
Nürnberg-Heidelberg	DE-DE	9	
Nürnberg-Linz	DE-AU	9	1.00
Paris-Metz	FR-FR	23	
Paris-Courtrai	FR-BE	6	0.26
Paris-Nancy	FR-FR	14	
Paris-Courtrai	FR-BE	6	0.43
Lyon-Nancy	FR-FR	6	
Lyon-Torino	FR-IT	4	0.67
Wurzburg-Erfurt	DE-DE	8	
Bremen-Groningen	DE-NL	3	0.38

Source: Thomas Cook (1992).

Before discussing the border effects we first have to pay some attention to transport modes chosen for business trips. The car is the main mode of domestic business trips (85 per cent) in The Netherlands. The train mainly takes the other 15 per cent. The share of the train increases with distance. The share of air transport is negligible given the small size of the country. In international business trips originating in The Netherlands the share of the car is still substantial (78 per cent). This indicates that most of the international business trips will have a destination a short distance away in neighbouring countries. Note that the travel time by car from the Dutch Randstad to large cities in Belgium (Antwerp, Brussels) and Germany (Ruhr area) varies from only 1.5 to 2.5 hours.

The effect of borders on the intensity of business trips of Dutch firms has been estimated using a spatial interaction model. In this model we include as explanatory variables:

-   mass indicators (gross domestic product, GDP) of regions of origin and destination,
-   travel distance between centres of gravity of the regions,
-   a dummy indicating when a national border is crossed.

To estimate the model we combine data on interregional domestic business trips in The Netherlands (132 flows between all pairs of 12 provinces) and data on trips between Dutch regions (4 clusters of provinces) and a number of European regions (66 pairs). Domestic travel data are collected regularly by the Dutch Central Bureau of Statistics. The international data were collected in a special survey by INRO-TNO (1990) for business trips by car crossing the Dutch-German border in 1990. The estimated results are presented in *Table 5.3*.

The conclusion of model version 1 (based on all 198 observations) is that distance decay is substantial in business trips by car. Other things equal the number of business trips between two cities at a distance of 100 km apart is 4 to 5 times as large compared to two cities at a distance of 200 km apart. The border effect is quite large according to the first model version: it means that crossing a border to another EU country reduces the number of trips by car to only 16 per cent (exp(-1.83)) of the number of trips one would expect without a border crossing. For a business trip to a country outside the EU one even finds a reduction factor of 5 per cent (exp(-3.07)).

The problem with model 1 is that it ignores mode choice of business travellers. The share of train and air will increase with distance. Hence, if we had data on the total number of travellers aggregated across all modes, we would find substantially higher figures for the number of trips to destinations in countries located further away; however, we do not have data on the number of international business travellers by train for combinations of regions. We solved this problem by confining our attention only to short distance international business trips. In the context of the present data set these are the trips to the German regions of Nieder Sachsen and Nord-Rhein Westfalia. We may safely assume that for these international destinations the car is still the dominant travel mode, just as it is for domestic business trips in The Netherlands. The results are presented in *Table 5.3* as model 2. Most parameters are not substantially affected by the change in data set, but the border effect certainly is. According to model 2, crossing the border to a German region implies a border factor of some 60 per cent (exp(-0.51)). This figure is much more modest than the result of model 1.

Table 5.3   Estimation of interaction model for interregional business trips from and in The Netherlands*

Explanatory variable	Model 1**	Model 2**
Constant	-10.7	-7.76
	(-6.86)	(-5.34)
Log GDP	0.84	0.78
(region of origin)	(14.33)	(13.99)
Log GDP	0.90	0.78
(region of destination)	(14.48)	(14.08)
Log distance	-2.25	-2.25
	(-21.04)	(-23.81)
Dummy EU	-1.83	-0.51
	(-7.67)	(-1.70)
Dummy other countries	-3.07	-
	(-9.76)	
R square	.884	.888
Number of observations	198	140

*   Dependent variable measured as log of number of trips
** t-values in parentheses

We may conclude that a careful analysis of border effects on business trips by car reveals that crossing the Dutch-German border leads to a reduction from a level of 100 (indexed) to about 60. This means that obstacles to interaction still exist between neighbouring regions that both have been part of the EU for decades. It is interesting to observe, however, that the obstacle impact of the border is smaller in this case of car based business trips than in the cases of rail and aviation presented in the preceding sections.

Border Effects on Local Border Crossing Transport (Road and Bus)

That borders function as obstacles for transport can also be observed from traffic intensities on and near borders. The general tendency is that flows on borders are much smaller than they are at some distance from the border. In this section we present some results for The Netherlands. Major express ways linking large cities like Amsterdam and Rotterdam with neighbouring countries display large differences in traffic intensities: near a large city they are very high, on the border they are much smaller. For example, in 1996 the A1 express way linking Amsterdam with Berlin had an intensity of about 140,000 cars per 24 hours near Amsterdam. The intensity at the German border some 160 km further away was only 14,000, a reduction to 10 per cent. Indeed the major difficulties in so-called hinterland connections of large cities do not appear near the border, but near the cities. Of course the large difference in intensity depends considerably on the high population density around the large cities leading to a high demand. In order to identify a border effect it is better to compare traffic intensities on borders with intensities near borders (say some 20 km away). The ratio between the two captures much better the obstacle effect of borders. We give the results for some highways in The Netherlands in *Table 5.4*.

The figures show that *on* the border the major highways have intensities that are clearly smaller (reduction factors of some 40 per cent) than *near* the border. Thus, the major use of international road transport corridors is for domestic purposes. There appears to be a difference, however, between trucks and passenger cars. We observe that the percentage of trucks *on* the border is clearly larger than *near* the border. This indicates that freight transport is more long distance oriented and less sensitive to borders than passenger transport.

Table 5.4   Obstacle effects of borders on some major Dutch highways (1996)

Highway	Border effect	Share of trucks on border	Share of trucks near border
A1	35%	43%	24%
A7	48%	25%	19%
A16, A58	37%	32%	19%

Source: AVV (1997).

Another way to study obstacle effects of borders is to consider border crossing public transport using busses. This gives an indication of the extent to which local economies across the border are integrated. We compared service levels of border crossing bus services with service levels at other places for the year 1993. We compared the frequency on the border with the frequency at a place some 10 km away from the border for a selection of bus routes. Some results are shown in *Table 5.5*. We observed a substantial difference in frequency of bus services on and near the border. Compared with a frequency of 100 in a standard border region, cross-border links on average achieve a score of some 35 to 40.

Table 5.5   Daily frequency of public transport busses on and near borders in The Netherlands

Border crossing point	Frequency per day on border	Frequency per day near border
Nieuw Schoonebeek	29	74
's-Heerenberg	42	61
Winterswijk	9	95
Putte	51	89
Luykgestel	27	67
Eede	27	104

Source: Schedules of public transport operators (1993).

Conclusions

With the ongoing process of economic integration in the EU certainly not all border related obstacles have been removed. This is no surprise given the various economic and non-economic dimensions of the obstacles surveyed in this chapter. The major bottleneck is in general not the lack of infrastructure for cross-border links, although along some borders there are indeed problems, such as in the Alpine countries. The major problems here relate to the high construction costs of infrastructure in mountain areas and to regulatory measures of national governments involved. The latter point underlines the importance of institutional aspects in the obstacle effects of borders.

Our analysis of cross-border transport services by various modes of collective transport reveals a double effect of borders. The first effect concerns the demand side: because demand for cross-border interaction is lower than for other destinations, the supply frequencies are lower. This supply effect will have an additional negative effect on cross-border interaction because of the lower frequencies. Thus, we observe the phenomenon that (demand related) *obstacles* to cross-border transport flows *create* additional (supply related) *obstacles*.

A policy implication of the statement that barrier effects of borders are mainly related to the *demand side* is that there is not much reason to invest large amounts of money in international links. Most of the problems in international transport relate to congestion near the large metropolitan areas, not to insufficient capacity in border regions. This would call for a careful analysis of Trans European Network proposals. In those cases where the *supply side* dominates the barrier aspects of borders, trans-national initiatives may be essential. The reason is that the benefits of removing the border will accrue to various countries. Given the national bias in project analysis, i.e. positive effects on economies in other countries are usually ignored, a trans-national perspective is needed. Asking transport users to pay toll to use trans-national infrastructure is a way to overcome this problem: toll revenue can be used to determine the willingness to pay of users from other countries.

Our conclusion that national borders exert a strong influence on cross border transport flows holds true for all spatial ranges. For example, for the *short range*, cross border *public bus* services have frequencies that are on average some 35-40 per cent of the level they are in other parts of border regions. For the *short to medium range* we find for *road transport* on average an obstacle factor of a similar magnitude. A tendency exists for freight traffic to be less sensitive to borders than passenger traffic. In addition, our analysis of business traffic shows that business trips are also less sensitive. Here we find a reduction to about 60 per cent of the normal non-border level. For *medium to long distance* we find for *rail transport* services that cross-border frequencies are reduced to some 44 per cent. The *long distance* connections are mainly served by *airlines*. We observe here a reduction to some 30 per cent for cross-border links. It is not impossible that this figure will increase as a consequence of increased competition in the European aviation market.

Our results for business trips (by car) and for freight transport by road imply that in the fields of trade and production the effects of barriers are more modest than in the other fields. This may be an indication that the development of openness in regional development has already proceeded further than is sometimes thought; however, the gap between domestic interaction of firms and cross-border interaction of firms remains significant.

Data problems are substantial in the field of border effect studies. Therefore some of the results reported here have been based on limited evidence. Nevertheless it is striking that the results are rather similar for most types of transport modes. With the increasing use of information technology the prospects for alleviating the data problem are favourable. The prospects would be even better when the increased use of information technology is accompanied by an increase in organisational efforts at the international level to achieve further standardisation of data bases.

We have identified at least two *paths for further research* on the theme of openness and borders. First, a point often overlooked in the analysis of border regions is that coastal regions may have a comparable lack of interaction with neighbours. The limited opportunities of coastal regions for interaction with other regions is a theme that deserves more attention in future research. The second path relates to the distinction between state variables and flow variables in the context of border related obstacles. We note that the concepts of interregional flows and interregional similarity discussed in the second section are closely related to the two basic types of regional concepts commonly used: homogeneous regions and functional regions. A systematic treatment of the spatial delimitations for both types of regions at the European level to find out the different roles of borders versus distance (proximity) is called for.

Acknowledgement

The author thanks Joost van Nierop for computational assistance, and Jaap Boonstra as well as Natalie Looy for data collection efforts.

References

Armstrong, H. and Taylor, J. (1993), *Regional Economics and Policy,* Philip Allan, Oxford.
AVV (Adviesdienst Verkeer en Vervoer) (1997), *Verkeersgegevens*, Rotterdam.
Boonstra, J. (1992), *Barrieres in verkeer en infrastructuur; een internationaaal perspectief,* Faculty of Economics, Free University, Amsterdam.
Capinieri, C. and Rietveld, P. (1997), *Networks in Transport and Communication; a Policy Approach.* Avebury, Aldershot.
Cattan, N. and Grasland, C. (1992), *'Migrations of population in Czechoslovakia; a comparison of political and spatial determinants of migration and the measurement of barriers',* Trinity Papers in Geography, no 8, Dublin.

Cook, T. (1992), *European timetable*, Thomas Cook Publishing, London.

Decroly, J.M. and Grasland, C. (1992), 'Frontieres, systemes politiques et fecondité en Europe', *Espace Populations Societes*, 2, pp. 81-118.

Donze, L. (1993), 'Barriers to communication in Swiss telephone flows', *Sistemi Urbani*, vol. 23, pp. 103-120.

Geenhuizen, M. van, B. van der Knaap, Nijkamp, P. (1996), 'Transborder European networking: shifts in corporate strategy?', *European Planning Studies*, vol. 4 (6), pp. 671-682.

Hofstede, G. (1980), *Cultures consequences; international differences in work related values*, Sage, Beverly Hills CA.

INRO-TNO (1990), *Internationaal zakelijk verkeer*, Delft.

Klaassen, L. H., Wagenaar, S. and Van der Weg A. (1972), 'Measuring psychological distance between the Flemings and the Walloons', *Journal of the Regional Science Association*, vol. 29, pp. 45-62.

Looy, N. (1994), *Grensregio's in een Europa zonder grenzen*, Masters thesis, Free University, Amsterdam.

Ratti, R. and Reichman, S. (eds) (1993), *Theory and Practice of Transborder Cooperation*, Helbing & Lichtenhahn, Basel.

Rietveld, P. and Janssen, L. (1990), 'Telephone calls and communication barriers; the case of the Netherlands', *The Annals of Regional Science*, vol. 24, pp. 307-318.

Rietveld, P. (1993), 'International transportation and communication networks in Europe; the role of barrier effects', *Transportation Planning and Technology*, vol. 17, pp. 311-317.

Rietveld, P. and Boonstra, J. (1995), 'On the supply of network infrastructure', *The Annals of Regional Science*, vol. 29, pp. 207-220.

Rietveld, P. (1994), 'Spatial impacts of transport infrastructure supply', *Transportation Research A*, vol. 28, pp. 329-341.

Rossera, F. (1990), 'Discontinuities and barriers in communications, the case of Swiss communities of different language', *The Annals of Regional Science*, vol. 24, pp. 319-336.

6 Openness: A Value in Itself? The Case of the Dutch-German Ems-Dollart Region

ENNE DE BOER

This chapter deals with public transport in border regions. The focus is on the Ems-Dollart Region (EDR) in the Dutch-German borderland, but the results tend to be more general in character. The EDR territories on either side of the border are similar in an economic sense. In such a situation, open borders are likely to have little positive impact on exchange and development. In public transport, openness is indicated by little regulation, large corporations, but no monopolies, and a large degree of integration. Public transport has been strongly changed by the respective national governments in the past decades. There has been decentralisation and deregulation, with continuing disparities between the two countries discouraging the creation of cross-border networks. The demand for cross-border public transport is quite modest, but this may increase if institutional differences between the two countries decrease considerably.

Introduction

In this contribution, the Dutch-German Ems-Dollart Region (EDR) is the subject analysed. It is an Euregion, and as such more a formal body than a coherent socio-economic border region. It is the most northern of the five Dutch-German Euregions, bordering the North Sea in the North, and the Rhine-Ems-IJssel Euregion in the South (*Figure 6.1*). The focus of analysis is on openness with regard to public transport. This means an emphasis is placed on infrastructure, organisational aspects, and supply and demand for passenger services according to schedules on both sides of the border and across the border. Taxi and touring services will not be included.

Regional public transport can be regarded as an indicator for socio-economic openness of borders and integration of the adjacent regions: a situation in which people cross the border for ordinary daily-life and business trips and not just for petrol and butter, or an excursion into a different world. The analysis of openness in this chapter, as manifest in the public transport system in the EDR, will be carried out against national facts and especially against the situation in other border regions.

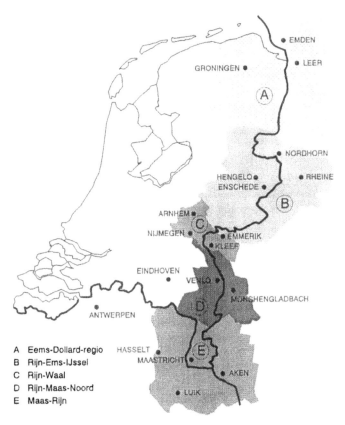

A   Eems-Dollard-regio
B   Rijn-Ems-IJssel
C   Rijn-Waal
D   Rijn-Maas-Noord
E   Maas-Rijn

Figure 6.1 Dutch-German Euregions

The data used are the product of different studies carried out partly for the EDR and partly for the East-Netherlands branch of the Dutch Ministry of Transport. Consequently, the figures presented here do not always cover the same area or the same period. The studies are based on a variety of approaches, i.e. observation, interviews and literature surveys.

The Ems-Dollart Region

As an Euregion, the Ems-Dollart Region (EDR) is a cross-border co-operative institution of regional and local governments. The EDR, with a size of 11.310 km^2, covers the area of seven regional governments. At the Dutch side these are the provinces (counties) of Drenthe and Groningen and at the German (more precisely the Bundesland of Niedersachsen) side the *Landkreise* of Aurich, Emsland, Leer and Wittmund, as well as the town of Emden (see *Figure 6.2*). The common state border is in fact much shorter than the coastline, i.e. 80 km and 180 km, respectively. The region was called after the tidal waters through which part of the national border runs today: the estuary of the German river Ems and the Dollart basin. These waters divide east and west, but in the past, well into the 19th century, the seaways constituted the most important connections. The EDR part of the Dutch-German border is old, dating from the 16th century. On the mainland a fairly straight borderline was drawn through the vast Bourtanger Moor, which was nearly impenetrable.

From an economic-geographic perspective, the Dutch and the German side are quite similar. The area is relatively sparsely populated: 148 inhabitants per km^2 against 442 for The Netherlands and 224 for Germany on the country level (EDR Operational Programme 1994-1999). The Dutch part of the region is more densely populated than the German part. The largest town in the Dutch part is Groningen with approximately 170,000 inhabitants. It has developed the role of a central town (market-place) in the province of Groningen and most of Drenthe over at least 500 years. Emden, which did not manage to develop central functions for the surrounding area, is the largest town at the German side with barely 50,000 inhabitants.

The fertile coastal plain and the cultivated moorlands made the region important for agriculture, except for the higher inland sands. Agriculture still is the most important land use, but it is declining. Even in the most agricultural area (the *Landkreis* Wittmund), it has only a share of 14 per cent in regional employment. Manufacturing industry has developed in areas alongside transport axes and in the region's seaports on the estuary: Emden (Germany) and Delfzijl (The Netherlands). The service sector is the most important sector, witnessing 29 per cent of jobs, and it is growing fast, though more on a government basis than on a commercial one. The economy is regarded as structurally weak: 9.9 per cent of the labour force is unemployed. Further, tourism is relatively well developed in both parts of the EDR, predominantly for longer stays during the summer season. The Eastfrisian (German) islands and the Drenthe woodland and heath are traditional holiday areas. There are, however, few cross visiting or foreign guests. This situation illustrates how much these areas have been integrated each in their respective national culture and economy.

Consequently, infrastructure leading to the nations' centres is well developed, but cross-border infrastructure much less so. The harbours of Delfzijl and Emden for instance, have inland waterways to Amsterdam and the Ruhrgebiet for ships with a maximum tonnage of 1,350 and 1,000, respectively.

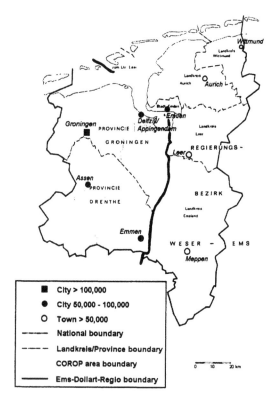

Figure 6.2 Provinces and Kreisen in the Ems-Dollart Region

Source: EDR, 1988.

Overlooking this region one wonders what function this border had and will have in the near future. In the past centuries, the border was in great part a natural barrier. From the second half of the 20th century this was much less a problem for the land-side, since the moorland had been cultivated. In the meantime the border had become a strongly institutional barrier, giving the commercial centres on both sides the opportunity to develop their hinterland relations without competition from each other, each getting development support from its own central government because of the peripheral position. The similarity of the areas on both sides, the modest quality of waterways into the hinterland, the small population density and the modest economic potentials, all these factors suggest that the opening of the border will have only a modest impact on cross-border interaction and integration. A further indication is that the organisations in the EDR are considered to be too weak for creating economic innovation. Some actors seek co-operation in larger regional centres like Oldenburg and Osnabrück, outside the EDR. Given this situation, the regional policy of the EDR focuses on the expansion of cross-border infrastructure and on tourism.

Public Transport on Each Side of the Border

In this section various characteristics of public transport in the Dutch part and in the German part of the EDR are analysed. The focus is on three issues of supply and demand of public transport, i.e. (1) market regulation, especially the uniformity, exclusiveness and continuity of services, (2) changes at the corporate level: scale and structure, geographical expansion and market orientation, and (3) market segments and integration. Openness is seen here as indicated by a low level of regulation, a large corporate scale, but no monopolies, and a large degree of integration. In addition, it is important to distinguish between internal and external openness: within the country and across the border. The two seem, however, to be related: a similar level of openness on both sides makes an integrative effect of an open border more likely. Of course, the dynamics of the internal openness require specific attention: are there tendencies that are converging or, to the contrary, diverging?

Regulation and deregulation

Public transport was until recently a strongly regulated market. The reasons for regulation are partly economic, partly social and partly safety oriented. The supply of services is subject to concessioning, in order to prevent counterproductive competition leaving only services to and between urban centres. Requirements can be made for services, fares and of course, technology. Public transport is not profitable as a rule and therefore, needs government subsidy, if not for services then at least for infrastructure.

The Netherlands has created a virtual state monopoly for rail transport outside urban concentrations. There were no railway companies like the German Bentheimer Eisenbahn AG, operating a freight line ending just across the border in Coevorden (Drenthe). In Germany parts of the rural networks were (stayed) in *Land*, *Kreis* or private hands, because the strict regulation of the national Deutsche Bundesbahn made their services relatively expensive. In The Netherlands, such connections have been abandoned as a rule. In line bus transport in The Netherlands outside cities of over 50,000 inhabitants, increasingly larger area concessions have been created since the 1930s. Today, apart from a few line concessions, only 10 area concessions are left. In recent years, the process has been accelerated by amalgamations of companies owned by a national holding, VSN. *Figure 6.3* and *Figure 6.4* show the situation in 1984 and 1998.

Subsidies for regional transport were exclusively granted by the national government. Accordingly, the national government could enforce a great deal of uniformity in the services. One type of ticket is in use for the entire country. Thus, a multi-zone ticket bought in Amsterdam can be used in Groningen.

In Germany, in contrast, the line concession is the principle outside urban areas. This implies for example, that the German part of the EDR counts twice as many concessionaires as the entire Netherlands (Ostfriesland, 1994). Ticketing is a strictly local matter as a rule, except in the areas of the transport alliances. These often serve a range of agglomerations and their suburban zones, like the Rhein-Ruhr Verkehrsverbund. Consequently, the system of subsidisation is a complex matter in Germany, with national tax measures, state subsidies on busses and infrastructure (through the municipality), and local operating subventions. Local authorities have a further stake in transport as providers of pupil transport and financiers of free rides for the handicapped.

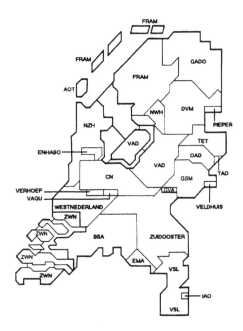

Figure 6.3 Area concessions in regional transport in 1984

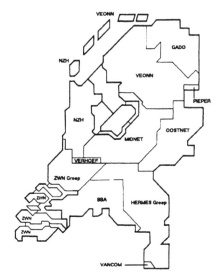

Figure 6.4 Area concessions in regional transport in 1998

Deregulation is under way in some respects, enforced by EU policy. In Germany the profitable trunk railway network is still under the control of the Deutsche Bahn (DB). It owns and manages the infrastructure of most regional networks. In principle, the tracks are now made accessible to other companies, that pay for use. Currently, this process is developing rapidly, but the great majority of the services is still offered by DB. In addition, the responsibility for the regional networks, i.e. for subventioning these, was delegated to the individual states at 1 January 1996. The necessary funds were also transferred from the *Bund* to the *Länder*. Several of the latter decentralised it even further to *Zweckverbände* , i.e. unions of authorities for a specific task, of counties and towns with a similar status. For example, Nordrhein-Westfalen decentralised to rather small *Zweckverbände*.

According to the new structure, regional authorities have to develop regional transport plans to co-ordinate trains and busses. This introduces a level of integrated planning hitherto unknown. Furthermore, some regional networks and services have been put up for tender. The contracts are not always awarded to DB in spite of the general price reductions it offers. In the German part of the EDR tendering has taken place for one railway line only, Esens – Sande, near Wittmund. The line concessions make tendering for bus services difficult here.

In The Netherlands there is no generally acknowledged need for regionalisation of parts of the railway network. Nevertheless, the principle that rail transport should be self-supporting has been adopted by the Ministry of Transport. According to this line the national government is seeking to reduce its subsidies for about 20 non-profitable lines. Efforts to transfer the responsibility for these lines have been effective so far only in the North. The province of Groningen will get the responsibility and subsidies for its secondary railway network by 2000. Overijssel, south of Drenthe was the first province to have a privatised railway line (Almelo - Mariënberg) operated by the regional bus operator Oostnet. Today, other operators than national NS (Nederlandse Spoorwegen) can get access to the profitable trunk network, but only Lovers Rail - a cruise operator - has gained permission to operate. Regarding regional bus transport, the intensive interference of the national government ended in January 1997, when both the authority and the subsidies were transferred to the provinces. Efforts to build geographically more functional entities like the German *Zweckverbände* (*Vervoerregio's*) failed. In fact, a planning structure has been given by the national government in which the provinces and municipalities have to develop their own transport plans. In addition, competition for area concessions was opened on a small scale and on an experimental basis. In 1994, the American based Vancom company acquired a temporary concession in the Maastricht area. Further tendering has been blocked temporarily, after a lawsuit initiated by the labour union, demanding a job guarantee for the personnel of the present concessionaires.

In The Netherlands, local authorities traditionally have the responsibility for pupil transport to primary and special schools. This is a less substantial obligation than in Germany. Since 1 April 1994 they have also had the responsibility for disabled transport. Nation-wide about 0,75 billion guilders (333 million Euro) per year will be spent on this transport, about three times as much as on pupil transport.

The previous developments no doubt are positive for openness from a national perspective. Increased regional responsibilities, structured by means of funding and a new planning structure are present on both sides of the border, as is the intention to introduce competition. There are, however, large differences in the original situation and in the character of the changes in the two border regions, which prevent a substantial increase in cross-border openness.

Change at the corporate level

In The Netherlands the National Railway Company NS, owned quite a number of local lines and rural tramways. When these were closed NS busses took over the services. When regional bus transport became unprofitable NS gradually bought these companies so that around 1970 national and regional transport were almost in one hand and subsidised by one agent: the national government. In the nineteen-seventies NS decided to have no further interest in unprofitable bus operations and a separate national holding was created with the national government as the only shareholder: VSN (United Dutch Regional Transport). In the end (about 1990) the conglomerate received subsidy of 60 to 70 per cent.

In Germany the national railway company was less dominating and so was its bus company, operating on former railway routes, as railway replacement transport. Yet it was of considerable importance after the inclusion of the Postbus, having line concessions all over the country. During the 1980s the Bahnbus company was made independent of DB and was split up into large regional companies like WEB (Weser Ems Bus) along The Netherlands - Niedersachsen boundary. These and local companies can almost do without operating subsidies by subcontracting, i.e.

hiring often marginal companies with ill paid labour, and by offering more restricted services than their Dutch counterparts. As a rule there are few evening services and few services on Sunday. In the *Landkreis* Leer for instance, all bus routes on the west bank of the Ems, operated by WEB and Auto-Fischer Leer, are without services on Sunday.

The Dutch government has tried to reduce bus subsidies by close control of the companies involved. In the end it realised that this policy only killed entrepreneurship. As a result, the companies were given more freedom and the duty to reduce loss to 50 per cent in about a decade, as well allowed a possibility to compete. As a reaction the national holding amalgamated a number of its companies and several bought a number of taxi firms. In this way the companies could face competition by their sheer size and by their more varied production capacity. Along with their touring branches they now offer cheap demand-responsive services, sometimes on a large scale. The Hermes Company, for instance, offers demand-responsive services all over its concession stretching from Vaals (across Aachen) to the central Dutch town of Utrecht. VEONN, operating in a great part of Drenthe, also offers this type of services under the epithet "Total Transport". Under pressure from the Ministry of Transport the holding is now being divided into two or more companies, to break its monopoly.

The railway company NS underwent the usual process of division into separate companies, within one holding, each bound to a budget and to generating income from other companies using their services, or from parties in transport or in exploitation of shops in station buildings or offices on NS property, etc. Further, facing the possibility of competition on unprofitable (subsidised) lines, NS created joint ventures with regional bus operators (VSN companies) to be able to offer efficient, integrated services. The newcomers, Lovers Rail and Vancom (bus), were associated in 1997 with foreign public transport providers, i.e. French CGEA (Compagnie Générale d'Entreprises Automobiles) and the British Arriva, respectively. The latter company acquired two of the VSN subsidiaries, GADO and VEONN, covering the Dutch part of the EDR.

The German picture is complex, with a growing number of regional railway companies. Sometimes these are bus operators that still had a freight railway branch, like the Bentheimer Eisenbahn or the Dürener Kreisbahn near the Aachen border. Within the EDR there are only a number of traditional railways: the Emsländische Eisenbahn (freight) and the Island railways, transporting summer tourists. The numerous German bus operators with only a few line concessions are likely to be engaged in touring activities and pupil transport. Another development can be observed in Nordrhein-Westfalen. The government forced its concessionaires into co-operations by threatening to withdraw its vehicle subventions. One such union is the WVG (Westfalische Verkehrgemeinschaft). It is normal for German bus operators to have touring activities, except for co-operations like the WVG: its participants have these. The taxi market in Germany is a matter of innumerable local concessionaires. In the border region EDR, one can observe the Landkreisen trying to modernise their bus services and their providers. Aurich had sort of a co-operation already, co-ordinating seven providers. In 1997 a regional "Verkehrverbund Ostfriesland" was established. Yet, the changes are small and slow.

The picture all-in all, is slightly confusing. The power of large national monopolists is being broken, in The Netherlands probably more strongly than in Germany, which can be seen as favourable for openness. The old national companies, however, are unlikely to cross the border for competition. Moreover, the developments in the Dutch and German parts of the EDR are so different, that it can hardly favour competition from across the border.

Market segments and integration

Public transport serves different types of users. The most important factor in serving different user categories is access to private means of transport: non-captive and captive riders. Non-captive riders are those with a choice. They will use public transport when it is (considered to be) better, i.e. faster, more reliable, cheaper, etc. Captive riders have different degrees of captivity. They may just temporarily not have a car available, or, the other extreme, not be able to walk. Someone who is able to cycle still has a considerable freedom of movement, dependent on the distance to activity places, the quality of the route and road (flatness, safety) and the weather. Traditional public transport, with fixed schedule, fixed route, fixed stops, can fulfil the needs of only particular types of captive riders. Passengers have to be able to walk independently to the (bus)stop, to board without help etc. and to find their way at the destination. Therefore, various special transport were developed to serve specific groups or activities, i.e. transport for the handicapped, for the elderly, transport to schools, hospitals and even to discotheques, the latter for safety reasons. In rural areas like the EDR, and certainly in its border zone, far away from Emden and Groningen, the only relevant types of professional special transport are pupil, handicapped and hospital transport (de Boer, 1991). The first two especially might in some degree be integrated into public transport.

Pupil transport was modestly developed in both countries until about 1970. A number of factors caused an unforeseen growth in later years. In Germany this was the creation of large school centres, in The Netherlands this was the intensive use of a range of special school types for children with learning difficulties, and in both countries this was the increased participation in an ever more specialised secondary and higher education. In Germany the cost of pupil transport in primary education amounted to about 2,5 billion Marks around 1990. The sorry state of public transport and of public finance were the reasons for a policy of integration, which was achieved almost completely. Rural bus lines are in fact 90 per cent pupil lines, which explains the absence of a Sunday service as mentioned in the previous section. In The Netherlands this type of integration was never that extreme, but due to increasing costs a situation was created in which the municipalities became responsible for the integration of transport to special schools into public transport, funded by the national government.

The most remarkable difference in policies for captive categories can be observed in the policy for transport of disabled people. In Germany qualified handicapped have free rides in existing, mostly traditional local transport systems. In smaller towns up to 50 per cent of local rides are made by this category. This situation explains the remarkable pressure for adapted vehicles in public transport. Low entry bus and tramway systems are far more common in Germany than in The Netherlands. On rural lines in the EDR, however, it is hardly found. This may have to do with the small scale of operations, which makes innovation difficult. Apart from concessionaire fares (general price reductions) there are no special provisions for the elderly. In The Netherlands low income disabled were until 1994 entitled to a "transport provision" provided by the social security administration, which was as a rule a financial provision. When under a new law the number of those entitled to a provision doubled by granting the elderly and the disabled equal rights, the municipalities were given the responsibility and the existing budget. Accordingly, they massively introduced an efficient system of collective door-to-door transport, to also be used by able elderly. Most municipalities in the EDR contracted either the GADO or the VEONN bus companies to take care of this, the latter one with a public system. Nation-wide the area had the highest coverage with this type of system (de Boer and Diepens, 1995).

Different captive categories are served best by public transport when this is adapted to their specific needs. Door-to-door transport or dial-a-ride with small vehicles is a good solution. Dial-a-ride systems have a strong tradition in Germany. The dial-a-ride taxi is a normal feature in public transport, but predominantly at marginal times in the (sub)urban fringes of towns. It is more a supplement to standard line transport than a provision to improve the mobility of the handicapped. In the EDR there is only a local system in the municipality of Rhauderfehn. In The Netherlands dial-a-ride systems were much less prominent. The form of a demand responsive fixed route operation was more common, i.e. a line bus that departs only when at least one passenger has telephoned before. There is a strong movement to integrate door-to-door systems, which sprang up in abundance under the disabled provisions law, into regular public transport. The first large system was introduced in 1997, in the region of the Achterhoek, and the second one in 1998 in the region of Oldambt. In the latter, the province and five municipalities are combining their means to provide line transport where possible, and demand-responsive transport where and for whom necessary. The province, responsible for public transport, used predominantly by pupils of secondary schools, spent about 2,5 million guilders (1,1 million Euro) in the old line system. The municipalities spent 1 million guilders (0,44 million Euro) on pupil transport to primary and special schools and had a budget of 3,5 million guilders (1,6 million Euro) for the handicapped.

In Germany *integration* seems to be almost perfect: pupil transport integrated into public transport, disabled people given free public transport; however, the modest bus schedules, the lack of door-to-door transport and lack of low entry busses suppress the demand of the handicapped and still hinders their social integration. In The Netherlands integration is developing in a more promising manner than in Germany, i.e. an integrated public transport with coherent line and door-to-door components providing especially for the handicapped with better opportunities for social integration.

Developing at a different pace, a widening border gap?

Looking back to the different aspects of public transport in the border regions, the following lines have become apparent. Market regulation has been strong in both countries. It is diminishing now, but more rapidly in The Netherlands than in Germany. In the EDR, this contrast between the two countries is clearly present. With regard to changes at the corporate level no clear-cut conclusions could be drawn. Developments in The Netherlands went more towards a monopoly than in Germany, but the latter country had, by contrast, a fragmented system of bus companies. On the national level, Germany with its many private railway companies shows a more positive development; however, there are no signs of this within the EDR. Furthermore, the integration of different market segments was better in Germany than in The Netherlands, but developments in the latter country are quite promising, especially in the Dutch part of the EDR.

The following conclusions seem to be justified. The situation in both countries in the past is roughly comparable in that the systems were relatively closed. The development is towards more openness within the systems, with different accents in both countries. In addition, the speed of change in the Dutch and the German parts of the EDR is different, leading to an increase of dissimilarities.

Public Transport Across the Border

Modest expectations

It has been mentioned previously that from an economic point of view, the Dutch and the German parts of the EDR have little to attract actors and activities from across the border. In addition, the previous section has shown that the organisation of public transport and the adjustments to new needs are also different. This situation seems typical for the disparities between the two societies in general, as caused under the influence of the respective national states and institutional systems. Given these circumstances, one might expect a modest size of cross-border public transport: the supply will suffer from institutional barriers, the demand from disinterest. In this section, cross-border public transport will be analysed on three levels, i.e. rail infrastructure, rail and bus services, and the demand for services. The EDR will be treated systematically against the background of other Dutch German border-regions.

Cross-border railways

The two parts of EDR are connected only by one railway line for passenger transport, i.e. the line from the city of Groningen, via Nieuweschans to Ihrhove on the German north-south main line along the river Ems. The

German part is of low quality. The maximum speed is only 60 km/h and at a few spots, the train has to slow down because of dangerous crossings (de Boer, von der Brelie, 1991). This sharply contrasts with the Dutch part: where on the section Groningen – Nieuweschans a speed of 120 km/h is allowed. The low quality on the German side is typical for a line the national company would like to give up. Operations on this line are a matter of co-operation between the national companies DB and NS. In this case, DB trains are used because the Dutch ones do not comply with German safety regulations, whereas the personnel is Dutch.

The railway system of both countries shows the effect of the border: cul de sac railways on both sides, ending for instance in Dutch Musselkanaal (disused). Some of these have never been connected. For example, the Dutch rail infrastructure reached the border at Ter Apel from Musselkanaal, but the Germans did not build their part (Waldorp, 1984). As a rule the dead ends were connected at some point in time to the network across the border, such as Enschede to German Gronau, and Bocholt and Borken to Dutch Winterswijk (*Figure 6.5*).

Several cross-border lines, including Groningen – Ihrhove, were built relatively early in the railway era. At those times, these lines were important for foreign trade, and competition between regions and railway companies created an abundance of connections. Groningen – Ihrhove was in fact the link from the harbour of Harlingen (province of Friesland) to the German hinterland (de Boer, 1988). The Dutch lines built after 1890 were usually economically marginal ones and were closed gradually, sometimes after only a few years of service. Of the fourteen connections for passenger transport across the German border in 1931 only five remain. Cross-border transport seems to have suffered more from mass motorization than inland transport. No doubt the formation of national railway companies was an additional cause, because these companies preferred to focus on a few lines for reasons of profit. Against this background, the Dutch and Belgian networks were disconnected at the border close to the city of Eindhoven, in spite of protests from this city, the fifth in population size in The Netherlands and headquarters of Philips (de Boer and de Haes, 1997). The remaining four cross-border lines are of better quality than Groningen – Ihrhove, only Heerlen – Herzogenrath is a single-track diesel line. The other ones are electrified and double track, with the exception of Venlo – Viersen.

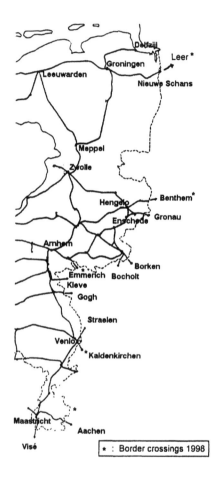

Figure 6.5 Railways for passenger transport across the German border in 1885 and 1998

Source: Jonckers Nieboer, 1938 (for 1885)

Level of services

One may distinguish various aspects of services in public transport. This section will focus on the time-related aspects:

- service period, the time between the first departure and last arrival, and

- number of departures and arrivals.

With regard to rail, one can observe two lines with passenger services in the early 1980s, that were later closed, i.e. the line from Nijmegen to Kleve and the one from Maastricht to Aachen (section Simpelveld – Aachen). Both date back to before 1890 and had a purely local function. The Maastricht-Aachen service was replaced with a service on the line Heerlen – Herzogenrath, which was used only as a cargo line before.

As far as the service period is concerned, it is clear that a 17-h period, standard in The Netherlands, is reached nowhere, but Heerlen – Aachen comes closest. The EDR line between Groningen and Leer, is the worst example, with less than 10 h in the direction of Germany. The first departure is often rather late, like on the Groningen – Leer line, at 8.23 h. This situation implies that the train cannot be used for commuting and educational trips. In fact local commuter trains have disappeared from the Hengelo – Rheine and Arnhem – Emmerich lines. A distinct improvement is however found on the Heerlen – Aachen line. The latter service is in fact the only one that operates on a regular hourly base. It has the highest train frequency of all cross-border connections. In general, one can say that the southern lines show most improvement. By contrast, Hengelo – Rheine, carrying German Interregio trains all the way from Amsterdam to Berlin, has dropped back to merely five train pairs per day, in spite of its designation for Eurocities in the Dutch national transport plan SVV II. Groningen – Leer is even worse, but it gained one train pair since 1983.

Two lines, the ones from Arnhem and Hengelo to Germany, base their service on truly international traffic, the relationship with Rhineland being stronger than that with Westfalen and Niedersachsen. In addition, the southern connections, from Venlo and from Heerlen, run through highly urbanised areas. Heerlen has a population of 90,000, Aachen one of 230,000. The "Heerlaken" conurbation numbers about 750,000 inhabitants. By contrast, the poorest service, Groningen – Leer, runs through sparsely populated area and was maintained in the past as a matter of principle, i.e. the detour over Hengelo was thought to be too large (de Boer et al., 1991).

There is quite a lot of bus transport across the border, such as the long distance "Euro" or "Inter" bus services of different companies, shuttles to holiday resorts, organised shopping trips to German towns like Oldenburg, Münster, Düsseldorf, etc. With regard to standard public transport, there is a considerable number of cross-border bus lines (*Table 6.1*). A few cities have more than one line across the national border: i.e. Bocholt/Winterswijk (2) and the Heerlen – Aachen city region (5). The rule seems to be that only regional centres close to the border can offer bus transport of any substance. Venlo, Bocholt, Nordhorn and Gronau (across Enschede) are above all attractive shopping centres.

In the EDR large (shopping) centres are relatively far from the border. The German towns of Leer, Papenburg and Meppen for instance are all on the east bank of the river Ems. This explains why, in 1990 there was only one line, Nieuweschans – Leer. In fact, this line was intended to support the poor service of the railway line, as rail replacement transport. The sum of the frequencies of train and bus is only 8 per day, whereas Nieuweschans has 23 trains per day in the direction of Groningen and Bunde, 4 km across the border, 17 busses per day to Leer. The additional two lines, connecting either small or distant towns, were the result of the EDR public transport program.

With a maximum of 12 daily departures on three routes the EDR has no doubt by far the worst cross-border public transport of all border regions. The difference with the inland public transport can be demonstrated easily by drawing a line at a distance of 5 km from the border, still east of the regional centres, and counting the number of bus routes crossing it, i.e. 12 or fourfold. The number of departures is about twenty-fold.

Demand for public transport

There is of course interaction between supply and demand in public transport; however, it is not possible to disentangle them. What can be observed is the actual use of systems that cross borders and passenger motives for crossing borders. There is, however, no recent and no systematic information on public transport use across the Dutch – German border. Consequently, we make use of data for 1985 on the Nordrhein-Westfalen border (Jansen, 1988). Only 1.3 per cent of those passengers passing the border took the train.

The railway crossings at Venlo, Arnhem and Hengelo had almost equal numbers of passengers: 430,000, 550,000 and 550,000, respectively. Of the 33,6 million motor vehicles crossing the border, 203,000 or 0.6 per cent were busses. The occupation rate of these busses was unknown.

**Table 6.1 Cross-border bus connections between Germany and The
Netherlands: service period and frequency (1996)**

Origin	Destination	Company (country), line number	Service period	Frequency (per day)
Nieuwe-schans	Leer	Weser-Ems-Busverkehr (DE)	09.15 – 17.15	4
Ter Apel	Haren	Levelink (DE)	08.50 – 17.45	3
Emmen	Meppen	Levelink (DE)	10.17 – 17.17	3
Ensche-de	Gronau	Oostnet (NL), nr. 3	08.10 – 17.10	9
Winters-wijk	Vreden	Oostnet (NL), buurtbus 196	07.50 – 18.45	12
Winters-wijk	Bocholt	Oostnet (NL), nr. (8)40	10.45 – 17.49	4
Doetin-chem	Emmerich	Oostnet (NL), nr. 24	07.12 – 16.42	10
Nijme-gen	Kleve	NIAG (DE), nr. 58	07.29 – 19.40	13
Venlo	Duisburg	NIAG (DE), nr. 929	07.38 – 17.05	6
Sittard	Heinsberg	KWH, nr. 36	05.21 – 18.48	8
Heerlen	Rimburg	Hermes (NL), nr. 28	07.24 – 19.30	13
Valken-burg	Herzogen-rath	Hermes (NL), nr. 47	06.48 – 16.26	12
Heerlen	Aachen	Hermes (NL), nr. 44	05.51 – 23.21	32
Kerkra-de	Aachen	ASEAG (DE), nr. 34	05.24 – 21.09	30
Maas-tricht	Aachen	VSN/Hermes (NL), Interliner 420	06.39 – 21.05	23
Vaals	Aachen	ASEAG (DE), nr. 15, 33, 65	06.03 – 22.59	66

Source: Published schedules and van der Borgt (1998).

With regard to travel motives the following results were obtained for inland and cross border travel (*Table 6.2*). In cross-border trips, the work and education motives are negligible, the shopping motive being dominant.

Table 6.2 Motives for national and cross-border traffic (percent share)

Motives	Germany	The Netherlands	Cross-Border
Commuting	24	17	3
Education	41	40	7
Shopping	11	13	63
Social	9	8	19
Recreation	*	4	3
Medical visits	7	-	4
Miscellaneous	8	8	1
Totals	100	100	100

* Included in miscellaneous

Source: Jansen (1988), adapted from Schulten (1984).

For the region of Aachen more recent data on public transport are available. The Dutch "Transport Region South Limburg" estimates a 2 per cent share for public transport across the border. More interesting are the results on travel motives. There is a remarkably higher share for work and education, i.e. 38 per cent, compared with the one for the Nordrhein-Westfalen border, which is 10 per cent. On the basis of the present scanty evidence one wonders whether there is a dividing impact of the border, i.e. whether flows across the border are smaller than on each side of it, and whether certain motives connected with public transport are influenced by the border. In the remaining part of this section, we present some more data on traffic and travel motives, with special attention for commuting, a motive that seems to be "suppressed" by the border. It needs to be mentioned that answering the questions above suffers from a lack of systematic data. National highway authorities do not register traffic outside their jurisdiction. Moreover they do not collect information on issues that are seen as not problematic: small traffic. We acquired, nevertheless, information from the respective highway authorities (North- and East-Netherlands) for a few highways.

Data concerning traffic flow on the A7 from the province of North-Holland through Friesland and Groningen to the border at Nieuweschans show a clear spatial variation, i.e. from about 5,900 vehicles per day to 69,400 (mean working day 1996). It is the contrast between the border at Nieuweschans and the Groningen city circular road. The small figure at Nieuweschans can not be explained by the empty land around it. The 32-km stretch from Den Oever (NH) to Zurich (F) lies on a dam in the water. Based on the distance to activity centres it should have had much less traffic than Nieuweschans, but there is considerably more, i.e. a total of 13,100 vehicles per day. This illustrates the border effect at Nieuweschans.

In *Table 6.3* travel motives for cross-border travel in the EDR are compared with those within Germany. The work motive, used in a broader sense than commuting, appears across the border to be less than half as important as in travel within Germany. The absence of travellers crossing the border with an educational motive is remarkable. The shopping and care motive(s) explain more than 50 per cent of travel. About a quarter of this is for filling up with petrol in Germany, where taxes are lower.

Table 6.3 Travel motives inside Germany (1989) and in cross-border travel in the EDR (1991) (% share)

Motives	*Germany*	*EDR border*
Work	26.8	12.8
Education	6.9	0.5
Shopping and care	27.4	51.3
Recreation	35.3	34.9
Other/unknown	3.6	0.5
Totals	100	100

Source: Von der Brelie and Wilkes (1992).

Data on working across the border were also available from social security, enabling a comparison for all border regions in The Netherlands (*Figure 6.6*). People working in Germany and living in The Netherlands are quite unevenly spread: about 60 per cent live in the province of Limburg. The numbers for Drenthe and Groningen are small, compared with other provinces and compared with commuting within the country. *Table 6.4* shows data for the five German EDR Kreisen, based on German health insurance registration. The latter source implies an underestimation, giving rise to a need for a cautious interpretation of the data. In 1994 only the (land) border Kreisen (Emst and Leer) had Dutch commuters. Their share in incoming commuting is very small, i.e. 2 per cent. Overall, the large number of incoming commuters can be ascribed to one employer: the Meyer shipyard in the town of Papenburg (Zanen et al., 1997).

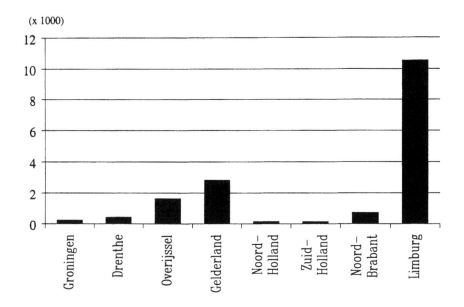

Figure 6.6 Cross-border commuting per province in The Netherlands

Source: Zanen et al., 1997.

There is a striking imbalance along the border in general: for each commuter from Germany there are ten from The Netherlands. This can be explained by two factors: a higher level of wages in Germany (attractive for the Dutch) and cheaper housing in The Netherlands, attractive for Germans who want to keep their German jobs. The latter is however, of little importance in the EDR: of the 287 people registered in 1994 54 were Germans living in The Netherlands (*Table 6.4*).

Table 6.4 **Commuters working in the German part of the EDR and insured somewhere else (1992/1994)**

	Emden	LK Aurich	LK Emst	LK Leer	LK Wittm.	German EDR
Incoming commuters						
1992	16,960	5,479	13,101	5,232	2,958	-
1994	15,170	6,031	12,592	5,286	3,087	15,551
Change (%)	-10.6	+10.1	+4.1	+1.0	+4.4	-
Of which from NL:						
1992	-	-	196	11	12	219
1994	-	-	261	26	-	287
Change (%)	-	-	+33.2	+136.4	-	+31.0
Of which foreigners (1994)	-	-	216	17	-	233
Insured employees (1994)	27,582	10,733	82,694	33,641	13,040	198,054

Source: Zanen et al. (1997).

With regard to another border region, the southern of the province of Limburg and adjacent Belgian and German regions, Breuer et al. (1993) discuss the phenomenon of imbalance. The imbalance in commuting between the Dutch and Belgian regions with Germany is largely due to the scientific and research potential of the Aachen region, including a university of technology for example. The relations with Aachen across the border are traditional, but the open borders invite people to combine the qualities of two countries for their daily life. In fact, Aachen is so close to the border that it has Belgian and Dutch suburbs, offering cheap housing for German migrants. Furthermore, in Belgium the taxation system is a factor of attraction. It explains the substantial commuting to Maastricht and environment, but the automobile and chemical industries north of Maastricht (Geleen) might also influence the pattern of commuting.

In general there are signs of a growth in commuting, due to open borders and better taxation arrangements for working across the border. A trend for growth can also be observed in the EDR (Zanen et al., 1997). In the Interreg-II Programme of the Euregion to the south of the EDR, growth of 200 per cent between 1983 and 1992 is indicated; however, in this particular case the growth can be largely ascribed to the establishment of Dutch firms just across the border. Strict Dutch planning control is likely to be the most important reason for this phenomenon.

The border between EU members has been opened in the sense that there is a free movement of people and commodities across the border; however, this movement is considerably smaller than within each of the countries. *Institutional barriers* or rather differences, seem to be the cause of this. As Zanen et al. state (1997): 'People don't look for work across the border, because they think everything is different there, and that is true indeed'. In addition, the national language constitutes an important barrier, in spite of the fact that the dialect is the same on both sides of the border. On the Dutch side of the Aachen border the dialect of towns like Kerkrade belongs sooner to the German domain than to the Dutch. Dialects however, are dying out gradually and education is monolingual. Another factor is the difference in school systems, i.e. the age at which obligatory education starts, the number of years spent in certain levels (primary, secondary), the character of (or school type in) secondary education, and the formal relation with employment (de Boer et al., 1991). Only mutual access to higher education has been arranged formally. Despite the latter situation, the university of Groningen, the only one in the EDR, had less than 10 German students in 1991. Early 1998 this was 65, out of a total of 17,750 students, and most of these were likely due to EU exchange programmes. On the other side, a considerable portion of cross-border travel is just created by institutional differences, especially the ones in taxation. This concerns also shopping, with petrol, alcohol and coffee as major examples.

Conclusions on cross-border transport

The supply of public transport across the border is generally poor, with the EDR in a sparsely populated border zone as the worst case. There are fewer railway lines, fewer bus routes, and fewer departures on those lines and routes, compared to the region of Aachen with a dense population on both sides of the border. Foreign operators are crossing the border only to the next town. The demand for public transport is considerably smaller than in inland transport. This is caused to a great extent by the virtual absence of travel motives like education and work, although especially the latter is growing fast in recent years. We may conclude that the border continues to be a barrier for these travel motives. There is no longer a formal barrier but a cleavage based on institutional differences, which makes the step across difficult. Much of the modest commuting that exists is based on a factor of attraction, i.e. higher wages in Germany and cheaper housing in The Netherlands.

Openness of EU Borders: Productive for the EDR?

The above analysis of cross-border interaction in the EDR has raised questions about using openness in view of the development of the region. The Dutch and German parts have little advantages to gain from integration, because there is little to attract people and activities from across the border. With regard to the specific perspective of this analysis, public transport, it is interesting to see that the European Community is putting on pressure to liberalise the sector. This implies that the market should be opened up for operators from across the borders. The process is underway in The Netherlands and Germany, be-it slowly and with different accents. In the EDR the differences are relatively large, which is not an inviting situation for foreign operators. Furthermore, the EDR has only very modest cross-border connections. The explanation is the institutional cleavage the border constitutes, i.e. concerning the official language, the administrative system etc. Therefore, important travel motives for public transport are absent.

Openness is no doubt essential for European integration. Our analysis shows, however, that it is not a sufficient condition for integration. In the EDR, the disadvantages of institutional differences are much bigger than the advantages enjoyed in the other country, and accordingly, openness will bring integration only very slowly. An integrated public transport market of some substance is not something for the near future.

References

AVV (Advies Dienst Verkeer en Vervoer) *Traffic data A12*, AVV, Rotterdam.
Boer, E. de (1998), 'Die nordniederländischen Eisenbahnen im Regionalschnellverkehr', in Akademie für Raumforschung und Landesplanung Hannover, *Regionalschnellverkehr in Norddeutschland und in den nördlichen Niederlanden*, Hannover, pp. 7-69.

Boer, E. de, and Brelie K. von der (1991), *Grenzüberschreitender öffentlicher Personennahverkehr, Entwicklung eines Konzeptes für die Ems-Dollart Region*, TU Delft, Delft.

Boer, E. de, Brelie, K. von der, and Wilkes, Chr. (1991), *Improving the railway Groningen – Leer – Oldenburg (– Bremen)*, Ems-Dollart Region.

Boer, E. de, and Diepens, J.M.H. (1995), 'WVG-vervoer tussen isolement en integratie', *Verkeerskunde Jaarboek 1995*, ANWB, The Hague, pp. 71–89.

Boer, E. de, and Haes H. de (1997), *Een railverbinding Eindhoven – Hasselt? Een verkenning naar een mogelijk alternatief voor uitbreiding van het wegennet*, TU Delft, Delft.

Boer, E. de, Nederveen, A.A.J. and Tacken, M. (1992), *Stelselmatige vergroting van woon-schoolafstanden in het voortgezet onderwijs*, Research Report, TU Delft.

Brelie, K. von der, and Wilkes, Chr. (1992), *Grenzüberschreitende Verkehre in der Ems-Dollart-Region, Untersuchung von Potentialen für den öffentlichen Verkehr*, Büro Seele, Aurich.

Breuer, H.W, Juchelka, R., and Vollings, A. (1993), *Die MHAL-Region und ihr öffentlicher Personenverkehr, Der grenzüberschreitende Bahn- und Busverkehr: aktuelle Situation und mögliche Entwicklungsansätze*, Aachen.

Borgt, R. van der (1998), *Grensoverschrijdend busverkeer*.
http://www.xs4all.nl/~rvdborgt/bus/grensbus.html

Eems-Dollard-Regio (1988), *Grensoverschrijdend actieprogramma Eems-Dollard-Regio 1988*.

Ems-Dollart-Region (1994), *Interreg II, Operationeel Programma 1994 – 1999*, Nieuweschans.

Jansen, P.G, and Rietmann, E. (1988), *Grenzüberschreitende Verkehrsverflechtungen im niederländisch/ nordrhein-westfälischen Grenzraum*, Deutsch-Niederlandische Raumordnungskommission – Unterkommission Süd, ILS, Dortmund.

Jonckers Nieboer, J.H. (1938), *Geschiedenis der Nederlandse Spoorwegen 1832 – 1938*.
Ostfriesland-Fahrplan 1994/95 (1994), Janssen Druck, Wittmund.

Pötsch, P. (1994), 'Grenzüberschreitender Zusammenarbeit im ÖPNV in der südlichen Oberrheinregion', *Eures Discussion Paper* dp-35, Freiburg i.B.

Schulten, P. (1984), *Der grenzüberschreitende öffentliche Personnahverkehr in der Euregio, Zusammenfassung der Ergebnisse*, Haaksbergen.

Waldorp, H. (1984), 'Ter Apel – Rijksgrens', *Op de Rails*, 52 (2), February 1984, pp. 48-52.

Zanen, T.J., Krüger, R., and Huebner, M. and Kröcher, U. (1997), *Een EURES-cross-border-project voor de Eems-Dollard regio – Haalbaarheidsstudie*, Groningen/Oldenburg.

7 The Determinants of Cross-Border Economic Relations. The Case of The Netherlands and Belgium

HENK VAN HOUTUM

In this chapter the position is taken that borders are not only barriers to cross-border interaction, but also relevant markers, at least for the inhabitants of a nation. In order to examine the economic influence of this latter aspect of borders the chapter focuses on the determinants of cross-border economic relationships of individual firms. The empirical study of the influence of borders on the internationalisation process of individual firms concerns the border regions of The Netherlands and Belgium. It was found that cross-border economic relations are significantly positively influenced by the personal willingness to enter cross-border relationships, be it economic or personal, and that entering into cross-border economic relationships depends heavily on the attitude towards the existence of the border and the perception of the difference in business conventions in the other country.

Introduction

One of the key issues of this volume is the openness of borders and the way regional actors may take advantage of this situation. Openness here refers to a free movement, to unhampered cross-border interaction (see also Ratti, 1993). Since regions neighbouring the borders are often peripheral, it seems reasonable for border regions to harmonise and "functionalise" the common space into an "active space" together with border regions on the other side of the border, which after all face the same problems and opportunities. A central thesis put forward in the field of regional economics and economic geography is that the existence of a border may hamper this active cross-border association. The border is seen as a barrier to interaction. Within this field, attention is mostly focused on exogenous obstacles or barriers when discussing cross-border interaction. It is argued that borders impede a smooth transfer or free movement of information and activities and should hence be seen as obstacles, obstructions to a higher level of integration and welfare. As such, borders cause non-linear shock-wise *discontinuities* in the cross-border flows of goods and services and the mobility of capital and labour by raising the accessibility costs.

This barrier effect of borders upon the mutual accessibility of European regions and cities has received a great deal of attention in regional economics and economic geography. Several scholars have attempted to quantify this physical impact of borders (see e.g. Bröcker, 1984; Rietveld and Janssen, 1990; Bruinsma, 1994, Plat and Raux, 1998). A large number of empirical studies have demonstrated the strong divisive effect of borders on interaction, cross-border relationships and integration (see van Houtum et al., 1996). Yet, most scholars argue that there is no simple relationship between cross-border interaction and the border. For instance, it is not clear to what extent a low level of cross-border interaction can be explained by a shortage of border crossing infrastructure (Plat and Raux, 1998). One may pose the question whether shortage of infrastructure is the result or the explanation. What is more, some scholars doubt whether the influence of the border can be traced back merely to its physical barrier effects (Bruinsma, 1994; Plat and Raux, 1998).

In this chapter the latter issue is examined in more detail while stressing another pressing theme connected with the explanation of obstacles in cross-border interaction and openness. A key argument in this chapter is that to understand the impact of borders on cross-border interaction and openness, one must take into account that *state* borders are barriers and that borders are at the same time relevant and socially wanted in the perception of the people of the *nation*. The latter aspect might also influence the degree of openness of countries and border regions.

The perception that borders are relevant can be traced back to legally and morally claimed and felt rights to protect and defend one's own country. In other terms, it is an expression of control of power over territory. This control function, the most essential and primary function of borders, is also economically important. It creates security and certainty within one's own domain, thereby enhancing the possibility of active socialisation within and identification with a certain space, and increasing trust and solidarity among the inhabitants of the country. In turn, this may have an important effect on the interaction between people and common welfare. What results is a trade-off between welfare via higher integration on the one hand and welfare via the trust in interaction encouraged by nationalism on the other hand. Still, within economics and economic geography little research has been devoted to these non-physical border effects on interaction. These endogenous obstacles and barriers to openness are explored in this chapter. Openness is herein viewed in terms of internationalisation of firms. Central in the discussion is the view that the extent of internationalisation is dependent on the willingness of the actors involved. Therefore, not the accessibility but the *intention* of individual actors is the focal point of attention in this analysis. Without the (latent) intentions of the entrepreneur, the start up of an international economic action will not even be sought. The question to be answered here is: what are the determinants of cross-border activities? In particular: is it the barrier effect of borders present despite the intention to cross borders, or is it the absent intention of the actors present despite a "barrierless" border that causes cross-border relationships still to be an exception rather than the rule?

With this in mind, the "active space" approach in border regions is explored theoretically in the next section. "Active space" is investigated here by using cross-border economic relationships between firms in border regions as an indicator. In this context, a set of hypotheses on the formation of these relationships will be formulated. This is followed by an answer to the question of whether the border is truly open or not, in other words, whether the border matters in the border regions of The Netherlands and Belgium, as indicated by the spread of economic relations of firms over the two countries. In the section that follows, the hypotheses are verified and the observed cross-border pattern is explained. The chapter concludes with a discussion of the merits of the approach used in this analysis and with proposals for future research.

What Explains the Process of Internationalisation?

In this section, the process of internationalisation is examined theoretically in terms of the number of cross-border economic relations of individual firms. Based on various theories hypotheses on the possible determinants of internationalisation are formulated. The determinants found in the theory can be summarised under six headings[1], i.e. the spatial growth of the firm, mental distance, the relationship preference, personal and professional networks in the foreign country, the degree of feeling at home in the culture of the neighbouring country, and border evaluation. Note that both the theoretical and the empirical analysis principally apply to small and medium-sized enterprises.

Spatial growth of the firm

The theory of the growth of a firm explains why and how firms develop over time (see e.g. Penrose, 1959). Håkanson has added the spatial component to this theory, by modelling the typical growth pattern of a firm in the process (stages) of internationalisation (Håkanson, 1979). Following this view, internationalisation can best be understood as a process of gradually widening circles in the water once the stone has been thrown in. Internationalisation can be seen as a stage model of decisions, that starts with the expansion of sales on the domestic market, followed by exports of the products, the creation of dyadic long-term international economic relationships with foreign firms and ends up with the establishment of a new firm in the foreign country. This evolutionary stage model is theoretically embedded in most internationalisation theories and has received strong empirical support in many studies (see e.g. Johanson and Wiedersheim-Paul, 1975; Bilkey and Tesar, 1977; Johanson and Vahlne, 1977; Cavusgil, 1984; Leonidou and Katsikeas, 1996). The development of the size of the firm and the development of the spatial spread of the activities of the firm are the indicators used here to describe this evolutionary internationalisation process. The following hypotheses can be formulated based on this concept (*Hypothesis 1*):

a   *Export percentage in the neighbouring country:* the higher the export percentage to the neighbouring country, the more economic relationships it will have in the neighbouring country.

---

1 Note that, by focusing on the individual actors, this chapter does not pay attention to the interaction with the other actor. It thereby abstracts from the step-wise process that internationalisation actually is (see van Houtum, 1998). The explanation given here can therefore only be partial.

b *The number of economic relationships in the home country:* the more economic relations in the home country, the more economic relationships it will have in the neighbouring country.

c *The number of active persons:* the greater the size of the company, the more economic relationships it will have in the neighbouring country.

d *The percentage of cross-border workers employed:* the higher the percentage of cross-border workers employed by the enterprise, the more economic relationships it will have in the neighbouring country.

e *Importance of the sector:* if the company belongs to the manufacturing industry, it will have more economic relationships in the neighbouring country; a construction company will have a smaller number of economic relations in the neighbouring country.

Mental distance

The next determinant involves the perception of entrepreneurs of the effect of cultural differences. Next to size characteristics of firms and the spatial spread of their activities, another crucial characteristic, maybe *the* determinant in the theory of the growth of firms, as Penrose (1959) saw it, is the "productive opportunity" of entrepreneurs. According to Penrose, the growth of firms should not be seen as a *deux ex machina*, but as a result of productive possibilities that entrepreneurs see and can take advantage of (Penrose, 1959). The perception and keenness of the entrepreneurs is also a crucial determinant in the internationalisation process. The decision makers' present knowledge as well as their perception of the profitability and the risk of internationalisation is then the issue. This view of the decision maker in the internationalisation process plays a key role in what can be called the Swedish evolutionary internationalisation model (see Johanson and Wiedersheim-Paul, 1975; Johanson and Vahlne, 1977). Central in this influential model is the argument that, starting from the assumption of bounded rationality (Simon, 1961), obstacles to internationalisation can be reduced through incremental decision-making and learning about the foreign markets and operations (Johanson and Wiedersheim-Paul, 1975). The interesting conclusion is that internationalisation starts with entering into countries that are psychologically close to one's own country, that is, those countries that are comparatively well-known and similar in business practices. One of the drawbacks of this theory, however, is that the measurement of the perception of obstacles and chances to internationalisation is not done at the level of individual entrepreneurs but strangely enough at the macro level, the level of countries (see van Houtum, 1998). Knowledge is assumed to be vested in the decision-making system of a society. As a result, the model does not deal with the individual decision-maker. What has resulted therefore, in the empirical tests of this model so far, is a comparison between the (cultural) differences of different countries instead of an individual estimation of and attitude towards these differences.

Another much cited internationalisation model in which perception plays a key role is that of Bilkey and Tesar (1977); this was explicitly set up for, and measured at, the level of individual entrepreneurs. This model included managerial attitudes on the effect of exporting for the firm as well as perceived barriers to exporting. Although the authors did not include the estimation of (the influence of) differences in business practices in their model, their conclusion resembles the Swedish internationalisation model in that learning effects are important. Therefore, these scholars also suggest that firms start with entering into psychologically "close" countries. Both, the Johanson et al. model and the Bilkey and Tesar model are much cited and used in the international business literature, as standards for the stage model theory of internationalisation (Cavusgil, 1980; Czinkota, 1982; Bradley, 1995; Crick, 1995).

The consequence of the assumption of bounded rationality, implying that an entrepreneur is physically incapable of gathering all the information needed or wanted, is that the entrepreneur lacks the knowledge to make an optimal choice. In this chapter, the thesis is put forward that entrepreneurs are not only physically bounded in their rationality, according to the previously discussed internationalisation theories, but also mentally, implying that their rationality and horizon of opportunities, may well be bounded by uncertainty-avoiding and conventional behaviour (van Houtum, 1999b). It is argued here that what determines the perception of entrepreneurs is the degree to which they are attached to the certainty created by a particular form of socially learned and accepted behaviour, and to breaking certain patterns and generating innovations in their production process and/or market orientation. It is expected that this trade-off between certainty and uncertainty determines, to a great extent, the perception towards entering into and developing economic relationships in the neighbouring country. This is called mental distance (van Houtum, 1998, 1999b). Mental distance is measured at the individual level and is defined as the estimation by entrepreneurs of the differences and the consequences of these differences in business conventions between a foreign country and the home country. Business conventions are defined as the socio-economic conditions for doing business, socio-cultural conditions (including language), and legal-administrative preconditions (Storper, 1997).

Entrepreneurs estimate the differences between business conventions and evaluate them with regard to the success of the relationship. The greater they perceive the differences in having relationships with entrepreneurs in the home country compared with having them in neighbouring countries, and the more negative their evaluation of these differences, the greater the mental distance is with regard to such relations in the neighbouring country. The expectation then is that the perception of great differences leads to refraining from establishing contacts and relationships with entrepreneurs in the neighbouring country. The reasoning behind this expectation is that great differences lead to greater adaptations and efforts to make the relationship in the neighbouring country to a comparable success. In other words, a greater investment is required, costing more mental effort, money, and time. As a consequence of these differences, there is greater uncertainty with regard to economic relationships in the neighbouring country. Entrepreneurs will wish to safeguard against this uncertainty, which leads to higher transaction costs and greater pressure upon the trust in one another that is required for the success of economic relationships.

The previous reasoning leads to *Hypothesis 2* concerning *mental distance*: the greater the entrepreneur perceives a mental distance between the home and neighbouring countries to be, the smaller the number of economic relationships the enterprise will have in that country.

The relationship preference

The third determinant concerns the strategic preference of the firm. Whereas the previous two determinants focused on the process of entering into a foreign country, regardless of the entry mode of the firm, this determinant explicitly deals with starting up economic relationships. One may suppose that some entrepreneurs are more open to contacts and/or more active in seeking out contacts than others. Likewise, one may imagine various types of preferences with regard to relationships (van Houtum, 1999a). The following preferences are taken into account: preference for projects with high profits despite higher risk; preference for steady, long-term relationships; preference for relations with a large network of business contacts and for business contacts and relations consciously established in the neighbouring country; preference for relations at short distances;

preference for knowledge concerning price/quality ratio of alternative partners; and, preference for relations in the home country.

The previous reasoning leads to *Hypothesis 3* concerning the *type of relationship preference:* the more the entrepreneur prefers an active search for relations and contacts in the neighbouring country, the more economic relationships the enterprise will have in that country.

Personal and professional networks in the foreign country

Networks are a key issue in regional economics and economic geography today. At the same time, there is much debate on the right interpretation and most of all, the right measurement of the influence of networks. Only few of the many network approaches available today deal explicitly with the importance of networks in relation to the internationalisation process. One of the most influential and elaborated concepts is put forward by Swedish scholars (Johanson and Mattsson (1987, 1988) and Håkansson and Johanson (1988)). These scholars provide an image of an enterprise that is socially embedded in, and relies upon, the relationships with its personal contacts and economic relations. Within this image, a firm ideally seeks for those relationships that will produce the necessary transactions as reliably well and economically as possible. Finding these relationships, however, is a complex task in which much uncertainty is involved. Especially in a context of internationalisation, in which optimal exchange partners cannot be easily identified and the market situation may be somewhat alien, a firm may be expected to rely on foreign network relations that are well-known, preferably personal and already existing, to enter the foreign market (Powell, 1990; Podolny, 1994). In other words, the international network theory suggests that firms use their existing contacts to create new relationships. The "strategic network identity" of these contacts then becomes a crucial issue (see Håkansson and Johanson, 1988). This observation raises the question of how well existing contacts are capable of linking the firm to attractive new relationships and what kind of contacts are best qualified for this function.

The indirect influence of networks on the process of internationalisation is measured by using two distinct types of indirect networks. The first type of network distinguished here is the personal network in the neighbouring country. This kind of network refers to the *number* of *personal* acquaintances of an entrepreneur in a certain region, e.g. friends, family members, or other personal acquaintances. They are, in any case, acquaintances that are not professionally involved in the enterprise. A social network that crosses the border may be considered advantageous to the development of economic relationships. The second type of network is the *professional* network in the neighbouring country. Professional acquaintances are persons that are associated with the company of an entrepreneur due to their profession, but that are not economic relationships with other firms. This category includes, for example, chambers of commerce, colleagues, and informal investors.

Based on the previous arguments *Hypothesis 4* concerning the *network of acquaintances* says: the more personal and professional acquaintances the entrepreneur has in the neighbouring country, the more economic relationships the enterprise will have in that country.

Feeling at home in the culture of the neighbouring country

Next, the degree to which an individual actor is capable of empathising and feeling affinity with the culture prevailing in the neighbouring country is expected to be of relevance to cross-border economic relations in that country (cf. Riedel, 1994). The inclusion of this determinant is based on the assumption that it not only important to understand the estimation and evaluation of the cultural differences in business practices, but that it also matters in the internationalisation process to what extent the decision maker feels at home in the culture of foreign country involved. The assumption then is a positive relationship between feeling at home in the foreign country and the size of the international economic involvement in that country. Following Harris, culture is defined as '...the total socially acquired life-style of a group of people including patterned, repetitive ways of thinking, feeling, and acting.' (Harris, 1993: 104). A distinction is made between the culture of the neighbouring country as perceived by the entrepreneur as private individual and as business person. This implies that feeling at home is differentiated into culture of living and culture of business. Accordingly, *Hypothesis 5* runs as follows: the more the entrepreneur feels at home in the living and working environment of the neighbouring country, the more economic relationships his/her company will have in that country.

Border evaluation

As explained above, in (regional) economics, the border is usually incorporated into the analytical model as a *barrier* to (spatial) activity. The role played by the actor's attitude towards the border as a barrier is often given minor attention. An analysis of border evaluation, however, might give us more insight into the direct influence of the border on the attitude of people. It is not the function, but the symbolical value of the border as a barrier that is then the issue (see e.g. Anderson and O'Dowd, 1999; Paasi, 1999). Scholars, especially in environmental psychology and human geography (e.g. Leimgruber, 1980, 1991; Riedel, 1994; Paasi, 1996) have pointed to the *relevance* of the border, mostly in the context of the basic human need of social identification and protection. People consider the border more or less important or valuable to their occupations. It is worthwhile to examine the degree to which the border is actually evaluated as relevant and what impact this evaluation has on international economic activities, especially at a time when this symbolical value of borders seems to be becoming more rather than less important, because of a growing time-space convergence.

In short, in the present analysis, both the aspect of the border as *barrier* and its *relevance* will be considered. A suitable method to measure this symbolism and the value attached to the concept and phenomenon of borders analytically, is to use attitudes expressed when evaluating the border as valid indicators (cf. Reynolds and Mc Nulty, 1968; Leimgruber, 1980, 1991; Riedel, 1994) (see *Appendix 1*). The assumption then is that entrepreneurs who regard the national border as irrelevant and not as a barrier will have more economic relationships.

Based on the previous arguments *Hypothesis 6* concerning *border evaluation* reads: the more an entrepreneur regards the border as a barrier, the smaller the number of economic relationships he/she will have in that country; and the less relevant the entrepreneur regards the border to be, the more economic relationships the enterprise will have in that country.

Does the Border Matter?

The central question in this section is whether a discontinuity effect can be ascertained at the state border between Belgium and The Netherlands. To answer this question an extensive survey was held among small and medium sized companies in three regions on either side of the border between The Netherlands and Belgium, to be exact in the province of Zeeland (Zeeland Flanders and Central and North Zeeland) in The Netherlands and the region Gent/Eeklo in Belgium (van Houtum, 1998) (*Figure 7.1*). An extensive questionnaire was used in the survey. All companies in the survey belonged to the construction industry, manufacturing industry, and wholesale sector. A total of 470 companies was analysed, of which 161 were in the region Zeeland Flanders, 219 in Central and North Zeeland, and 90 in the region Gent/Eeklo. All of them were asked to fill out the questionnaire for transactions taking place with the same partner on a regular basis.

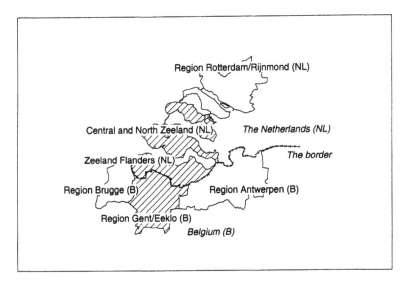

Figure 7.1 The research areas in The Netherlands and Belgium

It is worth noting that in the border regions involved the same language is spoken, although with a slightly different accent (Dutch and Flemish). The language as such therefore, cannot be seen a barrier to openness. What is more, since 1958 The Netherlands, Belgium and Luxembourg have formed an Economic Union, titled "Benelux", implying that all kinds of barriers for a free traffic of people and goods have been abolished and economic and financial policies are to some extent connected.

A first point of attention was the regional difference in the type of relations. It was found that national and international economic relationships are most often concerned with *supply and outsourcing* of production related activities, such as half-finished products and manufactured goods, for the firms in all three regions. *Sales market relationships*, such as with sales agents and market research organisations, rank in second place for the companies in the three research areas. *Business service relationships* are most often sought regionally or eventually somewhere else in the home country, but are generally less important than the market sales relationships. The number of *control relationships*[2] is the smallest for all firms. The probable reason for this phenomenon is that the relationship is not needed as much as sales market relationships. Besides it goes far beyond an exchange of commodities for money; it involves extensive co-operation with another company. Therefore, such a step is taken warily and relatively rarely.

Furthermore, it was found that the average number of relations per company in the home country and in the neighbouring country is unevenly distributed. The orientation is primarily national, and expands to the international level afterwards. It was found that in total, a number of 316 companies (67.2 per cent) in the response population have no relationship whatsoever in the neighbouring country as against 154 (32.8 per cent) with a relationship. The number of relationships in the home country is, on average, 4.6 times larger than in the neighbouring country and, on average, 23 times larger than in the directly neighbouring region of the neighbouring country. Moreover, it appeared that of the companies who do have economic connections in the neighbouring country, the majority has only one. These results vary strongly per sector: the manufacturing industry has by far the most frequent "one or more economic relations in the neighbouring country" (44.4 per cent), against 36.7 per cent in the wholesale sector and 17.5 per cent in construction.

The results differ strongly per region as well. *Figure 7.2* shows the distribution of the economic relationships of firms in the three research areas, i.e. Zeeland Flanders (ZF), Central and North Zeeland (CNZ), and Gent/Eeklo (G), over eight regions in The Netherlands and Belgium. The percentage of economic relationships as shown in the figure is computed by dividing the actual number of relationships in the different subregions by the total number of potentially possible relationships in the subregions given the amount of firms in the sample. The zero line in the figure can be regarded as the borderline between The Netherlands and Belgium. The horizontal axis

2 Control relationships are such economic relationships between enterprises that one may speak of a new (part) of the firm. A (partially) joint administration is kept. Examples are: joint venture, participation in other/own company, merger.

represents the percentage of economic relationships in the different regions. It should be noted that the figure concerns the total of the regional orientation differences for sales market relationships, service relationships, production process relationships, and control relationships.

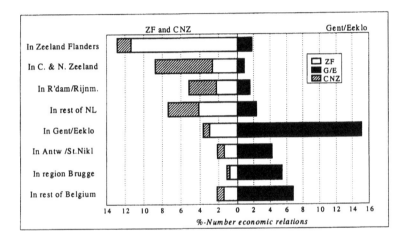

Figure 7.2 Number of economic relationships in The Netherlands and Belgium

The figure indicates that the greater majority of economic relationships is concerned with the home region. The proportion of national relationships clearly stands no comparison with the number of international economic relationships. In more detail, companies in Zeeland Flanders, relatively taken, have the greatest number of economic relationships in the neighbouring country (i.e. Belgium). They not only have significantly more supply and outsourcing relationships in Belgium than the companies in Central and North Zeeland, but also more of these relationships in Belgium than the companies in Gent/Eeklo have in The Netherlands. The estuary of the Westerschelde separating Zeeland Flanders and Central and North Zeeland apparently has an important dividing effect. Both Dutch regions are close to the border, but the Dutch region of Zeeland Flanders is separated from the rest of The Netherlands by the Westerschelde; a bridge or tunnel connection with the rest of The Netherlands is still missing. Accordingly, if one wants to avoid using a ferry, one has to travel via Belgium to reach

Zeeland Flanders over land (see *Figure 7.1*). The companies in Zeeland Flanders are therefore quite "close" to the neighbouring country.

Nevertheless, it is noteworthy that the companies in Zeeland Flanders have little economic interaction with the region of Brugge, while the physical distance between Zeeland Flanders and Brugge is equal to that between Zeeland Flanders and Gent/Eeklo. Apparently, the (functional) affinity with the region of Brugge is smaller. There is, however, interaction between firms in Gent/Eeklo and Terneuzen (Zeeland Flanders). The important channel between Gent and Terneuzen, linking the town of Gent to the Westerschelde, and hence the North Sea, and neutralising the barrier effect of the border, plays an important role in the explanation of this economic interaction (see also Allaert et al., 1991). In relative terms, companies in Zeeland Flanders have more relationships in Gent/Eeklo than in Central and North Zeeland. These differences in orientation should, however, not be overestimated. In total, including the companies in Zeeland Flanders, the effect of the state border on the geographical distribution of economic relationships is far greater than that of the 'water border' between Zeeland Flanders and the rest of The Netherlands. Hence, the institutional/political line between the two countries is stronger than the physical barriers to interaction. This is a striking result.

Belgium appears to be less important for companies in Central and North Zeeland; they consider the travel time too long or are more reluctant to enter into economic relationships in Belgium. The average number of relationships in the neighbouring country for the companies in the region is relatively low. Companies in Central and North Zeeland clearly enter more often into relationships in the region Rotterdam than in Zeeland Flanders or Belgium.

The companies in the region Gent/Eeklo have economic relations throughout The Netherlands. They are oriented more towards the rest of The Netherlands than towards Zeeland Flanders or Central and North Zeeland (*Figure 7.2*). Where companies in Zeeland Flanders seem to turn naturally towards Gent/Eeklo when entering into economic relationships in the neighbouring country, companies in Gent/Eeklo appear less eager to enter into economic relationships in Zeeland Flanders.

The above results lead to the following conclusion. There is no significant international co-operation between companies in the three research areas close to the Dutch-Belgium national border. Seen from the perspective of the three regions together, there is clearly a dividing impact of the state border on spatial economic relationships and from this perspective, the border does matter.

Hypotheses and Results

The fact that the border matters does not yet explain why this is the case. In other words, how can the divisive effect of the border be understood? To answer this question a multiple regression analysis is used to find an explanation for the number of economic relationships in the neighbouring country. In this way, the previously given hypotheses are set against the findings from the empirical research. The results are presented in *Table 7.1*.

In the first step the determinants were investigated using factor analysis. *Appendix 1* shows one example, i.e. the relationship preference. As a next step the items that emerged from the factor analysis were used in the multivariate analysis. No problems of collinearity were found to exist in the model. The complete model applied for the companies in all three regions and was found to explain the number of economic relations of the companies significantly[3].

A diversified picture emerges from the confrontation of theoretical hypotheses and empirical facts. The direction (positive and negative) of the significant variables conforms to the hypotheses; however, not all variables that were expected to play a crucial role appear to be significant in the explanation of the number of cross-border economic relations. *Appendix 2* shows the correlation between the independent variables and the number of cross-border economic relationships. A first remarkable observation is that the variable "number of professional acquaintances in the neighbouring country" does not significantly explain economic relations, while the variable "number of personal acquaintances" does.

Personal informal embeddedness in the society of the neighbouring country is apparently more important in explaining the number of cross-border economic relations than *professional* informal embeddedness. Another interesting observation is that whether the border is regarded as a barrier or not does not play a significant role in the number of economic relations. The perceived relevance of the border does, however, as predicted play a significant role in the number of cross-border economic relations.

---

3 $F(18/430)=13.21$, $p< 0.01$, adjusted R Square = 0.33.

Firms that regard the border as relevant, have less frequent and fewer cross-border economic relations. In addition, the dimensions of mental distance, in accordance with expectations, were significant and negative. The present results therefore, demonstrate that the perception and attitude of entrepreneurs vis-à-vis the border and the neighbouring country, as represented by mental distance, has a significant influence upon the number of cross-border economic relationship.

Table 7.1 Hypotheses versus empirical results

Content of hypotheses		Number of economic cross-border relations	
		Hypotheses	Results*
Feeling at home in culture of neighbouring country		+	- (n/s)
Mental distance	Expected negative effect of the relationship	-	- (s)
	Expected discrepancy with regard to business conventions	-	-- (s)
	Stringency in financial-economic conditions	-	-- (s)
Border evaluation	Barrier	-	+ (n/s)
	Irrelevance	+	+ (s)
Number of personal acquaintances		+	+++(s)
Number of professional acquaintances		+	+ (n/s)
Preference: networking		+	+ (s)
Preference: bold and well-informed search		+	- (n/s)
Preference: regionally/nationally bound		-	+ (n/s)
Age of the enterprise		+	- (n/s)
Size of the enterprise		+	- (n/s)
Export rate		+	+++(s)
Number of cross-border employees		+	- (n/s)
Number of economic relations in home country		+	+++(s)
Sector: manufacturing industry (yes/no)		+	++(s)
Sector: construction (yes/no)		-	- (n/s)

*n/s= not significant

Furthermore, the variable "relationship preference" can be subdivided into three categories (see *Table 7.1, Appendix 1*). Relationship preference characterised as "networking" appears to play a significant role in determining the number of cross-border economic relationships, which is according to expectations. Contrary to expectations, the other two types of relationship preference, "bold and well-informed" and "national/regional", are not found to be of significant importance. Among the variables concerning the growth of the firm, "export rate", "the number of economic relations in the home country", and the sector manufacturing industry are found again to be strongly positive and significant, thereby confirming the stage concept of internationalisation.

In sum, it was found that the number of economic relations in the home country, manufacturing as the sector, the export volume to the neighbouring country, the network characteristics of the entrepreneur, and a relatively small mental distance to the neighbouring country, are *decisive* factors in determining the size of the network of economic relations of a company in the neighbouring country. Apparently, the development path of the company and the entrepreneur's personal networks, perception and attitude are crucial factors in the explanation of the internationalisation pattern. The importance of the subjective evaluation of differences in doing business in the neighbouring country as opposed to the own country and of the concomitant consequences, is an innovative element in the explanation of the internationalisation pattern. It indicates that the rationality of the economic actors should not be overestimated and the role of perception and attitude should be incorporated.

Conclusions

In this chapter the aim was to clarify the influence of the national border, within the European Union, upon the development of international economic activities. To this purpose, the focus was on the number of cross-border economic relationships between companies in the border regions of The Netherlands and Belgium, i.e. Zeeland Flanders and Central and North Zeeland on the Dutch side of the border, and Gent/Eeklo on the Belgian side. Research hypotheses were formulated to explain the number of economic relationships. These hypotheses were put to the test using large-scale survey research.

The most important result found in the empirical analysis is that the networking of entrepreneurs as well as their attitude towards the neighbouring country matter. Despite the effects of globalisation and despite the efforts of the European Union and the Benelux Economic Union, the border is (still) a major issue in the development of cross-border economic relationships by firms in border regions of The Netherlands and Belgium. The abstract administrative border of the two countries, which has been demarcated in complicated and ambiguous circumstances, like wars and diplomatic trade-offs, and which is relatively young (officially since 1839), appears surprisingly hard to overcome. The national border seems to be rooted in the minds of people.

This research therefore, suggests that the traditional interpretation regarding borders as barriers in economic geography and regional economics is too narrow. A border of a country is not merely an obstruction, a hindrance to cross-border interaction. The border should also be seen as a relevant marker in a mental sense. The fact that borders exist is not without a reason. Despite the usual economics of scale arguments and political ambitions, it has to be realised that borders may serve other purposes than merely economic or political ones. National as well as regional communities are, and may explicitly want to be, distinguished (Cohen, 1993). Crossing the border is therefore, both a question of willingness or intention and breaking with the patterns of uncertainty-averse, routine behaviour. What matters is that the cross-border market is thereby divided in a spatial and in a mental sense. This leads to the interesting result that the actual feeling of "sameness" and recognition in terms of business conventions do not correspond with the small physical distance between firms on different sides of the border. Most importantly, this mental distance aspect of borders matters in managing openness. It has been demonstrated that the amount of mental distance has a proportionately negative effect on the number of cross-border economic relationships. The research results also suggest that it matters significantly what assumptions are being made on human economic behaviour. Entrepreneurs have a limited capacity for gathering the relevant information, and they also do not appear necessarily willing to do so. Entrepreneurs appear to possess certain perceptions and "belief sets" that need not necessarily align with what is economically realistic or desirable (van Houtum, 1999b). Uncertainty-avoiding behaviour is not determined by financial economic arguments alone. The difference between an action in the "here with one of us" and the "there with them" is for an important part connected to expectations rather than based on facts.

This result opens up a new stream of research questions. New questions that need to be answered are concerned with the way attitudes and expectations are formed in society, what role the border plays or has to play in this formation process, and why and how quickly the border is nested in the minds of people. These questions are under investigation in follow-up research. A suggestion for further research concerns a detailed analysis of the relation between the variables explaining cross-border economic relations, by way of Lisrel-analysis, such as between personal networks, feeling at home and mental distance. Another suggestion for future research is an analysis of the circumstances under which the border does not play (such) an imposing role as observed in the Dutch-Belgium border regions. Such an analysis would give better insight into the kind of (inter)activities that cause the additional challenges or uncertainties. Lastly, despite the extensive research on border regions in recent decades, it is still an open question, whether it matters for the way of entry, the speed, and the success of the internationalisation process where in a country firms are located. In other words: does the home region matter to the internationalisation process of firms? The often-cited and well-known analysis of Michael Porter, of the competitive advantage of nations, suggests that it does (Porter, 1990). His analysis however, does not concern individual firms but sectors (van Houtum and Boekema, 1995). It would be interesting to explore the applicability of Porter's concept on the evolutionary internationalisation process of individual small and medium sized enterprises. This would enrich the evolutionary theories on internationalisation, since these are mostly concerned with the geography of the country to be *entered* and not the geographic circumstances and conditions of the *home* country in which the internationalising firm is embedded.

The conclusion of this research is that internationalisation is a process that often starts in the home country. When a decision is made to initiate relationships with companies in other regions or even further away, the first step towards the neighbouring country has been made. In general therefore, international co-operation on a large scale is associated with relatively far-reaching national economic interweaving and a "border-crossing" perception. In short, the internationalisation process of small and medium sized firms is to a large extent a *mental* learning process.

References

Allaert, G. et al. (1991), *Een economische inventarisatie van de Kanaalzone Gent-Terneuzen. Een aanzet tot grensoverschrijdende samenwerking op economisch gebied.* Chambers of Commerce, Gent/Terneuzen.

Anderson, J. and O'Dowd, L. (1999), 'Borders, border regions and territoriality: contradictory meanings, changing significance', *Regional Studies*, vol. 33, pp. 593-604.

Bilkey W. and Tesar, G. (1977), 'The export behavior of smaller-sized Wisconsin manufacturing firms', *Journal of International Business Studies*, vol. 8(1), pp. 93-98.

Bradley, F.(1995), *International Marketing Strategy*, Prentice Hall, London.

Bröcker, J. (1984), 'How Do International Trade Barriers Affect Interregional Trade?', in A. E. Andersson, W. Isard and T. Puu (eds), *Regional and Industrial Theories*. Elsevier Science Publishers: North-Holland, Amsterdam, pp. 219-239.

Bruinsma, F. (1994). *De Invloed van Transportinfrastructuur op Ruimtelijke Patronen van Economische Activiteiten*, Nederlandse Geografische Studies, Utrecht/Amsterdam.

Cavusgil, S. (1980), 'On the internationalisation process of firms', *European Research*, November, pp. 273-281.

Cavusgil, S. (1984), 'Differences among exporting firms based on their degree of internationalization', *Journal of Business Research*, vol. 12, pp. 195-208.

Cohen, A.P. (1993), *The symbolic construction of community*, Routledge, London.

Crick, D. (1995), 'An investigation into the targeting of UK export assistance', *European Journal of Marketing*, vol. 58, pp. 1-21.

Czinkota, M. (1982), *Export development strategies: US promotion policy*, Praeger, New York.

Håkanson, L. (1979), 'Towards a theory of location and corporate growth', in I. Hamilton and G. Linge (eds), *Industrial Systems, Volume I*, John Wiley, Chichester, pp. 115-138.

Håkansson, H. and Johanson, J. (1988), 'Formal and informal cooperation strategies in international industrial networks', in F.J. Contractor and P. Lorange (eds), *Cooperative Strategies in International Business*, Lexington Books, Lexington MA, pp. 369-379.

Harris, M. (1993), *Culture, People, Nature: An Introduction to General Anthropology*, HarperCollins College Publishers, New York.

Houtum H. van, and Boekema, F. (1995), 'Regional economic competitiveness: Porter and beyond', in P. Beije and H. Nuys (eds), *The Dutch Diamond? The usefulness of Porter in analyzing small countries*, Garant, Leuven-Apeldoorn, pp. 185-218.

Houtum, H. van, (1998), *The Development of Cross-border Economic Relations; a theoretical and empirical study of the influence of the state border on the development of cross-border economic relations between firms in border regions of the Netherlands and Belgium*, CentER, Tilburg.

Houtum, H. van, (1999a), 'The action, affection, and cognition of entrepreneurs and cross-border economic relationships; the case of Zeeland-Gent/Eeklo', in G. Allaert and F. Boekema (eds), *Grensoverschrijdende activiteiten in beweging, onderzoek en beleid*, Van Gorcum, Assen, pp. 53-82.

Houtum, H. van (1999b), 'Internationalisation and mental borders', *Journal of Economic and Social Geography*, vol. 90, pp. 329-335.

Johanson, J. and Mattsson, L.G. (1987), 'Interorganizational Relations in Industrial Systems: A Network Approach Compared with the Transaction-Cost Approach', *International Studies of Management and Organization*, vol. XVII (1), pp. 34-48.

Johanson, J. and Mattsson, L.G. (1988), 'Internationalisation in Industrial Sysems - A Network Approach', in N. Hood and J.E. Vahlne (eds), *Strategies in Global Competition*, Croom Helm, New York, pp. 287-314.

Johanson, J. and Wiedersheim-Paul, F. (1975), 'The Internationalization of the Firm - Four Swedish Cases', *The Journal of Management Studies*, vol. XII, pp. 305-322.

Johanson, J. and Vahlne, J-E. (1977), 'The Internationalization Process of the Firm - A model of Knowledge Development and Increasing Foreign Market Commitments', *Journal of International Business*, vol. 8(1), pp. 23-32.

Johanson, J. and Vahlne, J-E. (1990), The Mechanism of Internationalization', *International Marketing Review*, vol. 7(4), pp. 11-24.

Leimgruber, W. (1980), Die Grenze als Forschungsobjekt der Geographie. *Regio Basiliensis: Hefte für Jurassische und Oberrheinische Landeskunde*, vol. 21, pp. 67-78.

Leimgruber, W. (1991), 'Boundary, Values and Identity: The Swiss-Italian Transborder Region', in D. Rumley and J. V. Minghi (eds), *The Geography of Border Landscapes*. Routledge, London.

Leonidou, L. and Katsikeas, C. (1996), 'The Export Development Process: An Integrative Review of Empirical Models', *Journal of International Business Studies*, Third Quarter, pp. 517-551.

Paasi, A. (1996), *Territories, Boundaries, and Consciousness: The Changing Geographies of the Finnish-Russian Border*, John Wiley, Chichester.

Paasi, A. (1999), 'Boundaries as social practice and discourse: The Finnish-Russian border', *Regional Studies*, vol. 33, pp. 669-680.

Penrose, E. (1959), *The theory of the growth of the firm*, Basic Blackwell, Oxford.

Plat, D. and Raux, C. (1998), 'Frontier impedance effects and the growth of international exchanges: an empirical analysis for France, *Papers in Regional Science: The Journal of the RSAI*, vol. 77(2), pp. 155-172.

Podolny, J. (1994), 'Market uncertainty and the social character of economic exchange', *Administrative Science Quarterly*, vol. 39, pp. 458-483.

Porter, M. (1990), *The competitive advantage of nations*, Mac Millan Press, New York.

Powell, W. (1990), 'Neither market nor hierarchy: network forms of organization', in B. Staw and L. Cummings (eds), *Research in organizational behavior*, 12, pp. 295-336.

Ratti, R. (1993), 'Spatial and Economic effects of Frontiers: Overview of Traditional and New Approaches and Theories of Border Area Development', in R. Ratti and S. Reichman (eds), *Theory and Practice of Transborder Cooperation*, Helbing & Lichtenhahn, Basel und Frankfurt am Main, pp. 23-53.

Reynolds, D.R. and McNulty, M.L. (1968), 'On the Analysis of Political Boundaries as Barriers: A Perceptual Approach', *East Lakes Geographer*, pp. 21-38.

Riedel, H. (1994), *Wahrnehmung von Grenzen und Grenzraumen: eine kulturpsychologisch-geographische Untersuchung im saarländisch-lothringischen Raum*. Geographisches Institut, Saarbrücken.

Rietveld, P. and Janssen, L. (1990), 'Telephone Calls and Communication Barriers, The Case of the Netherlands', *The Annals of Regional Science*, vol. 24, pp. 307-318.

Simon, H.A. (1961), *Administrative Behavior*, MacMillan, New York.

Storper, M. (1997), *The Regional World; Territorial Development in a Global Economy*, The Guilford Press, New York.

Appendix 1: Indicators and their dimensions

An attempt has been made to reduce the total number of items to a restrained number of dimensions using factor analysis (Principal Component Analysis, Rotation Varimax). Dimensions are groups of items under a common denominator. The items within these groups are closely related. The following variables are dealt with: relationship preference, mental distance, feeling at home in the culture of the neighbouring country, and border evaluation (see van Houtum, 1998). The first variable is given here as an example.

Dimensions of relationship preference

Factors and items	Factor loading	Dimensions
Factor 1		
Preference for steady long-term economic relationships	0.794	
Preference for economic relations with a broad contact network	0.737	Networking
Preference for conscious search for professional contacts and economic relations in the neighbouring country	0.719	
Factor 2		
Preference for knowledge concerning the price/quality ratio of alternative partners	0.744	Bold and well-informed
Preference for higher profit, despite higher risk	0.724	searching
Factor 3		
Preference for economic relations at short distance	0.823	Regional/
Preference for economic relations in the home country	0.712	national searching

Appendix 2: **Bivariate correlation between the independent variables and the number of economic relations in the neighbouring country**

Independent variables	Dependent variable: the number of cross-border economic relations in the neighbouring country	
	Coefficient 1)	Significance 2)
Number of professional acquaintances in the neighbouring country	0.19	0.02 **
Number of personal acquaintances in the neighbouring country	0.19	0.02 **
Type of relationship preference: networking	0.05	0.17 n/s
Type of relationship preference: bold and well-informed	0.01	0.38 n/s
Type of relationship preference: regionally/nationally bounded	-0.05	0.16 n/s
Feeling at home in the culture of the neighbouring country	0.08	0.05*
Mental distance: the expected negative effect of the relationship	-0.12	0.01**
Mental distance: the expected discrepancy in business conventions in the relationship	-0.14	0.00***
Mental distance: the stringency of the financial-economic conditions of the relationship	-0.08	0.05*
Evaluation of the state border: the state border is a barrier	0.00	0.50 n/s
Evaluation of the state border: the state border is irrelevant	0.12	0.00***

1) This expresses the direction (positive/negative) and the strength of the influence (Beta).
2) *** level of significance < 1 per cent; ** level of significance < 5 per cent; * level of significance < 10 per cent; n/s = not significant.

8 Cross-Border Economic Dynamics. Threats or Opportunities in the Swiss-Italian Labour Market

MARIO A. MAGGIONI AND ALBERTO BRAMANTI

This chapter contains an analysis of the structure and dynamics of the Ticino Canton labour market and, more generally, of the Insubrian economic system using an original population ecology derived class of models. The interactions between different types of workers in this border region are firstly modelled, then simulated, and finally compared with historical data series. The object of the analysis is further enlarged to take into account the role played by local firms. The concept of sustainability, with its three components: social, environmental and economic, is used to identify the conditions for long-run competitiveness and growth of a local system within an "active space" approach.

Introduction

In the slow, but gradual process towards the integration of the economic systems of the EU countries, great emphasis was given in the early phases to the free movement of goods (European Common Market). Even more decisive is the creation of a single currency area which anticipates an effective integration of all productive factors. In the near future European countries will start to experience a movement in the direction of a single labour market, although it is already acknowledged that equalising labour market conditions, salaries and training schemes is certainly a major challenge facing EU in the next decade.

It is certainly no mystery that one of the principal reasons for the conflict between the EU and the other European countries has been, and for many reasons continues to be, related to the free movement of people, especially workers. Switzerland is certainly very sensitive to this issue which continues to be a matter of debate although positive evaluations on the freedom of international movement of persons, in particular workers, have been recently expressed by some Swiss scholars.

> 'The possibility of freely recruiting personnel qualified to work in the different production sectors will make it possible to considerably improve macroeconomic productivity. Thanks to this increased productivity, Switzerland will not be 'submerged' by an excessive inflow of persons from EU countries since the labour demand will be limited.' (Rossi and Poretti, 1992: 7).

The focus of this chapter is on the labour market in the Ticino Canton and the aim is to show how the economic growth of the Canton cannot fail to take into consideration adequate labour market policies. Suitable attention must be paid to the phenomenon of cross-border workers which, although decreasing in terms of absolute numbers, continues to be extremely significant in the manufacturing sector and could become even more important in view of a slow, but gradual, process of qualitative up-grading of human capital thanks to the widening and diffusion of training and educational programmes.

The most interesting results obtained in the framework of training programmes, have been from an approach to the economic space of Ticino as an "active space" linked, through an "osmotic" border-contact zone (Ratti and Reichman, 1993), to the broader economic area including part of Northern Lombardy where it is easy to find important elements of synergy (Bramanti and Ratti, 1993; Maggioni and Bramanti, 1996).

Harmonisation processes and border regions

The intensity and pervasiveness of the processes of economic change are creating new "economic spaces" some of which overlap national borders (Guo, 1996). This phenomenon has recently been the object of much attention on the part of the EU and is characterised by about 10 thousand kilometres of territorial borders, with a good 60 per cent between member states. The border regions, therefore, cover 15 per cent of the European territory and contain about 10 per cent of its population. Although border regions are extremely different, with regard to the level of wealth and economic development, they all share the peculiar situation of being in the front line. They are the "pathfinders" in the transition process and therefore experience earlier, and more intensely than other territories, the challenges and opportunities related to the process of regulatory, legislative and procedural convergence that the EU provides for.

In a general sense, the problems of harmonisation and integration that the border regions have to face (Senn, 1993) can be linked to:

- the question of the cross-border mobility of workers which will be discussed further on, given its central importance to the Ticino case;
- the need for coherent territorial planning, in the absence of which a number of serious, "non-co-operation costs" will arise;
- the questions of an institutional nature which are most influenced by the "country effects", especially on personal services (training, assistance, health, etc.) and network services (transportation and communications);
- the questions related to production integration, including the creation of really cohesive cross-border economic spaces;
- the questions related to macroeconomic policies and, in particular, fiscal policies and exchange regulations.

These five categories are extremely important even within the Insubrian[1] cross-border area and become more important since this area represents an economic space which is potentially integrated but still marked by an extra-community border. The less urgent need to conform to the European regulatory and procedural convergence criteria, since Switzerland has not joined the European Economic Area, must therefore be counterbalanced now and in the future by greater and more decisive policy programming with regard to international co-operation.

In this framework the analytical concept of "regional active space" can be fruitfully used as it stresses a pro-active and flexible dimension of

1 The "Insubrian area", named after its first inhabitants, is made up of the Italian provinces of Varese, Como and Verbania and the Swiss Canton of Ticino (see *Figure 8.1*). It is an area of some two million inhabitants, strictly interconnected along the North-South axis to the metropolitan areas of Zurich and Milan. The area shows a high degree of economic autonomy and a rising awareness of its own potentiality (Bramanti and Ratti, 1993).

development where top-down leadership and financial resources are efficiently combined with bottom-up sensitivity and expertise to provide a local self-sustaining capacity. If such programming is accepted and coherently pursued, the border can act as a positive catalyst to foster growth in a programmed way, not related to individual convenience and short term economic conditions (Bramanti and Lampugnani, 1991).

Figure 8.1   The Insubrian economic system

The border and the Ticino Canton development model

The image of an "open region of Ticino" is gradually becoming a reality and is beginning to permeate the mentality of economic operators, public authorities and wider public opinion in Ticino.

> 'The commonly-held notion of the peripheral and isolated Ticino, blocked to the north by the natural barrier of the Alps and to the south by the national border, has gradually evolved towards other representations, mainly that of the model of integrated periphery and that of the model of emerging

peripheral development (somewhere between a local and global context) which lend themselves to original interpretations such as the 'active space' approach.' (Alberton, 1995: 86).

The future development of the Canton will certainly involve the development of an international services sector accompanied by a further improvement of the industrial sector. Its particular geographic and socio-cultural location (Ratti, 1995) can present a real advantage, provided the Canton can act as an intermediary and avail itself of its own growth possibilities, including among the most important factors, receptivity to new ideas and cross-border relations. The image which, according to the expectations of economic operators, the protagonists would like to give of themselves is one of quality as regards products and services, professionalism and creativity, and also of awareness of one's capacity to handle, in the best possible way, the opportunities which arise.

Any observation regarding the possession of specific "appeal" factors must not, however, make one lose sight of the fact that it is not the mere presence of such factors that constitutes the real crucial resource for development. The concept of "regional competitiveness" must therefore be interpreted as the capacity of the Ticino production system to attract and maintain, over time, companies with a steady (or growing) market share and, at the same time, maintain (or increase) the living standards of the population (Bramanti and Odifreddi, 1995; Maskell and Malmberg, 1997).

Any observation regarding the future development model of the Ticino Canton cannot leave out of consideration an analysis of its "open"[2] local labour market which is, to a large extent, linked to the larger economic space of Northern Lombardia and the Piedmont.

This chapter utilises a model of the dynamic relations existing between different types (or "populations") of workers considered within a broad economic and institutional context, i.e. the Insubrian labour market. The analytical framework chosen, an original approach derived from the population ecology's literature, makes it possible to model the development of an economic system through a stock-flow structure which shows the circuits of cumulative causation of the local development. These circuits determine, within a systemic framework, the emergence of three distinct types of sustainability viz. social, economic and environmental, which are

---

2 As regards the concept of "open labour market" see, among others, Antonelli and Maggioni (1997).

kept in balance and help achieve the long-term growth objective of the local economic system. This last observation, moreover, introduces the question of the competitiveness of the Ticino economy.

In this way, through successive variations, the logical system of the analysis is expanded to ascribe the right importance to some decisive, qualitative circuits viz. innovation, quality of the factors, investment in human capital and infrastructures, which alone can allow and will allow the Ticino economy to compete in the European market and, in the near future, to become a full member.

The use of a non-linear dynamic model introduces several effects of path-dependency and lock-in[3] in the analysis. Although the complexity of the system makes it impossible to draw a simple analytical model, it is, however, possible to sketch the framework of a dynamic system which can provide some indications regarding the different options to be pursued through appropriate policy measures. They include measures related to quality improvement of production factors (above all labour), and those related to an integrated management of the territory and a more co-operative management of the larger economic space of Northern Lombardia, with the Ticino area as the backbone, as a positive and regulatory hypothesis of many of the arguments developed in this chapter.

The idea of an "active space", as a coherent game of stimulus and answer played by different actors working within a territorial system, is particularly suitable to model local economic development. As a matter of fact the very challenge of sustainable development stands in proper co-ordination among the different factors underpinning the growth process in the territorial fabric. Partnership between actors seems to be a necessary condition for development, even if co-operation does not automatically imply symmetry or absence of power. The "active space" approach intends to put the following together:

- *Territorial span*; all firms and economic agents are deeply embedded within the territory in which they operate and in which they have historically developed. The territorial system has historically provided a

---

3 Non linear models of economic systems are heavily dependent on initial values of variables and the economy may easily end up in sub-optimal equilibria due to the role played by agents' past choices and increasing returns (see Arthur, 1994; Day, 1994).

mix of social and economical elements, routines and procedures that have contributed to the development of the system (path dependency).

- *Networking behaviour*; many firms and economic agents are informally linked together, giving origin to processes of collective learning and mutual processes of value-creation.
- *Governance structures*, relating to: the way of collective value-creation, resolution of conflicts among partners, production of "club-goods".

Increasingly, each business and each location will have to compete on a global stage and this implies developing a new economic awareness, building a common strategy and co-ordination with all the relevant actors. The management of human resources is undoubtedly a key factor for all the above mentioned implications and it is the only possible strategy to confront the creative destruction process deriving from innovation. Efficient labour markets are thus particularly decisive as a precondition for dynamic economic sustainable development.

All this seems to be dramatically important for the Swiss Ticino Canton where the trans-border labour market results in a greater flexibility and continuous adjustment to changing skill requirements, shifting demand and firms' relocation processes.

The results of this chapter closely correspond to the principal conclusions reached by numerous Swiss research studies regarding the impact of Switzerland joining the EU: the opportunities offered by integration, in fact, exceed the risks and the latter are by far fewer than those inherent in choosing a "solitary path" (Hauser and Bradke, 1992; Rossi and Poretti, 1992). The concept of "regional active space" would imply a positive role of Switzerland in shaping the process of gradual integration, as opposed to a "passive space" approach where the European institutional change is considered to be an exogenous, although important, variable in the definition of the Swiss model of development and growth.

A fragmented labour market

The Helvetian Confederation, which relied on the contribution and control of foreign workers during an important phase of its development, especially in manufacturing and services, pays great attention to foreign workers and shows understandably marked sensitivity in the border areas. Here the dimension of this phenomenon has become quite serious and is characterised by two types of foreign workers: resident and cross-border workers. As also shown in the seven regional referendum held between 1970 and 1990 regarding Swiss policy towards foreigners, the problem of foreign labour is of great concern to the Helvetian Confederation.

The abundance of foreign labour, especially in the Ticino border area, has triggered rapid development processes, mainly due to the low cost and high flexibility of this factor. The foreign component of the working population has always reacted as a "buffer" to the cyclical crises of labour demand, i.e. negative economic cycle, lowering instead of increasing the rate of Swiss unemployment. Indeed, especially in the Ticino area, the unemployment rates have been systematically lower than the European average and in particular the Italian rates (Rossi, 1987). The possibility of "exporting unemployment" has, however, implied an unpleasant and unexpected feed-back, namely some structural distortions of the labour market, by lowering the average quality of the labour supply, which have influenced the competitiveness of the Ticino production system in the medium-long term.

The institutional segmentation of the labour market implies a situation such that when an insider worker, i.e. Swiss citizen and resident foreigner, is fired, he/she becomes unemployed. When the same happens to an outsider worker, i.e. cross-border and seasonal worker, he/she is simply expelled from the Ticino labour market. This dynamic of "exporting unemployment" can be seen as a sort of free riding behaviour on behalf of Ticino with respect to Northern Lombardia. As far as the quality level of labour supply goes we know that a "risky situation", as is the situation of cross-border workers, does not facilitate investment processes in human capital. The Ticino labour market is therefore facing a trade-off between higher flexibility, with a lower quality level, and higher labour quality, with greater stability.

Therefore, since the 1990s the unidirectional "vicious circle" seems to have broken of exporting unemployment to Italy during periods of cyclical malaise and today a decrease in the demand for work coincides with a slight increase in the working population looking for work and, indirectly, in a rapid increase in unemployment[4].

---

4 The productivity of Ticino work - already below the Swiss average of about 11 percentage points - is even below the corresponding Italian percentage, especially in the border regions. This is not due so much to a production mix but to lower productivity in important sectors. The relatively large banking and transportation sector in the Canton have remained at a production level which is 16 per cent and 13 per cent lower than the Swiss

Theoretical Analysis: a Population-Ecology Approach

To study the connection between the labour market, sustainable development and the competitiveness of the Ticino economic system, that is, to study the "regional active space" (as defined in other chapters of this volume) applied to the specific case of the Insubrian area, we have chosen to adopt an approach derived from population ecology for the following main reasons[5]:

- It allows the analysis, within the same theoretical framework, of a multiplicity of interactions between different populations of workers: from co-operation to competition, from mutualism to predatory behaviour.
- It underlines the existence of simultaneous dynamic interaction within and between the workers' population and the co-evolution processes which are established between these populations and the macro-economic and institutional environment.
- The logistic model, applied to the employment growth process for a specific typology of workers, points out the structural features of a system and helps identify the objectives for economic policy measures. The *intrinsic growth rate r* is the maximum growth rate of the employment of a specific type of workers within an economy with infinite labour demand and where both competition effects on wages and congestion effects on productivity are irrelevant; the *carrying capacity K* is the total number of vacancies existing in the economy, when there is no segmentation, or the exogenously determined share of this total as defined by social, institutional and political factors; *coefficients of inter-population interaction* indicate the degree of complementarity and/or substitutability of the two types of workers.

This approach can be regarded as complementary to evolutionary economics which focuses on analysing the mechanisms that generate variety, often leaving aside the selective moment or describing it in

---

national sector average: all this is due to more than just the quality of the human capital utilised.

5 For a more detailed discussion of the advantages and drawbacks of the ecological approach see Maggioni (1994) and Gambarotto and Maggioni (1998).

simplistic terms as merely the result of a comparison between rough economic performance indicators such as profits (Maggioni, 1999).

The specific contribution presented in this chapter concerns the mechanisms for selecting variety based on the interaction between different populations, in our case, workers, and between them and the environment. "Populations", therefore, which represent a specific category of workers that is quite distinct from the other (populations) because of precise social, institutional and economic features. The interactions between two or more populations influence, and are indirectly influenced by, the institutional features of the local labour market. Such interactions help determine, together with other components (in the analysis initially considered exogenous and then gradually endogenised) the performance of the labour market itself and, more generally, the local economic system.

Many studies, both theoretical and empirical, which examine the Ticino labour market structure, indicate a considerable degree of fragmentation. *(Figure 8.2)*.

Years	1955	1965	1973	1979	1990
Swiss worker	61,350	69,246	65,856	66,451	93,892
Foreign worker:	20,617	58,912	72,177	64,857	83,367
Foreign resident	*15,363*	*41,903*	*41,189*	*37,163*	*43,115*
Cross-border worker	*5,254*	*17,009*	*30,958*	*27,689*	*40,252*

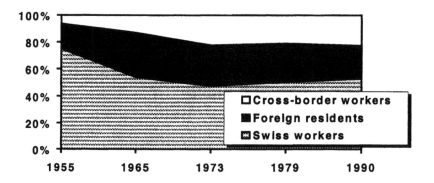

Figure 8.2   Long term dynamics in the Swiss Canton of Ticino labour market

Source: Bausch (1996).

The dichotomy between the jobs designated for foreigners and those reserved for indigenous workers continues to be great. The first characteristic of the market is, therefore, the separation of the labour market into two sub-markets: the first reserved for the Swiss and the second for foreigners. The first one offers qualified jobs particularly in the service sector; the second one does not take away any jobs from the internal workers but, on the contrary, it fills up the vacancies which the ongoing process of upgrading Swiss workers generates in the low edge of the labour market.

Even within the foreign workers the division between cross-border and resident workers depends on the changing Swiss labour legislation. The restrictive "Schwarzenbach law" (1968) regarding foreigners, for example, contributed to the rising role of cross-border workers in the late 1960s and early 1970s, and caused a strong competition between the two populations. For the Swiss firms the trade-off seems to be the one between a relatively higher level of skills (resident workers) versus greater flexibility (cross-border workers). Such a fragmentation influences the dynamics of the Ticino economy and therefore, is subject of analysis in the approach adopted here.

Inter-specific competition: cross-border workers versus foreign residents[6]

The internal dynamics of this peculiar sub-market can be analysed using a model of inter-specific competition between two types (or populations) of foreign workers in a common pool of low level exogenously determined vacancies. This situation can be thus modelled using a dynamic stock-flow model (see *Figure 8.3*) in which the development of the two populations of employed workers, cross-border (C) and foreign residents (F), are determined by the intra-specific indicators (r and K) and by the inter-specific competition (or substitution) coefficients ($Csub$ and $Fsub$)[7].

---

6 For a deeper analysis, see Maggioni and Bramanti (1996).

7 $dC\ dt$ and $dF\ dt$ in *Figure 8.3* refer to the variations in time of the employment of each type of workers, which are influenced by the above mentioned variables and parameters.

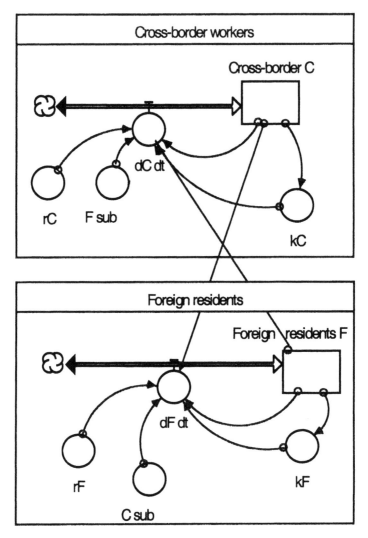

**Figure 8.3   Competition between cross-border and resident foreign
workers within the Ticino Canton labour market: a
stock-flow representation**

The two possible results of the expulsion of one category of workers
from the labour market and the co-existence of both categories can be
duplicated in the simulations (*Figure 8.4*).

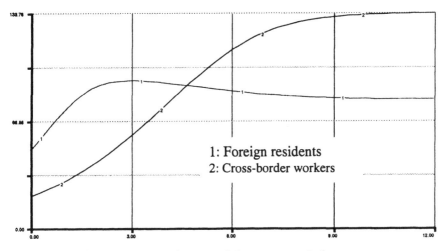

1: Foreign residents
2: Cross-border workers

A. A simulation of the co-existence of the two populations

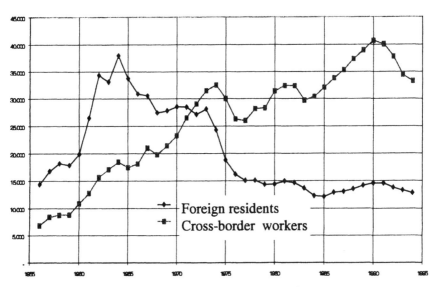

Foreign residents
Cross-border workers

B. Real data on the co-existence of the two populations

Figure 8.4   Competition process between cross-border and foreign resident workers: numerical simulation and historical evolution

It is important to point out that when the parameters of the system are given real economic values[8], the simulation (see *Figure 8.4a*) mimics the actual development of the cross-border worker population, seasonal workers and residents recorded in Ticino from 1956 to 1994[9] (shown in *Figure 8.4b*).

The rate of inter-population competition is therefore directly proportional to the degree of homogeneity of skills requirements of the population of firms and inversely proportional to the specificity of the two categories of foreign workers[10]. If we assume that firms in Ticino are all equal and have a uniform preference for a specific type of workers, then the most obvious result would be the survival of one, and only one, type of foreign workers, i.e. either the cross-border or the resident one. On the contrary, the data reveal that this is not the case. If this last observation is valid, and if a positive value is attributed to the simultaneous presence of two categories of foreign workers (since this allows a greater quantitative and qualitative flexibility of the local labour market and a greater resilience of the local economic system when facing exogenous shocks), the model seems to suggest a peculiar kind of public intervention which tends to increase, rather than decrease, the degree of local labour market segmentation for foreign workers.

In other words, if there are some objective features which cannot be eliminated and which discriminate against two categories of foreign workers, then it is preferable to have a policy which ensures the simultaneous presence of these two categories. On the contrary, when it is possible to solve the problem of heterogeneity, i.e. through specific investment, then policies designed to increase the efficiency of the local labour market by reducing the degree of segmentation are undoubtedly preferred.

---

8 The carrying capacity is assumed to be greater for cross-border workers than for foreign resident workers, due to the different impact of these two types of workers on the social sustainability of the system. The intrinsic rate of growth is higher for foreign residents, which are able to establish and exploit a more stable information and references network for job posting, and the substitution coefficient is higher in the case of cross-border agianst foreign residents than in the opposite case.

9 To be more precise, the simulation fits the data until 1990, then both populations are hit by an exogenous, regulatory shock.

10 Where specificity is related to the imperfect substitutability between foreign residents and cross-border workers. Obviously, this indicator is derived from qualitative features, such as training, skills, flexibility, union membership, of the two populations and the technical-organisational paradigm adopted by the local enterprises.

Inter-specific mutualism

It is equally possible, and of great interest, to model a system made up of two populations, in which each one benefits from the development of the other. Such a model describes, with a sufficient degree of approximation, the dynamics of the relations which exist between the total number of foreign and Swiss workers in the Ticino labour market.

The results obtained (see Maggioni and Bramanti, 1996) confirm the empirical evidence mentioned earlier and perfectly describe the dynamics of a local labour market in which the two categories of workers (Swiss and foreign) are not mutually substitute but, rather, display attributes of complementarity and synergy so that each type can increase its employment level even beyond the maximum threshold allowed (in isolation) by the local labour market. This is due to the presence of the other category of workers which fosters development, in a situation in which the "technological" constraint defines a certain optimum ratio between the two categories of workers[11]. The co-existence of the two categories therefore, allows the economy to employ a greater number of workers; but, at the same time, the effects of competition and inter-specific congestion prevent the system from undertaking an explosive path of growth.

Figure 8.5 shows a graphic demonstration of the structure and dynamics of the system. In this case, two logistic processes are shown which illustrate the growth of the population in isolation, by means of a well-known scheme: stock plus the two indicators r and K, to which have been added two coefficients of complementarity, *S comp* and *F comp,* which express the degree of synergy resulting from the joint employment of Swiss and foreign workers in the same economic system; because of the complementary nature of their individual skills and characteristics, the development over time of the two populations of employed F and S workers reaches values which are higher than those recorded by each category in isolation. The coefficient of complementarity refers to the technological characteristic of the labour organisation within firms. A possible example is the couple: F comp = 1/2; S comp = 2 which refers to a situation where the "optimal" job structure of a firm is 2 foreign workers for each Swiss.

---

11 One can think of a situation where the only production factor is labour of two qualities and the production function is a "Leontief type". In this situation the technical rate of substitution between the two inputs, i.e. foreign and Swiss workers, is always fixed and the shorter side of the labour market rations the use of the other input.

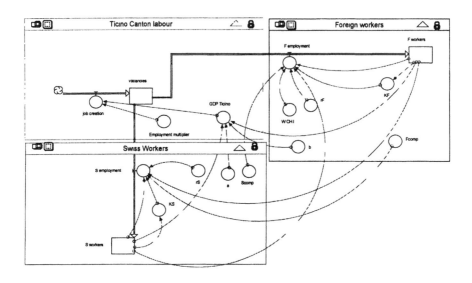

Figure 8.5   The Ticino Canton labour market: a stock-flow representation

This result can be further improved by introducing a measure of income level, proportional to the number of employed workers in each category, multiplied by its productivity: *a* for Swiss workers and *b* for foreign workers, and a mechanism to create vacancies by means of an "employment multiplier". The relative number of the two categories of workers cannot, however, be determined only by a technological relationship; it has, in fact, significant social implications. For this reason, an initial "social sustainability" indicator was introduced, calculated according to the ratio between the number of Swiss and foreign workers employed in the economic system (see *Figure 8.6*).

Sustainability: the frontier of development processes

Global sustainability is also becoming the *Leitmotif* of economic development. For advanced territorial productive systems, the challenge is just that of reproducing over time those elements which make it possible to maintain suitable levels of economic affluence and quality of life in the long run (Bramanti and Odifreddi, 1995). According to the argument developed in this chapter, the significance attributed to the three elements of sustainability considered important in the specific case of the Ticino economic system is the following.

The *social sustainability* of Ticino is certainly the easiest to grasp and is certainly the element which has been monitored the longest by the Canton policy makers in order not to alter in an excessively brusque way, and to control in the long-term, some strategic demographic ratios such as the ratio between the total number of foreigners and the resident population. The evident evolution of these ratios will also depend in a significant way on the naturalisation policy the Canton decides to adopt. In the model shown in *Figure 8.5* and developed in *Figure 8.6*, social sustainability is represented by the ratio between the number of foreign workers and the number of Swiss workers. The development, over time, of a band of oscillation in this ratio can, in fact, be set by law, whereas the variations of this indicator within this range are determined by the economic cycle of the local system. We are fully aware that, at the social scale, sustainability also implies considerations of equity, gender, education, civil rights, culture and other aspects of human development, but we confine ourselves to the interface of economic analysis, paying specific attention to the balancing of indigenous and foreign workers.

Ecological or *environmental sustainability* is also evident. The limitation of the physical space assigned to productive activities, given the physical morphology of the Ticino territory, and the increase in the pollution produced by enterprises and vehicle traffic, heavily related to the flow of commuters, has, in the last few years, reawakened the environmental sensitivity of the Canton administration. In particular, vehicle traffic seems to be a problem which can be solved with practical solutions aimed at increasing appropriate forms of public transportation and "forcing" personal habits towards a suitable modal-shift. Even if we know that the equation "more growth means more pollution", or, equivalently, anti-pollution objectives imply that growth should be restricted, is oversimplified (Grossman, 1995), in our model environmental dynamics are determined by the trade-off between the creation of pollution produced by the workers (greater for cross-border workers due to the greater average distance covered) and the costs of de-pollution which reduce the Canton's disposable income. Since increasingly higher percentages of the GDP must be set aside for environmental clean-up and water purification measures, this directly limits processes of job-creation and indirectly limits future income increases.

Finally, *economic sustainability* is linked to the broader question of the competitiveness of the enterprises in Ticino. This competitiveness depends on numerous elements including innovative processes which certainly play an important role. Processes which have a great deal to do with learning processes arising within the local labour market. All this brings with it the inevitable need for a serious re-qualification process and permanent training programmes for the labour force, in particular foreign and cross-border workers. It is therefore necessary to think in terms of narrowing the qualification gap between the different groups of workers present in the Ticino and improving their qualifications. In the model in *Figure 8.6*, economic sustainability is expressed using an indicator which measures the growth rate of the Ticino GDP as an initial approximation of the capacity of the Ticino economic system to maintain and increase its competitiveness. Evidently, the realism of the model is still unsatisfactory since it focuses exclusively on only a few circuits and presents only two populations of actors, Swiss workers and foreign workers.

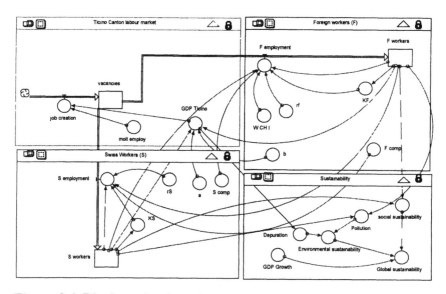

Figure 8.6 The introduction of sustainability in the Ticino Canton labour market

This shows the need of a better representation of the productive system, in particular, the introduction of a third economic population, namely, enterprises. The following section deals with this last point.

Towards a Dynamic Stock-Flow Model of the Economic System of Ticino

The preceding models showed a balance of sustainability in the long term, which already took into consideration the three kinds of sustainability, and which was basically linked to the interaction of the two types of workers by some feed-back mechanisms regarding stocks, i.e. number of workers, and income, which generates employment through some sort of multiplier.

The logic adopted in proceeding to formulate a dynamic model (stock-flows) of the labour market and, by extension, of the Ticino economy (see *Figure 8.6*) has enhanced the preceding results overcoming some of the structural limitations and introducing in the model another population, namely, the Ticino enterprises.

System competitiveness and sustainable growth

The previous model (see *Figure 8.6*) seems appropriate when analysing a past phase in the history of the development of the Ticino economy which has now come to an end. The main problem the economy is facing today, just like many other developed systems, is systematically summed up by the competitive challenge of international markets. To meet this challenge it is apparently no longer sufficient to have the advantage represented by the privileged access to a productive factor, i.e. cross-border and resident foreigners workers, which tends to be in "unlimited" supply and at a low cost. There are other elements of competitiveness and here the concept of regional competitiveness and business competitiveness comes into play and is closely related to the quality of the factors, i.e. human capital, financial capital, infrastructures, institutions, norms and routines, and to the mechanisms of generation/attraction of enterprises able to compete within international markets.

These observations, which have long been shared by a large group of regional economists, call for the introduction of enterprises as active agents in the model (see *Figure 8.7,* the Ticino region box). As a matter of fact, it is the enterprises which create the employment which generates income, and this is therefore the importance which must be attributed to the direct links which enterprises have with the dual labour market: in the Ticino case enterprises directly set in motion the mechanism of employment whereas, in the case of foreigners, they set in motion the job demand for foreigners[12].

A second point of importance is the concept of the creation/attraction mechanism contained in the model. This, in fact, implies a vision of local economic development which is both self-generating and competitive. A healthy productive system has to generate endogenous development and new firms, i.e. a creation mechanism, and at the same time should be sufficiently attractive to outside investments, i.e. an attraction mechanism. The model presented takes both these mechanisms into consideration in fostering the stock of enterprises operating in the area (Bramanti and Maglierina, 1995).

The workings of these two mechanisms, which are modelled like two-way flows, creation/failure, attraction/dispersal, bring us back to the qualitative part of the model which deals with that set of factors and phenomena which, proceeding from the quality of the factors, generates competitiveness for the system as a whole. The circuits can be easily identified in *Figure 8.7*: in addition to helping increase the GDP, the added value produced by the enterprises is invested in R and D and human capital. R and D generates innovation together with investments in training and human capital; innovation, in its turn, gives rise to the mechanism of the creation of new enterprises, complementary to the spin-off generated by the existing enterprises, and contributes to competitiveness. Finally, competitiveness is the combined result of innovation, infrastructures and human capital and, in its turn, heightens the attraction of the system.

12 A successful region must necessarily balance two complementary aspects of the development process (internal *vs.* external links, milieu *vs.* network, local identity *vs.* openness) to foster a process of sustainable growth and endogenous development. By sustainable growth we mean a growth process which can be replicated in the medium-long period, while we define endogenous development a process of development managed and controlled by local actors and institutions (Bramanti, 1995; Bramanti and Maggioni, 1996; Quadrio Curzio et al., 1997).

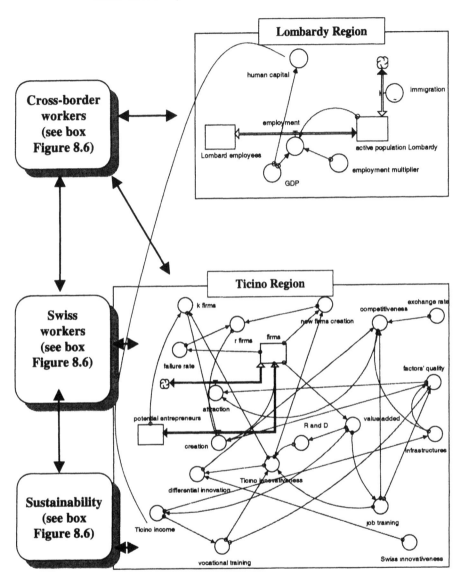

Figure 8.7 The Insubrian economic system: a stock-flow representation

Since the above-mentioned infrastructures all generate competitiveness, they are fostered by means of public investment (GDP) and help raise the quality of the factors together with the availability of credit, training, capital investment and the level of pollution. This last factor obviously acts in the opposite way with respect to the preceding factors.

The Insubrian labour market

As regards the worker population, it is particularly important to explain how, and in what way, the supply of cross-border workers originates. The mechanism assumed in the analysis is basically that of an unlimited offer and a controlled demand by the Ticino economic cycle and the cross-border labour market regulations. The driving force behind this market is therefore the available jobs for cross-border workers, attracted by the high wage differential; jobs which are directly linked to the labour demand of the enterprises, the employment rate of the Swiss and even the desired ratio (administrative control) between cross-border workers and Swiss workers.

The number of available jobs sets in motion cross-border employment which, in its turn, is controlled by a factor of temporal adjustment which depends on the wage differential (attraction) and the index of risk avoidance (aversion) since such jobs provide better wages but worse stability or job security. This high flexibility is represented in the model by the possibility of expelling cross-border workers whenever the number of workers exceeds the number of available jobs.

It is worthwhile to analyse the fundamental causes of this process. What are the mechanisms which make cross-border workers to leave the Northern Lombardia labour market? Here the idea is deliberately simplified and only takes into consideration the fact that the Ticino demand for cross-border workers is basically directed to already qualified workers and therefore "poaches" them from the market of employed workers in Lombardia[13]. This market is therefore enlarged by the stock of the working population (which therefore also includes the unemployed) which also encompasses the possible cross-border workers expelled from the Ticino labour market. A simple employment multiplier transforms the level of income into the number of vacancies in the northern Lombardia productive system (see the

13 This dynamics can be fruitfully analysed within a framework of repeated games, of the prisoner dilemma type. Once again the interested reader is referred to Maggioni and Bramanti (1996).

box Lombardia region in *Figure 8.7*). The working population in the area is not only enlarged by the demographic dynamics - which is, moreover, by now, almost completely stable - but also by the migration flows which, in the last years, have been very strong in the border areas.

For the sake of simplicity we do not show in the diagram all the boxes which fill as the complete model is built up. As a matter of fact what is going on within the Ticino labour market regarding the Swiss workers, has been already captured and shown in the relevant boxes of *Figure 8.6*.

The working population depends only on the demographic dynamics. The level of income generates vacancies by means of an analogous multiplier. Swiss and foreign workers, cross-border workers plus resident workers, once employed, generate income through a productivity index which is higher for Swiss workers, as a consequence of a greater supply of human capital. Moreover, the Swiss working population enlarges the stock of potential entrepreneurs through a mechanism of entrepreneurial selection which, in its turn, constitutes the link with the enterprise system. Swiss and foreign employed workers jointly produce a "social sustainability" indicator. It represents one of the three decisive indicators which contribute to the performances of the system in its entirety and, in particular, to the long-term growth objective.

The Role and Prospects for Institutional Collaboration: the Rise of a Regional Active Space

This chapter highlights the potentially significant role of trans-regional institutional co-operation which allows the exploitation of relevant external economies and fosters the sustainable development of the Insubrian economic system as a whole. The recent considerations made by the regional and industrial economy have underlined the increasingly important role assigned to the factor of space, i.e. territorial size, in co-determining the competitiveness of the system. Furthermore, attention is drawn to the necessity of a balance between the local and global elements, such as internal synergy and external relations, territorial identity and network relations, robustness and opening of the productive fabric. The necessity of a balance is particularly true in the case of a border region where the space which supports the economy intrinsically contains cross-border elements.

This is required by the need to make a general, long-term evaluation in considering the performance of a territorial system: we are only interested in sustainable growth. We could even argue that such sustainability acts as a filter, a selection mechanism and multiplier of the dynamic development of an area through a series of learning mechanisms which start-up feed-backs in the productive system, the government structures and the complex of policy instruments.

All the above leads to the underlining of the concept of "regional active space" as a way to interpret the problem of regional development, which is based on the interaction of openness, sustainability and the creative capacity of the specific territorial system. Applying this concept to the Ticino labour market, and to the broader problem of the competitiveness of such a system, seems particularly appropriate.

As far as openness is concerned, already today the presence of cross-border workers and the prospects of the free movement of workers in the EU mean that the Ticino labour market will be considerably influenced by "external" factors - although to a great extent "inside" the Insubrian Region considered as a whole. As applied to sustainability, it has been amply shown how the three elements are needed, i.e. social sustainability, concerning the problems connected to the identity of a people and the capacity to support the social burden, environmental sustainability, concerning the impact on the quality of life of residents and the attraction for outside elements both of which determine the quality of the Ticino productive factors, and economic sustainability. Finally, "regional active space", when applied to the creative capacity, shows its full potentiality since it increases the productivity and the competitiveness of all agents, i.e. firms, public institutions, and workers, and (indirectly) of the local economic system.

Although the model shown in *Figure 8.7* is rather sketchy, it contains numerous elements to link the fragmentation of the labour market to the competitiveness of the productive system, by means of learning processes, innovative activities and the up-grading of the industrial productive system. An initial major result is that the over-fragmentation of the Ticino labour market has so far been the result of the effects of the concept of border as a "barrier", or even more, as a "filter". This contribution has shown that in the presence of market fragmentation, since it is possible to maintain and reproduce it through administrative measures, a more rapid and balanced growth has been possible thanks to the simultaneous presence and utilisation of different populations of workers. If such a fragmentation disappears, or at least considerably decreases due to the freer movement of persons if Switzerland joins the EU, the strategies for sustainable growth will also radically change. It is at this point that the training programmes, in their broadest sense, and the institutional actors will both come into play to determine the new structural and dynamic scenarios of the Ticino labour market. In particular, it is important to point out that such a role seems to be decisive during the intermediate stage, which might even be significantly long, starting from an initial situation characterised by a highly fragmented market and strong entry barriers and moving to a final situation characterised by a market which is strongly integrated at the European level. Indeed, the process of European integration is the terrain on which the challenges of the next few years will be played out.

Acknowledgement

Although the whole chapter is the result of joint research, M.A. Maggioni is responsible for writing the sections "The theoretical analysis: a population-ecology approach" and "The Insubrian labour market", and A. Bramanti for "Introduction", "System competitiveness and sustainable growth" and "The role and prospects for institutional collaboration: the rise of a regional active space". We gratefully acknowledge financial support by MURST 40 per cent founds.

References

Arthur, W.B. (1994), *Increasing Returns and Path Dependency in the Economy*, University of Michigan Press, Ann Arbour.

Alberton, S. (1995), *Evoluzione economica e strutturale del cantone Ticino (1975-1995)*. *Ticino 2000. Piazza commerciale e finanziaria*, IRE, Bellinzona.

Antonelli, G. and Maggioni, M.A. (1997), 'Formazione, competenze e lavoro in contesti economici in rapida evoluzione', in P. Terna (ed), *La formazione e il lavoro al tempo delle reti telematiche*, Rosemberg & Sellier, Torino.

Bausch, L. (1996), *Frontalierato: problema o opportunità? Ufficio di Statistica, Aspetti statistici*, n. 11, Bellinzona.

Bramanti, A. (1995), 'Innovazione e ruolo dello spazio in una classe di modelli di sviluppo localizzato', in G. Gorla and O. Vito Colonna (eds), *Regioni e sviluppo: modelli, politiche e riforme*, Angeli, Milano.

Bramanti, A. and Lampugnani, G. (1991), 'Tessuto economico locale e rapporti di collaborazione nelle aree di frontiera', in F. Boscacci and G. Gorla (eds), *Economie locali in ambiente competitivo*, Angeli, Milano.

Bramanti, A. and Maggioni, M.A. (1996), 'Nuovi approcci per vecchi problemi: dove va lo sviluppo locale?', in G. Bazzigaluppi, A. Bramanti and S. Occelli (eds), *Le trasformazioni urbane e regionali tra locale e globale*, Angeli, Milano.

Bramanti, A. and Maggioni, M.A. (1997), 'Dinamiche transfrontaliere: predazione o cooperazione? Il caso del Canton ticino', in F. Boscacci and L. Senn (eds), *Montagna: area di integrazione. Modelli di sviluppo, risorse e opportunità*, Angeli, Milano.

Bramanti, A. and Odifreddi, D. (eds) (1995), *Lo sviluppo delle 'aree avanzate'. Apertura e identità nei sistemi economici territoriali: Varese come caso nazionale*, Angeli, Milano.

Bramanti, A. and Ratti, R. (1993), *Verso un'Europa delle regioni. La cooperazione economica transfrontaliera come opportunità e sfida*, Angeli, Milano.

Day, R.H. (1994), *Complex Economic Dynamics*, MIT Press, Cambridge MA.

Gambarotto, F. and Maggioni, M.A. (1998), 'Regional Development Policies in Changing Environment: an Ecological Approach', *Regional Studies*, vol. 32.1, pp. 49-61.

Grossman, G.M. (1995), 'Pollution and Growth: What Do we Know?', in I. Goldin and L.A. Winters (eds), *The Economics of Sustainable Development*, OECD-CEPR, Cambridge University Press, Cambridge.

Guo, R. (1996), *Border-Regional Economics*, Physica-Verlag, Heidelberg.

Hauser, H. and Bradke, S. (1992), *Traité sur l'EEE, Adhésion à la CE, Course en solitaire: Consequénces économiques pour la Suisse*, Verlag Rüegger, Chur-Zürich.

Maggioni, M.A. (1994), 'Modelli ecologici per l'analisi della dinamica industriale regionale', in F. Pasquini, T. Pompili and P. Secondini (eds), *Modelli d'analisi e di intervento per un nuovo regionalismo*, Angeli Milano.

Maggioni, M.A. (1999), *Clustering Dynamics and the Location of High-Tech Firms*, Ph.D. Thesis, University of Warwick, Coventry.

Maggioni, M.A. and Bramanti, A. (1996), '*Cross-Border Economic Dynamics, Threaths or Opportunities for Switzerland?*', Paper presented at the Annual Congress of the Swiss Society for Economics and Statistics, Lugano, 14-15 March, 1996.

Maskell, P. and Malmberg, A. (1997), 'Sapere localizzato e competitività industriale', in A. Bramanti and M.A. Maggioni (eds), *La dinamica dei sistemi produttivi territoriali: teorie, tecniche, politiche*, Angeli, Milano.

Quadrio Curzio, A., Fortis, M. and Maggioni, M.A. (eds) (1997), *I distretti economici delle alpi e delle prealpi centrali*, Credito Valtellinese, Sondrio.

Ratti, R. (1995), *Leggere la Svizzera. Saggio politico-economico sulle origini e sul divenire del modello elvetico*, Giampiero Casagrande Editore, Lugano.

Ratti, R. and Reichman, S. (eds) (1993), *Theory and Practice of Transborder Cooperation*, Helbing & Lichtenhan, Basel.

Rossi, M. and Poretti, R. (eds) (1992), *Lo spazio economico europeo in una regione di frontiera. Conseguenze per il Ticino dell'Accordo sullo SEE*, Istituto di Ricerche Economiche, Bellinzona.

Senn, L. (1993), 'Introduzione: verso un'Europa delle regioni', in A. Bramanti and R. Ratti (eds), *Verso un'Europa delle regioni. La cooperazione economica transfrontaliera come opportunità e sfida*, Angeli, Milano.

9 Which Role for Foreign Direct Investment? Active Space Development in Central and Eastern Europe

MARINA VAN GEENHUIZEN

Since the lifting of the Iron Curtain, significant flows of foreign investment have been directed to Central and Eastern Europe. There are, however, large differences in the use of the new openness by foreign investors, mainly connected to the divergent level of transition of the recipient economy. As far as regional "active space" development is concerned, the results give a rather pessimistic picture. So far, it appears that the contribution of foreign direct investment to regional innovation is limited to those cases where there was already an existing system of innovative suppliers in the region and where this system could be used within the investor's strategy. The preliminary observations indicate a need for a general policy to acquire and absorb new knowledge, as well as targeted policies to benefit from foreign direct investment in terms of spill over into the regional economy.

Introduction

Many governments in Central and Eastern Europe see foreign direct investment (FDI) as a catalyst for socio-economic improvement, and as an important source of alternative finance for the transformation of former state-owned enterprises in the absence of large domestic savings and limited access to international capital markets (cf. Svetlicic and Rojec, 1994; Meyer, 1995). In addition, FDI is considered to be a way of knowledge transfer to enhance the introduction of innovative products, new production processes, and management skills (e.g. Bell, 1997). The acquisition of western technology by Central and Eastern European firms has, however, seldom been forwarded as a policy issue in its own right, leaving FDI as an important vehicle for transfer (e.g. Radocevic, 1999).

FDI in transition economies has been extensively addressed in the literature in terms of size of flows, number of projects, location-specific barriers, and geographical distribution (e.g. Dunning, 1993; Meyer, 1995; van Geenhuizen and Nijkamp, 1996, 1998). Given the assumptions on knowledge transfer and innovation, there is now a recognition of the need for research on the way regional economies may benefit from investment impacts - through spill-over effects and active local learning - and become more competitive (e.g. Svetlicic and Rojec, 1994; Hooley et al., 1996; OECD, 1996a; Pavlínek and Smith, 1998; Smith and Ferenciková, 1998). The key questions here are concerned with the motivation of foreign investors to operate a subsidiary (joint venture) in Central and Eastern Europe within their broader strategy, the extent to which the investments create local forward and backward linkages - addressing the embeddedness issue - and the role of these investments in technology transfer and use of endogenous knowledge.

In addressing "active space" development in the regions of investment, the emphasis in this chapter is on institutional conditions that advance local learning in connection to the type of investment projects (van Geenhuizen and Ratti, 1998). The way in which the new openness since 1989 is used in terms of size of FDI on the country level is investigated first. Then this pattern of FDI is analysed regarding underlying factors, particularly the stage of transition or institutional proximity of recipient countries to the EU. In the next part, the peculiarities of the systems of innovation in Central and Eastern Europe are set out because these still determine important conditions for "active space" development. The entrepreneurial motives behind FDI and the broader corporate strategies are outlined next. Using various case studies from the literature, these motives and strategies are then linked to regional embeddedness of foreign subsidiaries and creative learning processes in the regional economy. The chapter is concluded with a few recommendations for policy action and future research.

Spatial Patterns of FDI

Foreign direct investment can be measured in different ways, with slightly different underlying conceptualisations (see, e.g. Meyer, 1995).

Accordingly, countries make different statistics available, such as statistics based on the cumulative foreign component of foreign investment enterprises and the cumulative balance-of-payments FDI inflows. This analysis makes use of the latter type of statistics.

With more than 13 billion US$, FDI inflow over the past seven years is largest in Hungary. The Czech Republic holds a second position, with Russia and Poland as third and fourth (*Table 9.1*). FDI stock in Central and Eastern Europe is heavily concentrated in countries adjacent to Western Europe, i.e. more than 80 per cent in total: Hungary 42 per cent, the Czech Republic 23 per cent and Poland 17 per cent. The per capita distribution, however, only partly underlines this concentration with Hungary, the Czech Republic, and also Estonia and Slovenia as the main recipient countries. Per capita figures indicate that FDI flows into Poland and Russia are relatively small in relation to their population size.

When considering the annual figures for 1995 and 1996, there seems to be a slight dip in 1996. This reflects particular underlying growth trends (EBRD, 1997). For example, various large infrastructure privatisations taking place in 1995 distort annual comparison, such as in Hungary (power sector) and Czech Republic (telecommunication). In 1996, Poland replaced Hungary as the leading recipient country, the latter holding that position since the lifting of the iron curtain in 1989. This change may be caused by approaching saturation in privatisation in countries that were in the forefront of the transformation in the past decade.

The countries in which the new openness is used to only a modest degree include Albania, Bulgaria, Fyr Macedonia, Belarus and Moldova (*Table 9.1*). The absolute size of FDI inflow here did not exceed 100 million US$ in 1996. Without exception these countries belong to the group where the economic transition is less advanced, leading to particular risks for investment. It can be concluded that the open borders of Central and Eastern Europe since 1989 have been used very differently regarding FDI. The factors underlying this differentiation are discussed in the next section.

Factors Underlying FDI Patterns

The spatial patterns of FDI partly reflect the influence of the state of transition (Meyer, 1995; Gorzelak, 1996; Lankes and Venables, 1997). The transition from a command to a market economy is the movement towards a new system for the generation and allocation of resources, with private production and well-functioning markets as corner-stones. A major component of this transition is the change and creation of institutions, including enterprises and legal structures (EBRD, 1996).

Table 9.1   Foreign direct investment in Central and Eastern Europe (CEE) and CIS

Country	Inflow * 1995	Inflow * 1996	Cumulative * 1989 –1996	Cumulative 1989-1996 per capita **
Albania	70	90	298	93
Bulgaria	82	100	425	51
Croatia	81	349	615	129
Czech Rep.	2,720	1,264	7,120	692
Estonia	199	110	735	477
Hungary	4,410	1,986	13,260	1,300
Latvia	165	230	644	258
Lithuania	72	152	285	76
Fyr Macedonia	13	39	76	38
Poland	1,134	2,741	5,398	140
Romania	404	210	1,186	52
Slovak Rep.	134	177	623	117
Slovenia	170	180	743	372
Totals CEE	*9,654*	*7,628*	*31,408*	*273*
Belarus	7	75	167	16
Moldova	73	56	161	37
Russia	2,021	2,040	5,843	40
Ukraine	300	500	1,270	25

* Million US$. Net inflows recorded in the balance of payment. Ex-Soviet republics in Asia excluded.
** US$.

Source: EBRD 1996, 1997.

It is now increasingly realised that transition is *not a linear* process, but a development with ups and downs, spurts and stops, regarding the different dimensions. For example, Bulgaria and Romania suffered serious macro-economic setbacks in 1997, but the crises triggered new governments that have begun to implement bold programs of stabilisation and structural reform (EBRD, 1997). In addition, transition does not follow one particular trajectory, but various different ones, dependent upon the interplay of generic macro-economic trends and local specificity's. It is, therefore, difficult to make generalisations valid for the entire area of Central and Eastern Europe.

The stage of transition is clearly different between countries according to indicators concerning enterprises, markets, financial institutions, and the legal system, but all have moved away from the initial stage in which primary emphasis was placed on reforms aimed at establishing markets and private ownership (*Table 9.2*). The more advanced stages are observed in the Visegrad countries (the Czech Republic, Hungary, Poland, and the Slovak Republic), Slovenia, and the Baltic States (Estonia, Latvia, and Lithuania). In this grouping Croatia is a borderline case. The countries in an advanced stage have pursued comprehensive market-oriented reform since the late 1980s or early 1990s. Gross Domestic Product (GDP) is generated mainly from the private sector, witness a share of 65 to 75 per cent in most cases. The most extensive privatisation programs have been implemented in the Czech Republic, Estonia and Hungary (EBRD, 1996). Having privatised most of their manufacturing enterprises, the two latter countries are now focusing on privatisation in the banking sector, infrastructure, and public utilities. In Latvia, Poland, Croatia and Slovenia large scale privatisation has lagged somewhat behind other areas of reform, although progress is now under way in all four countries. Poland's long-delayed mass privatisation program is now in the stage of implementation. By considering the location of countries in a more advanced stage of transition, it becomes clear that all of them border one or two EU states, except for the three Baltic states. It may be that a location close to the EU has allowed some interaction and institutional "infiltration" from the west during the years of the iron curtain, like in the case of Austria and Hungary.

The countries at less advanced stages of transition include Albania, Belarus, Bulgaria, Fyr Macedonia, Romania, Moldova, Russia and Ukraine (*Table 9.2*). They have all moved decisively to principles of market competition, but they are less advanced in enterprise restructuring (indicator C) and reform of financial institutions (particularly indicator G) than the Visegrad countries and the Baltic Republics. In 1997, Bulgaria and Romania restarted once lagging mass privatisation schemes, as well as direct sales of large enterprises. The private sector share in most of the countries listed above falls between 45 and 60 per cent of GDP.

Table 9.2   Progress in transition in Central and Eastern Europe and parts of CIS based on indicators (1997)*

Country	Private sector share in GDP (%)	Enterprises			Markets (trade)			Financial		
		A	B	C	D	E	F	G	H	I
Albania	75	2	4	2	3	4	2	2	2-	2
Bulgaria	50	3	3	2+	3	4	2	3-	2	3
Croatia	55	3	4+	3-	3	4	2	3-	2+	4
Czech. Rep.	75	4	4+	3	3	4+	3	3	3	4
Estonia	70	4	4+	3	3	4	3-	3+	3	4
Hungary	75	4	4+	3	3+	4+	3	4	3	4
Latvia	60	3	4	3-	3	4	3-	3	2+	3
Lithuania	70	3	4	3-	3	4	2+	3	2+	3
Fyr Macedonia	50	3	4	2	3	4	1	3	1	2
Poland	65	3+	4	3	3	4+	3	3	3+	4
Romania	60	3-	3	2	3	4	2	3-	2	3
Slovak Rep.	75	4	4+	3-	3	4	3	3-	2+	3
Slovenia	50	3+	4+	3-	3	4+	2	3	3	3
Belarus	20	1	2	1	3	1	2	1	2	2
Moldova	45	3	3	2	3	4	2	2	2	2
Russia	70	3+	4	2	3	4	2+	2+	3	3
Ukraine	50	2+	3	2	3	3	2	2	2	2

* Ex-Soviet republics in Asia excluded.
Indicators A-I stand for: A = Large-scale privatisation; B = Small-scale privatisation; C = Enterprise restructuring; D = Price liberalisation; E = Trade and foreign exchange system; F = Competition policy; G = Banking reform and interest rate liberalisation; H = Securities markets and non-bank financial institutions; I = Extensiveness and effectiveness of pledge law, bankruptcy and company law. Ratings from 1 to 4+ (with 4+ for most advanced economies).

Source: Adapted from EBRD 1997.

Russia holds a unique position, being alone for its sheer size. Problems of transition are different here from those in the Visegrad countries because private trade was eliminated a quarter of a century earlier, meaning that industrialisation took place almost from the beginning under the communist command economy, causing the absence of any roots (experience) in private ownership in manufacturing at the beginning of the transformation. Russia started the implementation of a reform scheme in 1992, involving price and trade liberalisation, small-scale privatisation, and unification of the exchange rate, however, fiscal and credit policies remained modest in their impact, and monetary policy was not able to stabilise the exchange rate of the rouble.

A major reason for FDI in Central and Eastern Europe is the rise of new consumer markets here (see e.g. EBRD, 1994). This would mean that countries with a relatively large domestic market, like Russia, Poland, Romania, and Ukraine, potentially attract more FDI than smaller countries. However, there are more factors involved. It is self-evident that conditions to invest in countries in advanced stages of transition are better than in the remaining countries. In fact, a study among western manufacturing companies revealing their perceived investment risks shows a close (negative) association between indicators on progress in transition and perceived country risk (a rank correlation coefficient of -0.89) (Lankes and Venables, 1997). Throughout all countries in this study, risk due to macroeconomic instability, and regulatory and legal risks turn out to be the most serious ones perceived.

Patterns of FDI at the country level also reflect the influence of geographical and cultural proximity. In the EU, Germany is by far the largest source of investment (Meyer, 1995; van Geenhuizen and Nijkamp, 1998). This partly explains the preference for providing FDI for countries like Hungary, Poland and the Czech Republic, based on short distances and a certain cultural similarity. The same holds for Austria, with important investments in adjacent Hungary, Czech Republic, Slovak Republic, and Slovenia (Alzinger and Winklhofer, 1998). Today, this situation seems changing. Germany is putting strong efforts into reunification and internal economic problems, at the same time that an important investing country has entered the field, i.e. the Republic of Korea. In 1996, for example, Korea became the largest investor in Poland with planned investments by Daewoo of 1,1 billion US$. The presence of Korea can also be observed in less popular transition economies, notably Romania (156 million US$ invested by Daewoo in Automobile Craiova) (UNECE, 1996).

In the remaining section we try to estimate the different border impacts on the inflow of FDI. In a "quasi-experimental approach" we compare pairs of recipient countries which are roughly similar in size of the domestic economy, but different in institutional distance in terms of transition stage. As we have seen above, the latter means also in most cases different in geographical distance to the EU. The results are given in *Table 9.3*. A reduction level of 6 per cent, for example, means a diminishing influence of a large institutional distance on cumulative FDI to 6 per cent. Of course, this is a rough estimate because some more factors may influence the size of FDI flow to a country, like duration of political stability and individual policies of recipient countries to attract FDI, causing some "noise" in the outcomes. We observe in all cases a reduction, but there is a large differentiation in the level, i.e. ranging from almost disappearance of FDI (1.3 per cent) to a reduction by approximately half (56.5 per cent).

Table 9.3   Reduction in cumulative FDI inflow in Central and Eastern Europe *

Pairs of recipient countries	Cumulative FDI in 1989-1996 (million US$)	Reduction level
Czech Rep.-Bulgaria	7,120 - 425	6.0%
Czech Rep.-Belarus	7,120 - 167	2.3%
Hungary-Bulgaria	13,260 - 425	3.2%
Hungary-Belarus	13,260 - 167	1.3%
Poland-Romania	5,398 - 1,186	22.0%
Poland-Ukraine	5,398 - 1,270	23.5%
Lithuania-Moldova	285 - 161	56.5%
Slovak Rep.-Moldova	623 - 161	25.8%
Slovenia-Moldova	743 - 161	21.7%
Latvia-Moldova	644 - 161	25.0%

*Albania, Fyr Macedonia and Croatia excluded.

Systems of Innovation

In the past several years attention has increased for the characteristics of innovation systems in transition countries, in view of prospects for economic growth (e.g. Dyker, 1997; Hutschenreiter et al., 2000). In western economies, innovation is a process taking place in companies and between companies in networks, such as with suppliers, customers, and with universities and other knowledge institutes (OECD, 1996b). Accordingly, innovation can be seen primarily as an interactive and socially embedded process, with a core of interpersonal relationships dominated by informal communication, and surrounded by formal co-operation using codified procedures and scientific language (e.g. Storper, 1996).

The current systems of innovation in Central and Eastern Europe are still more or less influenced by the features of the communist past (Dyker and Perrin, 1997). First, there were weak links between research institutes and companies, and concomitantly between research and the market. For example, in the former Soviet Union, science and technology were traditionally carried out in branch research institutes, being extensions of the appropriate ministries. These institutes responded to pressures from the hierarchy and were relatively isolated from the users of their findings (Egorov, 1996). In addition, in many countries the Academy system made a large contribution to basic and applied research, but again in relative isolation from companies. This isolation from the market has led to an almost absence of the notion of design which is essential in the reality of innovation in western economies (Dyker and Perrin, 1997). Particularly with regard to capital goods technology, there was neither feed back from users nor diffusion of user experiences to other users. In western economies large user-firms play a major role in interactively developing and modifying production equipment, as does the network of small specialised suppliers. Secondly, the scarce links between research institutes and companies were concerned with different concepts of technological progress compared to the Schumpeterian definition of innovation (cf. Egorov, 1996; Imre and Varszegi, 1996). Progress used to be defined in terms of either increase of unit capacities of existing equipment or re-invention (imitation) of western consumer goods. A third important feature at that time was the absence of notions of interactive technology transfer and innovation processes. Thus, the acquisition of new technology was seen as a one-way process while it was overlooked that the success in acquisition of (foreign) new technology depends also on the capacity to develop one's own technology (e.g. Bell, 1997).

Given the above situation, there is a need for R and D to be carried out by industry or actively connected with industry. With the collapse of many large industrial conglomerates and the absence of typical western institutions in the public and private sector, attention in policy is increasingly drawn to the development of small and medium sized enterprises (SMEs) (e.g. EBRD, 1995; Bernard, 1996; OECD, 1996a). SMEs are seen as important vehicles to introduce innovation and competition in local markets and to provide demonstration effects. They also allow flexibility in uncertain environments and provide re-employment opportunities for displaced employees from the state sector (OECD, 1996a). The numbers of SMEs are now rapidly increasing in most transition economies but there seems to be an emphasis on short-term (temporary) business requiring small capital investment, such as in trade and service activities. To date only small numbers of SMEs operate in manufacturing, for example, approximately 15 per cent of all SMEs in the Czech Republic, Estonia, and Lithuania (EBRD, 1995; OECD, 1996a).

With regard to the restructuring of the old R and D institutes, the current shift is from complex, technology-led projects towards simpler, market-oriented R and D and services like testing, quality control, and certification (Radosevic, 1995; Egorov, 1996). Accordingly, R and D institutes are now being changed into two directions, namely non-R and D and service organisations through privatisation and establishment of new companies (spin-off), and internal restructuring and the creation of new relationships between R and D in companies, independent institutes, and universities. The first direction seems to be fraught with difficulties, particularly the path of establishment of competitive companies as spin-offs from R and D institutes (e.g. Oakey et al., 1996). There is a danger that spin-off companies are launched from aims to compensate for funding shortfalls rather than from planning strategies including a careful selection of competitive products and markets. Furthermore, an impediment to the transformation of R and D institutes seems to be the brain drain. Low wages and lack of orders (budget cuts) have caused an outflow of research workers to other sectors. More importantly, there is a hidden brain drain involving a combination of formal maintenance of the workplace in a scientific institute and the spending of most working time by R and D workers on alternative activities, ranging from consultancy to plumbing. Developments like these are likely to erode the base of R and D institutes in some countries.

When adopting a regional focus, it becomes clear that in the communist past local production systems based on innovative products from own technology development could only develop in a limited number of cases. An example of such rare cases is the Skoda factory and its suppliers in former Czechoslovakia. Moreover, the critical institutional conditions for local learning as identified in specific regions in Western Europe and North America (e.g. Morgan, 1997) could not develop under communist rule. The self-organisation of regional communities and the development of initiatives aimed at societal progress (different from the official party system) in particular were often discouraged and sometimes even punished. Vertical structures or fragmentation prohibited the development of the horizontal co-operation that allows for open and interactive learning. This situation was connected with types of top-down planning dominated by fixed goals and rigidity rather than bottom-up planning based on responsibility of local actors and flexibility. Although major changes have occurred, it is reasonable to assume that the current institutional layer that supports the development of dynamic learning is still relatively thin. It can be concluded that ways to improve the conditions for regional "active space" development are still littered with stumbling blocks, a situation that cannot be changed overnight.

Foreign Investors and their Strategy

On a micro-level, foreign direct investment is the transfer by a firm of capital (and other resources) into a business venture abroad, aimed at acquiring control of the venture. Although motivation for FDI in transition economies tends to be mixed, important motives appear to be:

- cost-based, aimed at cheap inputs like low wages;
- market-driven, aimed at the penetration of new markets;
- resource-based, aimed at the exploitation of (scarce) natural resources;
- knowledge-driven, aimed at the use of specialist or cheap knowledge.

To date, market-driven motives seem to dominate inward FDI in Central and Eastern Europe, with cost factors as a second important motive (EBRD, 1994; Welfens and Jasinski, 1994; Meyer, 1995; Radosevic, 1997). Some authors have seen a gradual increase in the importance of cost-based FDI connected with declining standards of living and consumption in particular countries, and a slower market growth than expected. Accordingly, an increasing factor in investment decisions will be access to the relatively low-wage but often medium-skill workforce (e.g. Smith and Ferenciková, 1998). The type of subsidiaries involved, the broader strategic context of FDI decisions, and the consequences for learning are discussed in the remaining part of this section.

A useful typology of subsidiaries is given by Radosevic (1997) in which investment motives are combined with the technological deepening and integration of the factory within the (global) corporate network. The most primitive cost-driven FDI is concerned with *offshore factories*, common in sectors like the clothing industry and electronics. A concentration of this type can be found in the western Czech Republic, on the border with Germany. The firms are export-processing and vertically integrated into German subcontracting networks, based upon low-wage and low-skill labour. These investments compare with the US-Mexico maquiliadora type (Pavlínek and Smith, 1998). One step higher on the technology dimension are *sourcing factories*, very common in the automotive industry. General Motors in Hungary is an example within a broader strategy of the mother company to develop lower-cost sources for the supply of components. The same holds for the Daewoo Electronics operation in Poland. Furthermore, there is now an increasing number of *focused factories*, with VW-Skoda in the Czech Republic, Suzuki Magyar in Hungary and the Polish Fiat factory as examples (Radosevic, 1997). The sourcing here includes a broad range of components or entire products, a situation in which foreign investment may also be attracted by local suppliers. Focused factories are integrated at a relatively high level in global corporate networks.

Regarding market-driven motives, *trading companies* are certainly the most common type. In practice, trading companies are often coupled with assembly plants to avoid high tariff barriers. Most market-based investments aim at capturing additional new markets in Central and Eastern Europe but there are a few examples in which foreign investors have shut down the factories here to stop competition from the latter in Western European markets, eventually only leaving a trade organisation (e.g. Kiss, 1995; Bernard, 1996; Grayson and Bodily, 1996). A type of market-driven investment one step higher on the technology dimension is the *miniature replica*. These are emerging now in the form of take-overs of local firms, mainly in consumer goods (e.g. in food industry). With regard to a further category of investments, knowledge-driven investments, we may say that these are still rare, but there are some examples in Russia linked with the military sector working through subcontracting and joint-ventures (such as in aerospace and aviation). Resource-based investments are more common, most of them being of the *extractors* type. This type is found in Russia with its large reserves of oil and natural gas, and limited processing facilities, and also in Estonia with its vast amount of wood. The *processor* type of resource-based FDI is found in Central Europe mainly in the food processing industry (Radosevic, 1997).

So far, it is not possible to assess the relative importance of the above different foreign subsidiaries in transition economies. In investment studies a micro-perspective on subsidiaries has seldom been used in a systematic and comparative way, and this is the reason why the current evidence is still fragmentary. The preliminary evidence suggests limited prospects for dynamic learning and innovation in the subsidiaries involved, connected with offshore factories, sourcing factories, and trading companies, all without any (autonomous) R and D activity (Rojec, 1994; Bernard, 1996; Radosevic, 1997; Smith and Ferencikova, 1998). There are of course clearly positive aspects on a number of the above investments, i.e. the introduction of new technology and equipment critical for getting specific production from the ground - such as introducing state-of-the-art equipment and training at greenfield oil and gas attraction sites by Shell in Russia (Sharp and Barz, 1996) - and the upgrading of existing facilities to modern best practice standards. In the latter cases the investors provide know-how, manufacturing facilities, tools, components and licences for plant reconstruction, and introduce new corporate and management cultures. In addition, as previously indicated, there is a recent trend for increasing the number of focused factories, mainly in the automotive industry (Radosevic, 1997). Through their larger scope of activities and stronger integration in the global industry, the latter factories provide relatively good potentials for local dynamic learning both in the factory and the local supply chain.

Aside from the type of subsidiary, ownership relations may play a role in learning by domestic factories. In general, joint ventures offer good opportunities for the domestic partner to acquire western technology and derive new skills and management practices; however, there have been some important gradual take-overs of ownership shares of joint ventures by the foreign investor - such as in the Slovak Republic - thereby causing an increased isolation of the joint venture from its domestic partner (Smith and Ferenciková, 1998).

The picture indicated above is certainly not optimistic but it needs to be stressed that it is only a snapshot. On the medium term, the subsidiaries may integrate into global networks at higher levels, while increasing on site R and D activity (see also Gacs et al., 1999).

Regional Embeddedness

The regional dimension of FDI was often neglected in the first years of the transition because all the attention was attracted to problems of macro-economic stabilisation. Despite a lack of comparable regional data on FDI, there is sufficient indication for an ongoing trend of reinforcing existing regional disparities (e.g. Gorzelak, 1996; van Geenhuizen and Nijkamp,

1998; Hardy, 1998; Pavlínek and Smith, 1998). Thus, FDI inflow concentrates in the borderlands with the EU and in the large metropolitan regions. For example, Warsaw has received 35 per cent of all inflow into Poland and Bratislava 60 per cent of all inflow into the Slovak Republic. In contrast, many regions - often eastern parts – have received limited FDI or none.

Recently, the attention of researchers has shifted to looking at the regional impact of FDI in terms of embeddedness through market and supplier links, and causes of uneven embeddedness (Grabher, 1997; Radosevic, 1997; Hardy, 1998; Pavlínek and Smith, 1998; Smith and Ferenciková, 1998). The number of case studies dealing with impacts of FDI in terms of regional linkages is now rapidly increasing. Results from these studies point to both positive and negative developments (Kiss, 1995; Sharp and Barz, 1996; Pavlínek and Smith, 1998). Negative developments include the crowding out of traditional linkages with local suppliers of inputs (raw material, ingredients, components), such as found in parts of the Hungarian food industry, with these inputs being replaced by imports (Kiss, 1995). Positive developments occur if foreign investors act as organisers of a local supply base. For example, VW-Skoda in the Czech Republic has taken an active role in the establishment of its suppliers' network leading to a majority of VW-Skoda's purchases in the Czech Republic. Together with the introduction of a scheme for supply integration - requiring the suppliers to join the assembly line to install the parts themselves - opportunities have been created to increase local learning and innovation. The situation in the Slovak VW-Skoda investment (region of Bratislava) is completely different. Local supplier relationships are relatively few here, leading to a factory that is almost entirely disconnected from its regional environment (Pavlínek and Smith, 1998).

The difference in embeddedness between the VW Skoda factory in the Czech Republic and the one in the Slovak Republic can be explained as follows. First of all, the inherited production system was different. The Skoda factory in the Czech Republic had a tradition of skills and independent innovation with an established network of suppliers. In contrast, the Skoda factory in the Slovak Republic was a latecomer and became structured as a branch plant in the old system with limited autonomous production and R and D. Accordingly, the foreign investor has simply built upon an existing structure. Secondly, the existing supply structure matched with the investor's strategy, such as Just-in-Time production schemes. From a series of case studies it appears that regional supply structures are usually not created anew by foreign investors, whereas existing supply structures are truncated if investors cannot see advantages in using them. In the latter case, regional supply structures are replaced by internationalised systems of supply integration, leading to only modest opportunities for local learning and innovation (Pavlínek and Smith, 1998). So far, the causal analysis of regional embeddedness of FDI has revealed that there are two conditions under which a high regional embeddedness develops, i.e. an existing innovative supply system and a sufficient match between this system and the specific needs of the investors. It needs however to be mentioned that circumstances may change. For example, the favourable development around VW Skoda in the Czech Republic may come to an end now. The regional component suppliers here are increasingly involved in joint ventures with western firms, leading to the transfer of decision making and R and D from the Czech Republic to the mother companies abroad, thereby decreasing opportunities for local learning (Pavlínek and Smith, 1998).

Conclusion: Local Learning Connected to FDI

The preliminary evidence presented in this study indicates that to date FDI has had a modest impact on dynamic learning and innovation in transition economies. The acquisition of western technology and local absorption of this technology are no automatic by-products of FDI, thereby underlining the need for explicit and active learning and innovation policies that address problems in their own right. Policy action can follow two lines. The first line would focus on a general improvement of conditions to benefit from newly acquired technology, whereas the second line would include various targeted policies to attract innovative investments that develop a high level of regional embeddedness in terms of networks with innovative suppliers.

In order to benefit from newly acquired technology, there is a pressing need to continue to foster processes of technology transfer and absorption, such as by establishing liaison offices at higher educational institutes and universities, innovation centres, and technology centres focusing on generic technologies (such as micro-electronics) (e.g. Imre and Varszegi, 1996). Such centres are being established now in various countries, partially with the assistance of EU programs. Attention needs to be given to the operation of these centres and the development of active linkages with regional and national learning networks. An equally important but more difficult task is to teach domestic firms how to innovate and generate new technology, because acquisition and absorption of external technology is very much dependent on the technological level that is available. This covers the entire spectrum of learning mechanisms, from formal R and D to learning-by-doing using tacit knowledge. It involves the generation and absorption of new routines but also a further de-learning of traditional routines.

There are five policy ingredients to be proposed to regional and national governments regarding the attraction of FDI to foster innovative supply structures in a region. First, it is important to better satisfy specific corporate needs following from the more risky nature of innovative activities. For example, it is important to establish a safe system of intellectual property rights (such as patent protection), a condition that matters in the pharmaceutical and advanced chemicals industry (Sharp and Barz, 1996). Using generic tools to attract specific innovative industry, is a second ingredient. For example, the foreign investment agency Czechinvest is trying to attract innovative investments by offering incentives. A third policy ingredient is the improvement of skills of existing suppliers to satisfy the requirements of multinationals. This is mainly concerned to skills at the workplace level (western best practice) and management skills. A fourth but more difficult component would be to advance local self-organisation and local entrepreneurial initiatives that can interact with foreign investment projects (van Geenhuizen and Nijkamp, 1999). A good example in this respect is found in the region of Hungarian Székesfehérvár with a combination of a forward-looking local government, local entrepreneurial initiatives, and a skilled labour force (Business Central Europe, June 1998). A fifth, and again difficult ingredient, would be to pursue specific policy action to ensure that activities of foreign companies spill over into the local and wider economy. For example, a regulation policy may prevent the cutting of domestic supply chains. The previously discussed VW-Skoda investment in the Czech Republic seems to be a rare example in which the state was able to negotiate local content agreements (Pavlínek and Smith, 1998). Opportunities for negotiation may increase when existing suppliers are upgraded close to western standards.

With regard to regional (local) policies the following model may be used, be-it adapted to specific local circumstances: the establishment of industrial incubators and promotion of innovative product development while using the potentials of both local universities (or R and D institutes) and FDI. A particular tool is the "free science park", a combination of financial incentives to attract FDI and a knowledge transfer infrastructure connected with a university that supplies expertise to be commercialised. There are a number of initiatives that come close to this model. A careful monitoring of their operation is necessary to learn from their development.

It can be concluded that there are various interesting policy options to improve conditions for benefiting from FDI. The role of the above policies should, however, not be overestimated because some of them have a long lead-time, particularly the ones concerning institution building and upgrading of skills based on endogenous forces. Furthermore, there is still a lack of insights into the various ways processes work. Therefore, there is a need for solid comparative research to better underpin policy action. A first line would be to identify actual changes in existing FDI projects in terms of technology level, integration in global corporate networks, and regional embeddedness. Such research aims to assess whether the same investments remain consistently disembedded or turn out to improve over time. A second research path would be to elaborate causal analysis of regional (dis)embeddedness and local learning, in order to identify which important factors need to be influenced by policy action in favour of a move to "active space" development.

References

Alzinger, W. and Winklhofer, R. (1998), 'General Patterns of Austria's FDI in Central and Eastern Europe and a Case Study', *Journal of International Relations and Development*, 1, pp. 65-83.

Bell, M. (1997), 'Technology transfer to transition countries: are there lessons from the experience of the post-war industrializing countries?', in D.A. Dyker (ed), *The Technology of Transition*, Central European University Press, Budapest, pp. 63-94.

Bernard, K.N. (1996), 'Eastern Europe: A Source of Additional Competition or of New Opportunities?', in B. Fynes and S. Ennis (eds), *Competing from the Periphery, Core Issues in International Business*, The Dryden Press, London, pp. 439-474.

Business Central Europe, June 1998.

Dunning, J.H. (1993), 'The Prospects for Foreign Investment in Eastern Europe', in P. Artisien, M. Rojec and M. Svetlicic (eds), *Foreign Investment in Central and Eastern Europe*, ST. Martin's Press, New York, pp. 16-34.

Dyker, D.A. and Perrin, J. (1997), 'Technology policy and industrial objectives in the context of economic transition', in D.A. Dyker (ed), *The Technology of Transition*, Central European University Press, Budapest, pp. 3-19.

EBRD (European Bank for Reconstruction and Development) (1994), *Transition Report, Foreign Direct Investment*, EBRD, London.

EBRD (European Bank for Reconstruction and Development) (1995), *Transition Report, Investment and Enterprise Development*, EBRD, London.

EBRD (European Bank for Reconstruction and Development) (1996), *Transition Report, Infrastructure and Savings*, EBRD, London.

EBRD (European Bank for Reconstruction and Development) (1997), *Transition Report, Enterprise Performance and Growth*, EBRD, London.

Egorov, I. (1996), 'Trends in transforming R and D potential in Russia and Ukraine in the early 1990s', *Science and Public Policy*, vol. 23 (4), pp. 202-214.

Gacs, J., Holzmann, R. and Wyzan, M.L. (eds) (1999), *The Mixed Blessing of Financial Inflows*, Edward Elgar, Cheltenham.

Geenhuizen. M. van, and Nijkamp, P. (1996), 'Foreign Investment and Regional Development Scenarios in Eastern Europe', *Current Politics and Economics of Europe*, 4, pp. 1-12.

Geenhuizen, M. van, and Nijkamp, P. (1998), 'Potentials for East-West Integration: The Case of Foreign Direct Investment', *Environment and Planning C*, 16, pp. 105-120.

Geenhuizen, M. van, and Nijkamp, P. (1999), 'Regional Policy beyond 2000: Learning as Device', *European Spatial Research Policy*, vol. 6 (2), pp. 5-20.

Geenhuizen, M. van, and Ratti, R. (1998), 'Managing Openness in Transport and Regional Development: An "Active Space Approach", in K. Button, P. Nijkamp and H. Priemus (eds), *Transport Networks in Europe*, Edward Elgar, Cheltenham, pp. 84-102.

Gorzelak, G. (1996), *The Regional Dimension of Transformation in Central Europe*, Jessica Kingsley, London.

Grabher, G. (1997), 'Adaptation at the Cost of Adaptability? Restructuring the Eastern German Economy', in G. Grabher and D. Stark (eds), *Restructuring Networks in Postsocialism: Legacies, Linkages and Localities*, Oxford University Press, Oxford, pp. 107-134.

Grayson, L.E. and Bodily. S.E. (1996), *Integration into the world economy: companies in transition in the Czech Republic, Slovakia, and Hungary*, IIASA, Laxenburg.

Hutschenreiter, G., Knell, M. and Radosevic, S. (eds) (2000), *Restructuring Innovation Systems in Eastern Europe*, Edward Elgar, Cheltenham.

Hardy, J. (1998), 'Cathedrals in the Desert? Transnationals, Corporate Strategy and Locality in Wroclaw', *Regional Studies*, 32, pp. 639-652.

Hooley, G., Cox, T., Shipley, D., Fahy, J., Beracs, J. and Kolos, K. (1996), 'Foreign direct investment in Hungary: Resource acquisition and domestic competitive advantage', *Journal of International Business Studies*, Fourth Quarter, pp. 683-709.

Imre, J. and Varszegi, G. (1996), 'The Transformation of Science and Technology in Hungary', in A. Kuklinsky (ed), *Production of Knowledge and the Dignity of Science*, EUROREG, Warsaw, pp. 231-242.

Kiss, J. (1995), 'Privatization and Foreign Capital in the Hungarian Food Industry', *Eastern European Economics*, vol. 33 (4), pp. 24-37.

Lankes, H-P. and Venables, A.J. (1997), 'Foreign direct investment in economic transition: the changing pattern of investments', *The Economics of Transition*, 5, pp. 331-347.

Meyer, K.E. (1995), 'Foreign direct investment in the early years of economic transition: a survey', *Economics of Transition*, vol. 3 (3), pp. 301-320.

Morgan, K. (1997), 'The learning region: institutions, innovation and regional renewal', *Regional Studies*, 31, pp. 491-503.

Oakey, R.P., Hare, P.G. and Balazs, K. (1996), 'Strategies for the exploitation of intelligence capital: Evidence from Hungarian Research Institutes', *R and D Management*, vol. 26 (1), pp. 67-82.

OECD (1996a), *Small Firms as Foreign Investors:Case Studies from Transition Economies*, OECD, Paris.

OECD (1996b), *Science, Technology and Industry Outlook 1996*, OECD, Paris.

Pavlínek, P. and Smith, A. (1998), 'Internationalization and Embeddedness in East-Central European Transition: The Contrasting Geographies of Inward Investment in the Czech and Slovak Republics', *Regional Studies*, 32, pp. 619-638.

Radosevic, S. (1995), 'Science and technology capabilities in economies in transition: effects and prospects', *Economics of Transition*, vol. 3 (4), pp. 459-478.

Radosevic, S. (1997), 'Technology transfer in global competition: the case of economies in transition', D.A. Dyker (ed), *The Technology of Transition*, Central European University Press, Budapest, pp. 126-158.

Radosevic, S. (1999), *International Technology Transfer and Catch-up in Economic Development*, Edward Elgar, Cheltenham.

Rojec, M. (1994), *Foreign Direct Investment in Slovenia*, University of Ljubljana, Ljubljana.

Sharp, M. and Barz, M. (1996), 'Multinational Companies and the Development and Diffusion of New Technologies in Eastern and Central Europe and the Former Soviet Union', in A. Kuklinsky (ed), *Production of Knowledge and the Dignity of Science*, EUROREG, Warsaw, pp. 191-207.

Smith, A. and Ferencíková, S. (1998), 'Inward investment, Regional transformations and uneven development in Eastern and Central Europe', *European Urban and Regional Studies*, 5, pp. 155-173.

Storper, M. (1996), 'Innovation as Collective Action: Conventions, Products and Technologies', *Industrial and Corporate Change*, vol. 5 (3), pp. 761-790.

Svetlicic, M. and Rojec, M. (1994), 'Foreign Direct Investment and the Transformation of Central European Economies', *Management International Review*, vol. 34 (4), pp. 293-312.

UNECE (United Nations Economic Commission for Europe) (1997), *East-West Investment News* (various issues), United Nations, Geneva.

Welfens, P.J.J. and Jasinski, P. (1994), *Privatization and Foreign Direct Investments in Transforming Economies*, Dartmouth, Brookfield VT.

10 In Search of Complementarities. Initiation of Interaction with Russian Karelia

HEIKKI ESKELINEN

This chapter gives an analysis of the opportunities and barriers for co-operation between the Russian Republic of Karelia and Finnish regions on the other side of the border. The findings are overall dichotomous. On the one hand, the region has developed an institutional structure that allows for cross-border initiatives like in many normal European border regions. On the other hand, there are no signs that the region can evolve to a functional region. The prospective Karelia Euregion is not only very large, with different functional interests in the different parts, but the necessary complementarities in the regional economy are hard to find.

Introduction

The complementarity of the partners' resources is an important precondition for successful cross-border co-operation. This setting can be analysed both from a static and a dynamic perspective. The latter is essential for an assessment of long term development: are geographically contiguous partners able jointly to create unique competitive resources by means of cross-border activities? If this is the case, the resulting functional interdependencies lead to the formation of integrated trans-boundary structures.

The political and economic change in Eastern Europe and the partial opening of the East/West divide in Europe have permitted direct connections and co-operative relationships between border regions. Direct trans-boundary links, which were illegal only a decade ago, are now supported by specific programmes. This has clearly been a dramatic change: room for regionality has been created even across this divisive border (Christiansen and Joenniemi, 1999: 11). In fact, these developments have given rise to a kind of laboratory situation for the analysis of, for instance, how cross-border links emerge and co-operative practices are created, and what their potential is for the regions which have been functionally separated from each other for a long time.

Although each border region is a unique case, certain structural factors clearly condition the development and patterns of cross-border interaction and co-operation along the East/West divide. These include, among others, the economic potential, settlement pattern, infrastructure networks, historical and cultural links of the neighbouring regions, as well as the length of their separated development and their roles in relation to the ongoing European Union integration process.

This chapter consists of an investigation of the first steps towards interaction and co-operation in a peripheral area in the European North, between Finland and the Russian Federation, across the western border of the Karelian Republic (Russian Karelia). This region lies between two potentially major regions in Europe, the Baltic region and the Barents region (see *Figure 10.1*). The analysis deals with both conscious policy programmes to promote cross-border co-operation and spontaneous developments in the same direction. More specifically, the focus is, firstly, on the roles of actors at different spatial levels in this process, and secondly, on the key issues conditioning the rocky road of cross-border co-operation initiatives towards genuine trans-boundary regionalism.

The border region under consideration here accounts for more than half of the Finnish-Russian border, which since 1995 has also been the only land border between the European Union and the Russian Federation. Thus, it is natural to use the experiences of the EU's internal and external borders as a point of reference in the assessment of how cross-border co-operative practices have been compatible with a European regime.

The Border between the Baltic Sea and the Barents Sea

The Finnish-Russian border, which is more than 1200 kilometres long, runs from the Baltic coast almost up to the Barents Sea (see *Figure 10.1*). The border, which has shifted many times in the past, has historically been an interface and a battlefield between eastern and western cultures and politico-economic spheres of influence in northern Europe. Yet it did not represent a categorical dividing line and an effective barrier to interaction until after the October Revolution and Finland's independence in 1917 (see, e.g. Laine, 1994; Paasi, 1995).

Figure 10.1 Finnish-Russian border area

Obviously, the closed border has had negative repercussions on the development of the regions adjacent to it. In Finland, the eastern part of the country has suffered from a lack of contacts to its historical centres of gravity. This problem was accentuated as a consequence of the Second World War, when a part of eastern Finland was annexed by the Soviet Union. The population of the ceded territory, more than 400,000 people, fled to Finland, and they were replaced by a population from various parts of the Soviet Union. As a result, functional regions were split, important infrastructure networks were cut off, and the region on the eastern side of the border was effectively disconnected from its previous history. The potential for a border dispute, the Karelian issue, has been a standing topic in public discussion in Finland in the 1990s with the realignment of Finnish-Russian relations due to new circumstances.

Direct interaction across the border between individual partners was practically non-existent during the Soviet era; the exceptions were arranged and controlled in terms of the bilateral agreements between the two countries (for an interesting example: see Tikkanen and Käkönen, 1997). Foreigners were allowed to visit only a few places on the eastern side of the border, where activities were subordinated to security and military considerations. Crossing points for international passenger and goods traffic were located in the southern corner of the border area, on the routes to St. Petersburg, which is by far the largest centre in the border region.

Thus, the construction of direct links between the Finnish-Russian border regions had to begin almost from scratch after these links gradually became possible during the perestroika years in the Soviet Union. With regard to some basic preconditions for developing cross-border co-operation, that is, factors such as existing networks of infrastructure links, a sufficient supply of public and private services, and the compatibility of institutions (see van der Veen, 1993), the conclusion is straightforward: prospects are much bleaker than across any internal borders, and most external borders, in the EU. Firstly, the regions in eastern Finland and north-western Russia are sparsely populated, and their cross-border infrastructure networks have not been developed. Secondly, these regions have been, for a long time, typical peripheries in their own spacio-economic systems, and they have followed very different development trajectories ever since the October Revolution. Thirdly, the introduction of a European cross-border co-operation regime is in this case conditioned by the fact that the Russian Federation will not become a EU member in the foreseeable future, and thus motives for strengthening links across the border do not include pre-integration.

Notwithstanding the severe constraints mentioned above, interactions and co-operation across the Finnish-Russian border have emerged in many forms in the 1990s. Clearly, various actors in the border area have been sufficiently well-motivated to create contacts for a number of purposes, including security, regional development, industrial competitiveness, cultural and humanitarian considerations, and so on. This pursuit of trans-boundary regionalism is linked to simultaneous changes in socio-economic development doctrines, and in relationships between central and regional levels of governance and administration on both sides of the border.

Towards Co-operative Practices: Policy Initiatives

The first wave of direct contacts between partners across the Finnish-Russian border emerged spontaneously as soon as travel restrictions were relaxed in the USSR in the late 1980s. Individuals, civic associations and local organisations, such as municipalities and schools in the Finnish border region, created links across the divide, and the central governments had to react to this grass-roots activity as a fait accompli. To set the rules of the new practice, the governments of Finland and Russia signed treaties concerning neighbourhood relations, trade, and so-called near-region co-operation in 1992. The last-mentioned treaty was the first of its kind in Russia, and it implied that not only Finland but also the Russian Federation accepted the development of direct interaction and co-operation between border regions and, at least in principle, gave their support to such a strategy. In this context, the most important political precondition for improving economic, cultural and other connections across the border was Finland's decision not to advance territorial claims against the Russian Federation. This was also an important prerequisite for the country's membership in the European Union in 1995.

Before Finland's EU membership, cross-border co-operation was primarily a bilateral matter between the two countries, on one hand, and actors in the border regions themselves, on the other. In Finland, the issue of cross-border co-operation was first placed on the policy-making agenda mainly in two contexts, in domestic regional policy and as near region co-operation.

Lines of policy action in Finland

As part of the anticipated, and later realised, membership of the EU, *domestic regional policies* have undergone a major upheaval in Finland in the 1990s (see, e.g. Eskelinen, Kokkonen and Virkkala, 1997). In principle, changes in institutions and strategies have emphasised the role of cross-border co-operation. Firstly, responsibility for regional development was transferred to regional councils, which are bottom-up organisations based on municipalities, whereas in the past this responsibility lay with top-down regional organisations of the central government. Secondly, regional development strategies were reordered to harmonise with EU practice in terms of programmes, and the promotion of border region development was formulated as one such programme. However, as regional councils have no right to levy taxes, they did not have, before the implementation of EU Interreg programmes in 1996, sources of funding ear-marked for promoting cross-border co-operation. In any case, regional councils and municipalities were from the very outset active in organising local initiatives and unofficial consultative bodies, for instance, between eastern Finland and the Republic of Karelia, which later also proved to be of practical importance in the preparation of actual co-operative projects.

The state, for its part, has invested in the improvement of transport connections and border-crossing facilities. The rapid growth of trade flows between Finland and Russia - after the collapse of earlier bilateral trade in the early 1990s - and transit traffic have been the most important reason for these measures, and upgrading interaction at local and regional levels between the neighbouring border regions only an additional argument.

The *near region co-operation* with Russia, which is funded from the Finnish state budget, focuses on the Karelian Republic, the Murmansk and Leningrad regions, as well as the city of St. Petersburg (see *Figure 10.1*). Since 1996, the Ministry of Foreign Affairs has taken a more decisive role as the co-ordinator of this activity, and emphasis has been placed on linking it with the respective policies of various international organisations, the EU in particular. Regional working groups, which include representatives from government ministries and regional organisations, have been given the responsibility to co-ordinate measures.

According to the strategy revised in 1996, the reduction of environmental risks, the promotion of stability, security and welfare, and the improvement of preconditions for economic co-operation are the ultimate aims of the Finnish near-region co-operation policy. In practice, various environmental and educational projects have been the most important instruments towards these aims in the neighbouring regions in Russia. Finland allocated approximately 510 million FIM (excl. loan guarantees) of national budget funding to the near-region co-operation with Russia in 1990-97 (Ulkoasiainministeriö, 1998). Russian Karelia received more than one-fourth of this funding and was also a recipient in several multiregional measures.

From a Finnish perspective, near-region co-operation has faced problems similar to those often encountered by recipient and donor countries in development aid programmes. A weak commitment by Russian partners has been a problem in some cases, and expectations have also been contradictory in similar ways as to those in the EU Tacis programmes: the focus in the West is on technical assistance whereas direct investments are called for in the East. Yet, contradictions are not limited to the implementation of projects in Russia, they are also evident in the allocation of funding to actors in Finland. Large projects in certain priority areas are the declared purpose of the state-funded near-region co-operation, but this is bound to conflict with local needs and plans, as small local and regional partners are easily sidelined.

Towards the West on the fringe of the Russian Federation

In recent years, the relevance of regions (oblast, respublika, krays, okrug) as political and economic actors has been accentuated in the Russian Federation. This tendency has received special attention in border areas, including Finland's four neighbouring regions (St. Petersburg, Leningrad, Russian Karelia and Murmansk).

In contrast to many fringe areas in Russia, the regions bordering Finland have not attempted to sever their ties with the Federation, but have tried to utilise their geographical position to initiate direct ties with foreign countries, especially with the neighbouring regions (e.g., Bradshaw and Lynn, 1996). Russian Karelia is an example of a region which has been attempting to redefine and reconstruct its role as a kind of national unit within the context of the Russian Federation. This form of regionalisation derives from its institutional position in the Soviet ethno-federalist system, where there was a link between a titular group (in this case: Karelians) and its administrative territory (until 1991: the Karelian ASSR), even if the actual role of the minority was politically and culturally marginal (see Lynn and Fryer, 1998). The development of direct economic and political links with foreign countries, with Finland and its border regions in particular, has been an important instrument in this kind of nation- or region-building process. For the same purpose, the Republic of Karelia is also a member of the Barents Euro-Arctic Region (see Bröms et. al., 1994). However, an important obstacle to the implementation of various activities in this multinational and multiregional setting is the fact that the bulk of its neighbouring region in eastern Finland has been excluded from the Barents co-operation.

Legislative changes in the Russian Federation have been the necessary preconditions for creating direct links between Russian Karelia and foreign countries. Many of them have been based on case-specific agreements between the Republic and the Federation, and thus have contributed to the increasing variation in the rules of the game concerning foreign firms and other organisations in Russia. However, the impact of these changes has remained limited due to the unclear division of power and responsibilities between the Republic and the Federation (see Eskelinen et al., 1997). In addition to legislative changes, direct links have been supported by the opening of crossing points for international traffic without case-specific authorisation. This occurred on the western border of the Karelian Republic in Vartius in 1991 and in Niirala in 1993 (see *Figure 10.1*).

From the perspective of the Karelian Republic, in the final analysis the success of cross-border co-operation has to be evaluated on the basis of its contribution to the alleviation of the severe structural problems during the Russian political and economic transformation.

New partner: the European Union

Cross-border co-operation as such is not a new phenomenon. For decades, many border regions in Europe have been involved in various forms of local co-operative arrangements and joint paradiplomatic activities, which have since the 1980s been streamlined to fit into the European Union regime. A common characteristic of this organised cross-border activity has been tension between the domains and authority of different actors; the established realms of "high (foreign) politics" and "low (local) politics" are challenged in cross-border co-operation. Although the EU cross-border co-operation regime was originally designed for the internal borders between member states, more recently it has also been applied, with some adjustments, to the Union's external borders (see, e.g. Scott, 1999).

Obviously, the motives for cross-border co-operation on the external borders of the European Union, as well as its targets and arrangements, differ from those on its internal borders. A major difference stems from the fact that co-operation across external borders has in several cases, including the Finnish-Russian one, implications for international relations. This setting is especially intricate in the case of multinational co-operation regions such as the Barents region and Baltic region in northern Europe, where actors on different spatial scales are involved.

The Interreg initiative is the EU's most concrete instrument for supporting cross-border co-operation. On the Finnish-Russian border it was introduced at the end of 1996, when the first projects were given funding (see Segercrantz et al., 1998). This activity is organised through three programmes: South East Finland (towards St. Petersburg and the Leningrad region), Karelia (the Republic of Karelia), and the Barents (the Murmansk region). These three programmes, which will receive a total of MECU 33.9 from the EU during the period up to 1999, have introduced a new stage in co-operation across the Finnish-Russian border.

Firstly, they emphasise the role of regional actors, because the regional councils have been given responsibility for the administrative tasks of the programmes. In the case of the Karelian programme, for instance, three regional councils (North Karelia, Kainuu and Northern Ostrobothnia) share this responsibility. This institutional solution has, in fact, created a new co-operative region in Finland, defined on the basis of neighbourhood: the Republic of Karelia borders these three regions (see the shaded area in *Figure 10.1*). On the other hand, the division of labour between the three different Interreg programmes on the Finnish-Russian border is not without problems, especially because potential functional connections between St Petersburg and more northern areas in eastern Finland were not given due consideration in their preparation.

Secondly, Interreg funding, although its use is limited to the territory of the EU, Finland in this case, has clearly provided additional resources for cross-border initiatives. In fact, it can be regarded as a policy innovation in Finland because no corresponding funding mechanism has been developed for purposes of domestic regional policies or near-region co-operation. In addition, the fairly comprehensive programming activity involved in the Interreg programme has already contributed to a more careful selection of policy priorities. Thirdly, the need to create procedures for reconciling different programmes has become a more acute policy concern. Currently, the EU has no mechanism which would secure links between the Interreg and Tacis CBC projects, which are under way on both sides of the border.

The tension between the central government and the regions in the implementation of the Interreg projects and the technical problems of cross-border programmes (e.g. between Interreg and Tacis) have contributed to the fact that the regional councils in the Karelia Interreg region together with the Republic of Karelia have launched preparations for an Euregion (Cronberg, 1998). They seek to create an integrated system of decision-making and administration for promoting cross-border co-operation. This is clearly a very ambitious target, and it will in practice put EU policies towards external borders and Russia to a serious test. The relevant strategic decisions have to be made in the preparatory stages of the next Interreg programme period.

Overall, the Interreg programme has opened the way for regional cross-border co-operation that was previously dominated by the state, although the financial competence of the regional councils is still limited and they struggle to wrest more competence from the state, and even from their member municipalities. Interreg programme management - not unlike the whole EU structural policy administration - seems to have become an arena of "contested governance" (Lloyd and Meegan, 1996), with competition not only vertically, between the EU, the state and the regions, but also horizontally between actors at a regional level, or specifically the ministries.

Trends, Repercussions and Prospects of Cross-Border Flows

Cross-border co-operation is usually promoted on functional grounds in the sense that it is based on the problems and opportunities faced by the partners. Thus, the results are conditioned by the suitability of the functional linkages concerned and their potential: the existence of complementary assets tends to contribute to the construction of interdependencies, whereas regions with completely different resource bases and competing goals find it difficult to initiate successful co-operation. Furthermore, the possibilities of developing co-operative practices obviously depend on relations between local co-operative arrangements and national and multinational ones, and cross-border initiatives can be used for purposes of extending the scope of decision-making at the regional and local level (see, e.g. Keating, 1998).

As surveyed above, the development of cross-border co-operation programmes across the Finnish-Russian border has been intertwined with tendencies towards regionalisation on both sides of the border, at least in the sense that the implementation of cross-border initiatives would not have been possible without the active involvement of regional and local partners. However, in the final analysis co-operation programmes and other cross-border initiatives have to be evaluated on the basis of the dynamics of actual exchanges and their contribution to socio-economic development in a broad sense. In the following, transport and economic relations across the border are surveyed, and evaluated in terms of spatial development patterns.

Transport

Passenger traffic between Eastern Finland and Russian Karelia increased rapidly in the 1990s. In addition to the international crossing points, Niirala in the southern part of the region and Vartius in the northern part, see *Figure 10.1*, temporary crossing points are used for special purposes such as the import of roundwood. In total, approximately 800,000 people used these crossing points in 1997, which accounts for about one-fifth of the passenger traffic between Finland and Russia: see *Table 10.1*.

Table 10.1 Passengers between Eastern Finland and the Karelian Republic in 1990-97, and goods traffic in 1994 and 1995

	Niirala	Vartius	Border (total)
Passengers (all border crossings, 1000 persons)			
1990	75	16	962
1991	105	32	1328
1992	149	62	1309
1993	191	84	1631
1994	202	98	2026
1995	442	181	3714
1996	504	195	4017
1997	493	209	4084
Goods traffic by road and by rail, 1000 tons			
Export[*] 1994	79	7	2129
Export 1995	108	4	2519
Import[**] 1994	2298	1490	9200
Import 1995	3236	1839	14019
Transit 1994	862	24	7081
Transit 1995	569	35	5031

[*] Export from Finland.
[**] Import to Finland.

Sources: Statistics of Frontier Guards of Southeast Finland, North Karelia, Kainuu and Lapland; Statistics of Finland's National Board of Customs and Ministry of Transport 1996.

Trade and direct investments

From the point of view of Russian Karelia, the first and foremost issue in cross-border co-operation concerns the development of trade and investment flows with foreign countries, and especially neighbouring regions, in the years during the collapse of the regional economy. Clearly, foreign trade volumes have increased. In 1991, the republic's exports outside the CIS countries were 24 million US dollars, and in 1997 505 million. Growth in imports has been smaller; in 1997 imports amounted to 144 million dollars. The most important trading partner of the Karelian Republic is Finland. In exports, its share was about 31 per cent, and in imports about 21 per cent in 1997. The respective shares of the EU (incl. Finland) are 76 per cent and 59 per cent (Goskomstat Respubliki Karelija, 1998).

As far as foreign direct investments are concerned, a total of 427 joint ventures or firms with 100 per cent foreign ownership were registered in the Karelian Republic between 1990-97. Of these firms, less than one-third could be considered to be operative at the end of 1997. Although the relative share of foreign economic activity is slightly higher on average in the Karelian Republic than in the Russian Federation in terms of the number of firms, the typical foreign investment is very small, and therefore the volume of investments per capita is not higher in the Karelian Republic than in Russia as a whole. The small size of investments made in Russian Karelia can be explained by the fact that the major share of the investments, around 60 per cent, have been made by small firms and individual businessmen from neighbouring Finland.

Indisputably, the above figures demonstrate that the process of opening to the west has had some impact. Yet a closer look clearly indicates that the positive expectations concerning the potential of cross-border exchanges in the early 1990s have thus far remained overblown visions in the crisis-stricken Russian Karelia. Firstly, the export flows consist of a few staple commodities (e.g. timber), and three-digit growth figures should be seen against a very low starting level. Secondly, although most of the foreign firms have imported some new technology and upgraded the skills of employees for its utilisation, the diffusion process has proceeded slowly; spill-over and spin-off firms have remained almost non-existent, and in contrast to the targets of official policies, the overstaffing of enterprises is bound to lead to redundancies in the case of nearly all investments. Thirdly, the total absence of foreign investors in the paper and pulp industries, which form the backbone of the economy in Russian Karelia, is the most explicit manifestation of the failure of internationalisation strategy thus far.

In Eastern Finland, the expectations concerning increased cross-border exchanges have also been slow in their realisation. Due to the sheer size difference of the two regional economies, the Karelian Republic cannot currently be an important trading partner for eastern Finland, and thus its role in the export-led growth strategy of the latter has remained minimal in the 1990s. For instance, the share of the entire Russian Federation in the imports of the easternmost region in Finland (North Karelia) is only 100 million FIM, that is, about 2.5 to 3 per cent. Nevertheless, the geographical proximity of Russian Karelia has obviously been reflected in the formation of some small-scale economic relations across the border. Some companies in eastern Finland have subcontracted assembly work to Russian Karelia or participated in investments in tourism. Overall, it has been estimated that the economic significance of passenger traffic, tourism and services on the Russian near regions is of the same (limited) magnitude in eastern Finland as the trade of goods (see Eskelinen, Haapanen and Izotov, 1997).

Adaptation to openness

With regard to cross-border interaction, positive visions entail the rise of a trans-national region where the resources of two neighbouring countries and regions are intertwined in the sense that the functional integration of economic, and other, activities contributes to the creation of new resources and competitive advantages. This depends, here and everywhere, on the competitiveness and learning capacity of firms, public organisations and other relevant actors (see van Geenhuizen and Ratti, 1998).

In practice, the legacy of peripherality implies that border regions might not be capable of forming a trans-national region, but remain passive corridor regions. Transport routes run through them, but the border regions concerned are not capable of serving as seedbeds for interrelated economic activities. There is also a third potential path of development. In some isolated border regions, the prospects offered by the opening might not imply the growth of any competitive economic activities, not even transport. Therefore, cross-border links remain occasional and unimportant, and frontier regions might be utilised for quite different purposes, for instance, they might be defined as nature protection regions. All these three prospective paths have received attention in blueprints for cross-border co-operation between Russian Karelia and eastern Finland. With regard to the formation of integrated cross-border structures, the lack of resources on both sides, and functional differences between the two partners have set severe constraints. The above empirical observations on transport and trade flows illustrate this very clearly.

On the western side of the border, in Finland, infrastructure facilities are up-to-date, and an increasing number of economic activities are being linked to global product competition (see, e.g. Maskell et. al., 1998). Even if the record of the 1990s is not at all good in terms of employment, the metamorphosis with regard to the problem of peripherality seems to be proceeding as a result of changing competitive conditions[1]. As the sectors which have obtained their competitive edge from low-cost labour, such as the clothing industry, have already lost their importance in Finland, demand for subcontracting based on cheap labour remains limited. The new technology sectors, for their part, do not find competitive resources on the eastern side of the border, at any rate not in the Karelian Republic (see Eskelinen et al., 1999).

There is one important exception to the setting of limited joint interests and complementarities outlined above: the operational environment of forest-based industries is different from that of other industrial sectors. To a significant degree, the capacity of pulp and paper industries in eastern Finland has relied on imports of roundwood from Russia (Niemeläinen and Vanharanta, 1996). This raises a serious challenge to forest industries in Russian Karelia due to the fact that increased exports of raw material tend to raise the price and make technologically backward combines even more weak in competition. Yet, forest industry companies on both sides of the border also have common interests; in particular, foreign customers should remain convinced that forests are maintained on an ecologically sustainable basis, both on the western and eastern sides of the border. This challenge, which probably puts the learning capacities of the two partners to the most

1 In the Nordic peripheries, eastern Finland being a typical example, the key problem has traditionally been accessibility: distances have implied real costs, and thus undermined competitiveness. Nowadays, the economic legacy of peripherality makes itself felt, to an increasing degree, in the structural characteristics of the operational environment: are small urban regions, scattered far away from each other, capable of supporting learning and endogenous growth, and thus acting as seedbeds for economic growth, which is increasingly based on non-price competition? In terms of settlement patterns, this challenge implies that the traditional centre/hinterland pattern should develop into a more nodal model, in which even small urban settlements would have, quite irrespective of geographical distances, functional ties in an international economy (see Eskelinen and Snickars, 1995).

demanding test, does not primarily concern the replanting of cut forests, but the protection of biodiversity (cf. Lehtinen, 1994). If a shared strategy for action can be found, the next task would be to find investors for the modernisation of forest-based industries in the Karelian Republic. These investments would not stop imports of raw material to Finland, but they would contribute to an increased political and social legitimacy of this trade in Russian Karelia.

At the moment, a more permanent pattern of daily connections has emerged only in the vicinity of the international crossing points, Niirala and Vartius. The prospects for these local (mini-scale) trans-national regions-in-the-making will be conditioned by the roles of the respective border crossings in international transport networks to and from Russia. Clearly, the construction and competitiveness of transport corridors across the Finnish-Russian border have to be evaluated in connection with the development of links between north-western Russia and western Europe. International traffic demand has to be warranted in advance to justify the required large investments, but this potential only exists for one route or for a few routes. On the other hand, each locality would need satisfactory infrastructure links across the border to establish day-to-day co-operative practices along the lengthy border. From an international perspective, the potentially most important railway route from the Karelian Republic probably runs from Russia's northern resource regions through the Vartius crossing point (see *Figure 10.1*) towards the harbours on the northern coast of the Gulf of Bothnia on the Baltic.

In the long-run, the formation of corridor and local trans-national regions will be reflected in the settlement pattern of the border regions. The potential repercussions of spacio-economic developments along the Finnish-Russian border basically seem to be bipartite. On the one hand, new cross-border connections expand the sphere of influence of some centres on the western side of the border, and can thus temporarily support their traditional roles in the division of labour, based on raw material supplies and, to some extent, low labour costs. On the other hand, centres on the eastern side of the border should be able to initiate a process of transformation into nodal centres, by creating contacts with different business partners in an international economy (cf. Eskelinen and Snickars, 1995). However, resources for development of this kind are essentially smaller in Russian Karelia than in St. Petersburg.

Conclusions

The Finnish-Russian border is still guarded as carefully as it was during the Soviet era and a visa is still needed to cross it. In contrast, the other institutions and practices of interaction and co-operation between border regions have changed dramatically. The process of shifting towards proximity-based relations between actors in the border regions was initiated at the grass-roots level in the late 1980s; a political framework for it was established in bilateral agreements between the two governments in 1992, and from 1995 the European Union has also become involved in promoting cross-border connections. At the regional level, informal information exchange between neighbours has been developed in seeking a joint political strategy to create integrated administrative structures spanning the border.

The evolution of cross-border co-operative practices and policies is intertwined with the processes of regionalisation in several ways. On the one hand, to construct them regional actors must gain, or be entrusted with, the decision-making capacities required for building external links with foreign partners. On the other hand, the aims of cross-border co-operation is to compensate for the negative influences of borders, and in a long run to develop functional trans-boundary regions. In the case discussed in this chapter, the partners are highly asymmetrical, and thus the regionalisation processes related to cross-border co-operation have also evolved along different lines.

In Russia, border regions have tried to create direct links with foreign partners to escape the economic crisis in the Federation as the latter's capacity to control decision-making in the regions and redistribute resources to them has largely waned. In the case of the Karelian Republic, for instance, the search for a kind of "national sovereignty" can be interpreted as a continuation of its earlier role as an administrative region nominally devoted to an ethnic minority. In practice, the vagueness of relations between the Federation and the Republic throughout the 1990s has been a serious impediment to the formulation of any coherent policy in utilising cross-border links for assuaging the crisis.

In Finland, a partial regionalisation of the domestic regional policy institutions was a precondition for EU membership. The Interreg programme has been a step forward in this respect, as it is administered by regional councils, even if national government ministries have the final say in decisions on additional domestic funding. The near-region co-operation, which was instituted earlier, has been a centrally controlled foreign policy activity. Against this background, the Interreg initiative has been an important policy innovation in Finland, and it has also led to co-operation between regional councils in constellations on a new functional basis.

Currently, the regional councils of the Karelia Interreg region in Finland and the Karelian Republic in the Russian Federation aim at establishing cross-border co-operation on a more permanent basis. In a wider perspective, this Euregion strategy attempts to fill the empty space, a "no sea's land" between the two major prospective trans-national regions in northern Europe, the Baltic region and the Barents region. Yet, whatever the merits of the Euregion Karelia strategy, there is no guarantee that the region could evolve towards a functional region. Even its identification as a region is an ambitious goal: a major obstacle is the fact that the prospective Karelia Euregion is very large, and its different parts at present have different functional links and interests. With the potential exception of the forest sector, the necessary complementarities of the partners are hard to find.

Overall, the findings concerning developments along the Finnish-Russian border are paradoxically dichotomous. On the one hand, the whole cross-border co-operation regime has been turned upside down in the space of a decade, and the region, to an increasing degree, resembles a normal European border region. On the other hand, cross-border co-operative activities have not been able to make a visible contribution to the fulfilment of the socio-economic development targets usually put forward as their primary motivation, and the prospects for this do not seem promising. In the long run, a fundamental strategic question will concern the extent to which these peripheries are able to link their mutual co-operation to structures and lines of action on different spatial levels, from the local to the multinational.

References

Bradshaw, M.J. and Lynn, N.J. (1996), 'The Russian Far East: Russia's Wild East', *PSBF Briefing*, No. 9 November 1996, The Royal Institute of International Affairs, London.

Bröms, P., Eriksson, J. and Svensson, B. (1994), *Reconstructing Survival. Evolving Perspectives on Euro-Arctic Politics*, Fritzes, Stockholm.

Christiansen, T. and Joenniemi, P. (1999), 'Politics on The Edge: On the Restructuring of Borders in the North of Europe', in H. Eskelinen, I. Liikanen and J. Oksa (eds), *Curtains of Iron and Gold. Reconstructing Borders and Scales of Interaction*, Ashgate, Aldershot, 89-115.

Cronberg, T. (1998), *Euroalueen kehittäminen Interreg Karjalan ja Karjalan tasavallan välille*, Pohjois-Karjalan liitto, muistio 17.8. 1998, Joensuu.

Eskelinen, H., Haapanen, E. and Druzhinin, P. (1999), 'Where Russia Meets the EU. Across the Divide in the Karelian Borderlands', in H. Eskelinen, I. Liikanen and J. Oksa (eds), *Curtains of Iron and Gold. Reconstructing Borders and Scales of Interaction*, Ashgate, Aldershot, pp. 329-346.

Eskelinen, H., Haapanen, E. and Izotov, A. (1997), *The Emergence of Foreign Economic Activity in Russian Karelia*, Publications of the Karelian Institute 119, University of Joensuu.

Eskelinen, H., Kokkonen, M. and Virkkala, S. (1997), 'Appraisal of the Finnish Objective 2 Programme: Reflections on the EU Approach to Regional Policy', *Regional Studies*, vol. 31(2), pp. 167-172.

Eskelinen, H. and Snickars, F. (eds) (1995), *Competitive European Peripheries*, Springer Verlag, Berlin.

Geenhuizen, M. van, and Ratti R. (1998), 'Managing Openness in Transport and Regional Development. An Active Space Approach', in K. Button, P. Nijkamp, and H. Priemus (eds), *Transport Network in Europe: Concepts, Analysis, and Policies*, Edward Elgar, Cheltenham, pp. 84-102.

Goskomstat Respubliki Karelija (1998), *Statisticheskii bjulleten* No1, No2, Petrozavodsk.

Keating, M. (1998), *The New Regionalism in Western Europe. Territorial Restructuring and Political Change*, Edward Elgar, Cheltenham, Northampton MA.

Laine, A. (1994), 'Karelia Between Two Socio-Cultural Systems', in H. Eskelinen, J. Oksa and D. Austin (eds), *Russian Karelia in Search of a New Role*, University of Joensuu, Karelian Institute, Joensuu, pp. 13-25.

Lehtinen, A. (1994), 'Neocolonialism in the Viena Karelia', in H. Eskelinen, J. Oksa and D. Austin (eds), *Russian Karelia in Search of a New Role*, University of Joensuu, Karelian Institute, Joensuu, pp. 147-159.

Lloyd, P. and Meegan, R. (1996) 'Contested Governance: European Exposure in the English Regions', *European Planning Studies*, 4 (1), pp. 75-98.

Lynn, N. J. and Fryer, P. (1998) 'National-territorial change in the republics of the Russian North', *Political Geography*, vol. 17 (5), pp. 567-588.

Maskell, P., Eskelinen, H., Hannibalsson, I., Malmberg, A. and Vatne, E. (1998), *Competitiveness, Localised Learning and Regional Development. Specialisation and Prosperity in Small Open Economies*, Routledge, London.

Niemeläinen, H. and Vanharanta, H. (1996), *Investment Opportunities for Forest Product Industries in North Karelia*, University of Joensuu, Dep. of Economics, Joensuu.

Paasi, A. (1995), *Territories, Boundaries and Consciousness: The Changing Geographies of the Finnish-Russian Border*, Wiley, Chichester.

Scott, J. W. (1999), 'Evolving Regimes for Local Transboundary Co-operation. The German-Polish Experience", in H. Eskelinen, I. Liikanen and J. Oksa (eds), *Curtains of Iron and Gold. Reconstructing Borders and Scales of Interaction*, Ashgate, Aldershot, pp. 179-193.

Segercrantz, W., Kokkonen, M., Eskelinen, H., Mäkelä, K. and Segercrantz, T. (1998), *Interreg II A Karjala-, Kaakkois-Suomi ja Etelä-Suomen rannikkoseutu B ohjelmien väliarviointi. Väliarvioinnin loppuraportti*, VTT, Karelian Institute, Joensuu.

Tikkanen, V. and Käkönen, J. (1997), 'The Evolution of Cooperation in the Kuhmo-Kostamuksha Region of the Finnish-Russian Border', in P. Ganster, A. Sweedler, J. Scott and W-D Eberwein (eds), *Borders and Border Regions in Europe and North America*, San Diego State University Press, pp. 163-175.

Ulkoasiainministeriö (1998), *Tausta-aineistoa lähialueyhteistyöstä vuodelta 1997 ja 1998*, Kauppapoliittinen osasto, Keski- ja Itä-Euroopan toimintaohjelmat, KPO-23, Helsinki.

Veen, A. van der (1993), 'Theory and Practice of Cross-border Co-operation of Local Governments: the Case of EUREGIO between Germany and the Netherlands', in R. Cappelin and P.W.J. Batey (eds), *Regional Networks, Border Regions and European Integration*, Pion, London, pp. 89-95.

11 Cross-Border Co-operation in the German-Polish Border Area

STEFAN KRÄTKE

Cross-border co-operation and regional development at the eastern boundary of the European Union, particularly in the area of the German-Polish border region, are the subject of this chapter. The main focus is on the regional economic integration in the German-Polish border area. On the Polish side an expansion of low wage industries is frequently expected as a result of western firms' strategies which relocate production to reduce costs; however, the other widespread strategy of opening up new markets is likely to "overlook" the border region. Cross-border regional integration is also impeded by communication barriers and a highly competitive regional economic environment. The last part of the chapter deals with the prospects of a "special enterprise zone" on the Polish side of the border area and with approaches to promote regionally based cross-border co-operation.

Border Regions and the Diversity of Regional Development Patterns

This chapter deals with the characteristics of cross-border co-operation and regional development in the area of the German-Polish border regions, focussing on the problems of regional economic integration, particularly the emerging regional system of production and regulation in this area. Border regions are characterised by the fact that they contain economically, politically and culturally separated territories. Yet they also function as areas of contact and mutual infiltration. In the economic sphere, border regions frequently provide opportunities for taking advantage of economic gaps. On the basis of regional specialisation and competence, complementary structures can be exploited and converted to trans-border co-operation between enterprises; however, the evolution of an integrated economy in border areas can be obstructed to varying degrees, depending on the level of asymmetry between economic, legal and political systems on either side of the border. Due to the collision of dissimilar economic and socio-cultural systems borders can act as communication barriers which result in insecurities and high transaction costs in the sphere of economic co-operation. Even in a scenario of "open borders", there is no guarantee that a process of cross-border regional integration is taking place. The separating effect of international borders proves to be highly significant.

German-Polish border regions differ from Western European border regions in three central aspects. Firstly, the *external border* of the European Community cuts through German-Polish border regions. Border areas within the boundaries of the EU are situated within an economic territory of relative uniformity, providing significantly better conditions for integrated regional development. The external boundaries of the EU impose major restrictions on the freedom of movement for people, labour and the trans-border flow of capital and commodities. Secondly, the two regions that meet at the German-Polish border are both currently subject to an extensive process of socio-economic transformation. The border areas on the German side, formerly part of eastern Germany, are faced with the problems and consequences of radical social change after becoming part of the Federal Republic. Regions on *both* sides of the border are at present in a situation of high labour market pressure and uncertain economic development prospects. Thirdly, the German-Polish border regions, unlike most border areas within the EU, can be classified as belonging to those border regions which are affected by a *sharp divide* in income and wage levels between neighbouring areas. According to the Institute of the German Economy, the German-Polish divide in incomes for 1993, calculated as the average wages per hour in the manufacturing sector, comes to a ratio of about 10:1 (Ribhegge, 1996). The wage divide may have a significant impact on restructuring the spatial division of labour in the border area.

Contemporary regional research has emphasised that despite growing supra-regional and international interlinking of production and markets, the regions' specific structures and capacities should not be underestimated. Regions have specific potential for development and varying capacities for renewal (Pyke and Sengenberger, 1992; Amin and Thrift, 1994; Storper, 1995; Asheim and Dunford, 1997). Key importance is now being attributed to the regions' institutional resources. In the context of this changing focus of regional analysis, three points can be taken as indications of the capacity of regional economies to renew and compete: firstly, the existence of an institutional variety of enterprises (with the status of independent enterprises), secondly, the existence of co-operative relationships and negotiating structures within the regional economy, and thirdly, the existence of functioning communication structures and a close network of institutions to promote innovation, transfer of knowledge and co-operation. The region's *institutional resources*, i.e. its socio-economic patterns of interaction and organisation are furthermore influenced by specific socio-cultural characteristics of the regional actors, which are shaped by the developmental history of a region and its specific economic environment (Krätke, Heeg and Stein, 1997).

A crucial aspect of improving development chances of border regions lies in the strengthening of trans-border co-operation and the *regional integration* of border areas (Cappellin, 1993; Ratti and Reichman, 1993). The problem of asymmetry in legal and political systems of regulation on both sides of the border has mostly dominated in border area research (cf Schabhüser, 1993; Beck, 1997). Attempts at solving this problem have frequently been made by creating new institutional forms of political co-operation in the vicinity of the border. The formation of Euregions particularly can be seen as an institutionalised form of trans-border co-operation between neighbouring regions aiming at a partial eradication of developmental blocks and contributing to a joint resolution of trans-border problems. With the establishment of Euregions in the German-Polish border area, trans-border co-operation structures have been created on a political/institutional level which contribute to a positive climate for co-operation between all actors in the border region. On the basis of experiences of existing Euregions, four Euregions, comprising the whole of the German-Polish border area, have been constituted since 1990; however, trans-border co-operation should not be reduced to co-operation of local state organisations in the region. The constitution of an Euregion is a process of political-administrative region building, which by itself does not create an economically and socially integrated trans-border region. Trans-border co-operation in terms of regional inter-firm linkages is of equally great importance and should be considered in dealing with the developmental capacities of a border region. A border region can only first really be considered integrated, if the regional systems of production and regulation on both sides of the border are *connected*, the intensity and quality of these trans-border linkages is another question.

Border regions at the eastern boundary of the European Union are a special group within the East-Central European region, which today (as a whole) are subject to a far-reaching process of socio-economic transformation. New patterns of regional development in the transformation process arise out of the interplay of the regions' endogenous potential, i.e. physical structures, institutional resources and innovative capacities, and the strategic orientation of western enterprises in terms of the regional distribution of their investment activities in East-Central Europe. Many believe that foreign direct investment activities, such as the purchase of existing firms by western corporations, joint ventures, and the establishment of new production locations by western firms will fuel regional economic growth. The activities of trans-national firms are expected to bring in additional capital, more jobs, the transfer of advanced technology and management skills. Foreign investment activities of western enterprises can, however, be based on different strategies, which carry different modes of regional development. Those enterprises, that set primacy on cost minimisation and "cheap" production factors, are going to use East-Central European regions as low-wage production sites using out-sourcing, joint ventures and relocation of mostly low level functions of production. In contrast, enterprises that value high quality production, technological advances and constant innovation activity are going to include East-Central European regions in their location strategy primarily for the purpose of opening up new markets. Thus it is not ruled-out that trans-national enterprises will grant strategic privileges to selected production locations in East-Central Europe, i.e. to revaluate a region in the direction of setting up a specialised centre for technologically advanced production within their Pan-European or global locational network. We should be aware of the *diversity* of possible regional development paths and modes of integration of East-Central European regions into the socio-economic space of a new Europe: the restructuring of the regional system in East-Central Europe will produce no uniform mode of regional development, but rather a mosaic of quite different regional development patterns (Krätke, Heeg and Stein, 1997).

Regional Economic Restructuring in the German-Polish Border Area

The particular geographical situation of the German-Polish border region can be considered to be ambivalent: this border area is frequently characterised as a geographical interface between "East" and "West" which might offer fairly good development prospects; however, geographical proximity to the Central and eastern European countries by itself is, for the German part of the border area at least, no guarantee for the attraction of investment in the area. New economic links between German and Polish firms do not necessarily stimulate the location of industrial activity *within the border zone* of both countries. The German part of the border region in the mid Oder area is situated in the close vicinity of a significant metropolitan centre of the Federal Republic: Berlin. The expected positive radiation effects of Berlin, however, have up to now been concentrated on the close surroundings (fringe area) of the metropolis. Most of the jobs created by new investments in Brandenburg are concentrated in the districts to the immediate south, south-west and west of Berlin (Krätke, 1996), so that Berlin detracts possible investment from the border area.

The Polish part of the border region in the mid Oder and Neisse area is not far from Poznan, one of the large dynamic urban centres of Poland. Since 1990, these agglomeration centres of the Polish regional system in particular have proved to be most attractive to firms who base their choice of location on the accessibility of markets, the supply of qualified labour and "institutional" density, including potential suppliers and services. Therefore, the agglomeration of Poznan particularly will exercise a considerable "pull", detracting interest from the Polish part of the border region. We have also to take into account that major road and railway transit connections between Eastern and Western Europe traverse the mid Oder and Neisse border area. The Frankfurt/Oder - Swiecko crossing on the motorway alone deals with approximately 50 per cent of all heavy goods road traffic to Poland and the CIS countries. At present the *lack* of permeability of the German-Polish border is one of the major development blocks experienced in the border region: the increase in transit traffic has put so much strain on border crossings that drivers of vehicles with heavy goods must frequently endure up to 40 hours of waiting. These conditions also greatly hinder local and *regional* trans-border traffic and the regional economy.

The situation is complicated further by the specific "border regime" in force at the external borders of the European Union. This makes itself felt in foreign trade and customs duties, in turn affecting the permeability of the border for goods, and also the openness for people. However, with an agreement between the EU and Poland, the Czech Republic, Slovakia, Hungary, Romania, Bulgaria and the Baltic States aimed at expanding foreign trade with the EU through harmonising customs regulations, the relocation of labour-intensive production activities by Western European firms to these countries has been considerably encouraged, since companies obtain privileged treatment regarding customs duties (and taxes) on basic materials which are temporarily imported to be processed and later re-exported to the EU countries as more advanced products. These regulations, however, are valid for the associated countries' total area, they do not imply a privileged treatment of production activities within *border regions*.

The process of *economic restructuring* in the border region has led to plant closures and job losses on both sides of the border. The *German* side particularly has experienced massive de-industrialisation: in all of Brandenburg, the number of manufacturing jobs fell from 1990 to 1995 by roughly 75 per cent. In the administrative region of the Chamber of Industry and Commerce of Frankfurt (Oder), which covers the East-Brandenburg districts of the border region, in 1997 only 29 out of every 1000 residents were employed in industry compared to the 86 out of every 1000 residents in the West German Bundesländer (IHK, 1997b). Thus the number of industrial jobs per inhabitant is now roughly one third of the respective figure for West Germany. New economic activities and newly founded firms can only compensate for the job losses in industry to a very limited extent. On the *Polish* side, in the Voivodship Gorzow (northern part of the mid Oder and Neisse border area) the total number of jobs fell from 1990 to 1993 by 12 per cent. Since 1993, an increase of jobs has been registered in the private sector. In the Voivodship Zielona Gora (southern part of the mid Oder and Neisse border area) the total number of jobs fell from 1990 to 1993 by 14 per cent. After 1993, an increase in manufacturing jobs was registered (Krätke, Heeg and Stein, 1997). Compared to the massive de-industrialisation on the German side, the Polish side, as far as the development of industrial employment is concerned, shows *relative stability*.

Economic restructuring on the border region's *German* side has led to the formation of new economic linkages, the most important are the new *supra-regional* links with firms located in *West-German* regions, while intra-regional linkages between enterprises in the manufacturing sector remain weak and undeveloped. According to a survey of 300 manufacturing firms in the Oder region (on the German side), the supplies for the production of regional firms are predominantly bought in the *West-German* Bundesländer (Zidek, 1995): three quarters of the firms reported that their suppliers were located in West Germany, and a further significant number in the rest of Western Europe. The number of suppliers located in the region of East-Brandenburg or in Berlin was below 15 per cent. The main reason for this lack of intra-regional links is the integration of most privatised regional firms in the production networks of western firms. The 1997 business report of the regional Chamber of Industry and Commerce in Frankfurt-Oder (IHK, 1997) emphasises that the inter-firm co-operation *within* East Brandenburg is quite weak. The German Institute for Economic Research (DIW) also draws attention on the deficient *functional* structure in the enterprise sector: the economy of Brandenburg '... is made up, for the most part, of companies owned by West-German or foreign firms. As a result, companies in the Land usually only perform executive functions within the existing product lines, while central business functions, such as research and development, distribution and marketing, are located in the West.' (DIW, 1997: 108). Thus on the German side, there remains only a weak base for developing cross-border inter-firm networks of *regionally based* enterprises.

Regarding the border region's economic restructuring, we have furthermore to consider external cross-border investment activities in the area: as regards the spatial division of labour pattern in the extended pan-European economic space, it is usually assumed that a given region will specialise in those activities which make use of its comparatively "cheap" production factors. Therefore, it is often assumed that the industrial regions of East Central Europe will specialise in *labour intensive* production, while Western European industrial regions will concentrate on more scientifically and technologically advanced production (Kröger et al., 1994). In contrast to these expectations, the regional distribution of foreign direct investment in *Poland* reveals a more differentiated pattern of development: first of all, the regions with large urban agglomerations have the greatest concentration of foreign investment. Secondly, most of the regions in Western Poland have only a "mid level" concentration of foreign investment.

The regional distribution of joint ventures and subsidiaries of foreign firms (*see Figure 11.1*) can be interpreted in terms of *two location patterns*. A first pattern emerges in the concentration of foreign investment activity in the regions of large urban agglomerations. Typical of this pattern are locations established for the purposes of opening up a new market. According to a 1996 survey conducted by the German-Polish Society for Economic Promotion, out of the German firms investing in joint ventures or subsidiary companies in Poland, 69 per cent cited the Polish market and the EU-association of Poland as the reason for their investment, and only 21 per cent cited wage cost as their reason. Those economic activities of western firms, which seek to open up the *Polish* market, tend to be located in large urban areas rather than in the border regions of the western part of Poland. Secondly, the weaker concentration of foreign investment activity in the regions of western Poland can be interpreted in terms of another location pattern: in this pattern, new *supra-regional* chains of production in terms of export processing activities emerge, which are based on cost-cutting strategies. As mentioned above, there are presently favourable conditions for exploiting labour cost differentials by western firms. This type of trans-border relationship stimulates the development of *one side* of the border region, while the other side of the border is overlooked. The more this type of trans-border co-operation spreads, the greater the threat of a *divided* development in the border region.

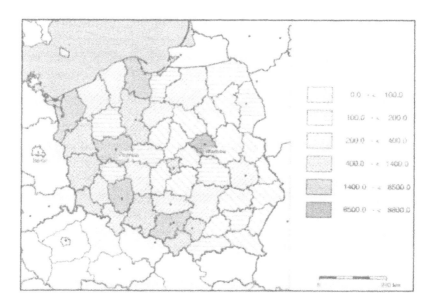

Figure 11.1 Distribution of firms with foreign capital in Poland in 1996
(joint ventures and subsidiaries of foreign firms)

The development of the German-Polish border region towards an *integrated* trans-border economy could in particular be promoted by trans-border co-operation between *regional* firms. Regionally based inter-firm linkages might increase the firms' competitiveness on both sides of the border region; however, up to now the examples of trans-border co-operation between *regional* firms have been of little quantitative significance (Krätke, Heeg and Stein, 1997) when compared to other types of German-Polish inter-firm co-operation.

The traditional industrial branches of the mid Oder and Neisse region are (on both sides) textile and clothing and wood processing and furniture making (see *Figure 11.2*). In the area of South-Eastern Brandenburg, particularly around Cottbus and the towns of Guben and Spremberg, there was a concentration of textile and clothing industry. In the nineteen-fifties, industrial policy on the German (GDR) side promoted the creation of large industrial complexes at locations outside the Berlin region. In this context, a large semi-conductor factory in Frankfurt (Oder) became an industrial centre for micro-electronics, and the EKO iron processing combine (Eisenhüttenstadt) founded a new steel industry centre. Of these two industrial complexes, only the EKO plant has survived. On the Polish side of the Oder region, the voivodship's capital Gorzów in particular expanded its metal and chemical industries. Manufacturing plants for furniture and silk were also constructed in Gorzów. Further industrial activity centres are to be found in e.g. the border town of Slubice, with its textile and clothing and furniture making. In the voivodship of Zielona Gora there are also many manufacturing plants for the textile and furniture industries, for example, in the towns of Zary and Zielona Gora. As far as employment structures are concerned, the German-Polish border area has been developed as an industrial region, however, it does not present a homogenous economy, but rather a region with fragmented and territorially uneven distributed industrial activities.

The low grade of regionally integrated cross-border co-operation in this area is reflected in the findings of an empirical survey on the regional economic linkages of the area's traditional industries in 1997 (Krätke, Heeg and Stein, 1997). A total of 40 firms, from both sides of the border, responded to our questionnaire concerning the spatial structure of their supplies and the current state of the concomitant cross-border linkages. Of the firms included on the German side, the wood processing and furniture industry shows only negligible cross-border interaction, and the textile and clothing industry has a low grade of cross-border links. Of the firms included on the Polish side, the wood processing and furniture industry shows a considerably high degree of regional economic interaction *inside the Polish area,* but there are no cross-border linkages to the German side. It is only the textile and clothing industry which has a degree of cross-border regional linkage worth mentioning.

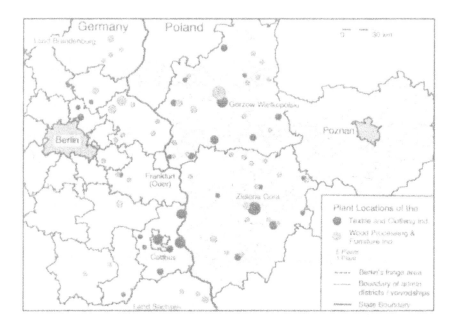

**Figure 11.2 Location pattern of traditional industries
in the German-Polish border area (1997)**

However, the geographically far-reaching linkages to firms in EU countries
are much more important, and this indicates that the textile and clothing
industry on the Polish side of this border region is (to a certain degree)
being incorporated in trans-national production networks by performing
low wage export processing activities.

Figure 11.3 shows the regional structure of supplies in two particular
industries (textile and clothing and furniture industry) by differentiating
between (a) firms located on the German side of the border region and (b)
firms located on the Polish side of the region. The bars indicate the shares
of supplies received by the border region's firms from different regions,
e.g. firms of the textile and clothing industry on the border region's
German side received nearly 10 per cent of their supplies from firms
located in the West-Polish border area, which indicates a modest cross-
border economic interaction. Firms of the furniture industry on the border
region's *Polish* side received almost 50 per cent of their supplies from
firms located in the West-Polish border area, which indicates that there is a

considerable degree of *regional* economic interaction within the *Polish* part of the border region.

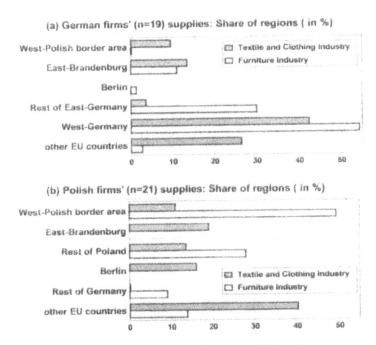

Figure 11.3 Regional structure of supplies in traditional industries of the German-Polish border area (in 1997)

Regional economic restructuring in the German-Polish border area is influenced by "non-economic" factors, particularly the socio-cultural environment of the region's economy. In border regions with a pronounced socio-cultural divide between its neighbouring parts there are special barriers to communication and co-operation between economic players to be overcome. In such border areas, the development of inter-firm networking might be hindered by a lack of trust between actors on both sides of the border. This lack of trust can only be overcome over a long period of time through a positive interactive learning process.

As regards the German-Polish border region, there was a massive *population transfer* in this area after the Second World War. Those who settled in this region believed for many years that their stay was to be temporary. After the Federal Republic of Germany formally recognised the Oder-Neisse boundary between the two countries in 1970, such uncertainties began to subside; however, for a long time, the border between the GDR and Poland was closed, and this prevented interaction between the region's inhabitants and disturbed the possibility of developing contact between the neighbours (Lisiecki, 1996). Although there has been greater contact between the people of both sides of the border since the opening of the border in the 1990s, Germans and Poles in this region have not yet realised that they might have *mutual* advantages in an integrated region which transcends national borders. Polish sociologists have examined the attitudes of the residents in the border area to find out whether traditional prejudices and negative attitudes towards the neighbouring nationality have been reinforced or have changed (Lisiecki, 1996). The findings indicate that in contrast to the situation in Western European border regions the mental social distance between German and Polish residents of the border area is still considerably high.

On the German side, reservations about the Poles are based on hardened resentments (among the region's inhabitants) and strong fear of competition. The "open border" has currently created competition between neighbouring regions, which did not previously exist. Racist attitudes and the fear of competition has become a breeding ground for the repeatedly reported attacks on foreigners, including Polish students, tourists and business travellers. These incidents are a widely recognised negative milieu factor in the German part of the border region. In a number of towns, Neo-Nazi groups have succeeded in creating "nationally liberated zones" where they control the accessibility of public space and where they dominate the political socialisation of young people (Schröder, 1997). Such widespread racist attitudes and the publicly apparent activity of Neo-Nazi groups are today damaging the region's chances for development.

New Developments in the Border Region's Economy and Institutional Infrastructure

Up to now, the new economic ties between Germany and Poland have not led to the emergence of a *regionally integrated* trans-border economy (see also Krätke, 1999). A large share of these new links are a cross-border linkage between firms in *West-Germany* and the border region's *Polish* side, and the hitherto existing co-operation between German and Polish enterprises is on a "qualitatively low level": its main direction has been the manufacturing of supplies or semi-finished goods in Poland by the Polish subsidiaries of German firms or by German-Polish joint ventures for the German market. This type of co-operation simply makes use of wage advantages for employers in Poland. Until 1997, however, it was open to doubt whether low wage assembly plants would have a chance of setting up in large numbers in the Western-Polish border regions as long as the Polish government did not grant them privileged conditions. In the years preceding 1997, the Polish government set up six "special enterprise zones" aiming at the attraction of foreign direct investment, which was predominantly located in some "interior" regions of the country. In October 1997, the Polish government set up eleven new "special enterprise zones" throughout the country. One of these is the special enterprise zone Kostrzyn-Slubice in the Western Polish border region (near the cities of Kostrzyn and Slubice), established for a twenty year period (*Figure 11.4*).

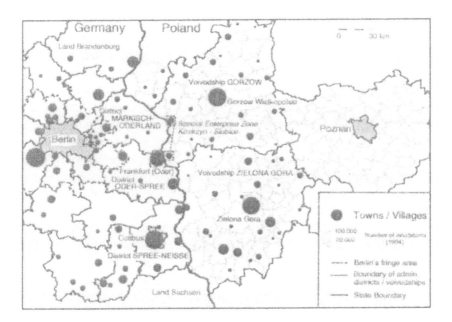

Figure 11.4   Location of the special enterprise zone Kostrzyn-Slubice in the mid Oder and Neisse area (broken line)

Western firms that invest a certain minimum amount of capital or create a certain minimum amount of new jobs are offered several tax exemptions for a ten year period, increased rates of depreciation and simplified procedures in the field of administrative approvals. Local state representatives hope to create up to 10,000 workplaces in the special enterprise zone Kostrzyn-Slubice through attracting investment by western firms, in particular export processing activities to supply the Berlin-Brandenburg market. Up to the end of 1998, however, only one branch plant of a foreign firm had been established in the special enterprise zone, and this one did not come from Berlin or from West-Germany, but from Croatia. It is still an open question whether the expectations of the new special enterprise zone prove to be realistic in terms of job numbers; additionally, the quality of workplaces and technological level of the economic activities in the special enterprise zone are open to doubt and should be carefully evaluated.

The politics of special enterprise zones are not only aimed at creating new workplaces, the hope for an advancement of the working populations' level of training and the technological level of the host country's economy also plays a role. The most famous example of a special enterprise zone in a border location is the Mexican "Border Industrialization Programme" in the US-Mexican border region. This area shows also *another side of the coin* in such a development strategy: we have firstly to consider the amount of public financial funds spent to attract foreign direct investment to a special enterprise zone; secondly, we have to consider the overall low quality of workplaces in export processing factories which predominantly supply low skilled assembly jobs. Furthermore, a concentration of export processing factories does not contribute to the development of a competitive *regional* economy, since branch plants, as established in special enterprise zones, regularly have only minor inter-firm links to the local economy of the surrounding region. Regional economies, which are dominated by a branch plant structure, are characterised by regional economic *disintegration*. The maquiladora factories in the US-Mexican border region purchase less than 2 per cent of their inputs from regional suppliers (Brannon et al., 1994). The maquiladoras' lack of regional economic embeddedness is also impeding the transfer of advanced technology to the host country's enterprise sector and regional economy. Export processing factories in special enterprise zones are primarily incorporated into the global production network of trans-national firms. Regarding the international experience, it cannot be taken for granted that the establishment of special enterprise zones is a promising approach for developing the *regional* economies of the host country; it might also lead to the spread of export processing industries with a quite low grade of regional economic integration.

The *alternative* to this particular mode of regional economic development might be described as a concept of qualitative change and improvement of the border region's economy: border regions can advance from poor to improved competitiveness by extending manufacturing and design skills in their existing industries and by developing innovative systems of production. A qualitative improvement of this type includes the promotion of cross-border co-operation on a *qualitatively* high level. Here, it might be emphasised that a significant number of successful regional economies have been based on *regionally* integrated production networks in "traditional" industries (Pyke and Sengenberger, 1992; Ernste and Meier, 1992). The border region's developmental policy can basically be orientated towards developing trans-border networks of regional firms and towards improving regional industrial competence. In a number of European regions, attempts at local economic regeneration have concentrated on such an approach, and this leads to the question as to whether the textile and clothing industry and the furniture industry of the German-Polish border region can be qualitatively improved with new production concepts and trans-border co-operation.

In the German-Polish border region, the south Brandenburg textile industry in the Lausitz area (part of the Euregion Spree-Neiße-Bobr) has made some steps towards industrial revitalisation on the basis of new production concepts: the industrial district of Lausitz with the towns of Guben, Forst and Spremberg was the traditional home of the textile industry in Brandenburg; after 1990 it was hit by a dramatic collapse and a 90 per cent loss of jobs in the textile industry. In the last years, however, textile production activities in the district of Lausitz have began to stabilise on a different base. A number of new manufacturers have set up business to produce cloth in Forst, Guben and Spremberg and have won a place in the market by supplying *high quality* textiles. The strategic shift from mass production and cheap exports in favour of high quality products in the district of Lausitz has opened up new perspectives. The manufacturers are not only building on existing industrial competence in the region, but are promoting its further development with a new production concept. The shift in emphasis towards high quality products is an attempt to secure promising future markets for textiles. These approaches, however, cannot compensate for the dramatic drop in employment figures in the region's textile and clothing industry. As indicated above, a certain degree of cross-border inter-firm co-operation among German manufacturers and Polish firms on the other side of the border has developed. In most cases, we are dealing here with a "low quality" mode of co-operation which is based on using the Polish side for low wage export-processing activities; however, there have also been some approaches to developing a cross-border inter-firm network of textile and clothing manufacturers which is based on a more "equal partnership". For example, the "textile project Guben" established, with the help of the Guben innovation and technology centre and the German-Polish society for economic promotion, a "high quality" co-operative inter-firm network of six regionally based firms on both sides of the border to organise jointly the design, manufacturing, and marketing for clothing for women. Success of this type of initiatives is not guaranteed, however, because the firms need to find a place in the market.

Thus we can observe activities for regional economic regeneration in the German-Polish border area which are backed by a certain institutional environment of the regional economy and seek to promote cross-border inter-firm co-operation. The most important example is the work of the German-Polish Society for the Promotion of the Economy. This institution offers various consulting services for small and medium-sized firms trying to develop production and distribution networks. The organisation gives particular assistance to firms that are looking for partners on the other side of the border and are interested in establishing trans-border business contacts for co-operative ventures. In addition to the acquisition of external investments, the activities of this institution include the organisation of German-Polish business meetings in Gorzow and the organisation of a "co-operation stock exchange" for the purpose of bringing possible partners together. These activities and services might be taken as examples or starting-points for a regional development policy which pursues the creation of a dense network of establishments for the promotion of innovation, knowledge transfer, and co-operative relationships in the region, in order to increase the industrial competence and competitiveness of firms on *both* sides of the border region. Such activities can be understood as approaches to improving the institutional resources of a regional economy which transcends the nation states' boundary.

Conclusion

New economic links between Western European and East Central European firms do not necessarily lead to the formation of trans-border *integrated* economic regions. Instead, there is a tendency towards the development of new economic links, such as those between enterprises in West-Germany and Poland, which *overlook* the border region. The development prospects for the German-Polish border region in a new Europe are determined by the type of trans-border economic relations ("high quality" versus "low quality" co-operation) that will dominate in the future. A regional development policy which is based on the persistence of the border region's economic division, particularly the wage divide, cannot be regarded as a forward looking strategy. A strategy of qualitative change and improvement, including cross-border co-operation between *regional* firms might be an alternative path to regional development in the border area. This basically involves the strengthening of supportive institutions in the regional economy which are capable of promoting innovative structures on *both* sides of the border, since socio-economically integrated border regions might, in the long run, have a better standing within the European fabric of regions. As regards the future scenario of more and more "open borders" in Europe, the regions' economic positioning is increasingly determined by their innovative capacities, i.e. the regions' specific institutional resources and production system.

References

Amin, A. and Thrift, N. (eds) (1994), *Globalization, Institutions, and Regional Development in Europe*, Blackwell, Oxford.

Asheim, B. and Dunford, M. (1997), 'Regional Futures', *Regional Studies*, 31 (5), pp. 445-455.

Beck, J. (1997), *Netzwerke in der transnationalen Regionalpolitik: Rahmenbedingungen, Funktionsweise, Folgen*, Nomos, Baden-Baden.

Brannon, F. and James, D. and Lucker, G.W. (1994), 'Generating and sustaining backward linkages between Maquiladoras and Local Suppliers in Northern Mexico', *World Development*, 22 (12), pp. 214-230.

Cappellin, R. (1993), 'Interregional Co-operation in Europe: an Introduction', in R. Cappellin and P.W.J. Batey (eds), *Regional Networks, Border Regions and European Integration*, Pion, London, pp. 1-21.

DIW (1997), 'Überlegungen zu den künftigen Leitlinien für die Wirtschaftspolitik Brandenburgs', *DIW-Wochenbericht*, 64 (6), pp. 105-109.

Ernste, H. and Meier, V. (eds) (1992), *Regional development and contemporary industrial response. Extending flexible specialisation*, Belhaven Press, London.

IHK (Chamber of Industry and Commerce) Frankfurt/Oder (1997a), *Konjunkturbericht 2, Halbjahr 1996*, IHK, Frankfurt (Oder).

IHK (Chamber of Industry and Commerce) Frankfurt/Oder (1997b), *Jahresbericht 1997*, IHK, Frankfurt (Oder).

Krätke, S. (1996), 'Where East meets West - the German-Polish Border Region in Transformation', *European Planning Studies*, 4 (6), pp. 647-669.

Kräkte, S. (1999), 'Regional Integration or Fragmentation? The German-Polish Border Region in a new Europe', *Regional Studies*, vol. 33 (7), pp. 631-641.

Krätke, S., Heeg, S. and Stein, R. (1997), *Regionen im Umbruch, Probleme der Regionalentwicklung an den Grenzen zwischen "Ost" und "West"*, Campus, Frankfurt-M./New York.

Kröger, F., Hasselwander, S., Henkel, C.B., Hesch, R., Trixl, E.-M. and Turowski, O. (1994), *Duale Restrukturierung, Wettbewerbsfähig durch west-östliche Arbeitsteilung*, Schaffer-Poeschel, Stuttgart.

Lisiecki, S. (1996), 'Die offene Grenze - Wandlungen im Bewußtsein der Grenzbewohner', in S. Lisiecki (ed), *Die offene Grenze*, Verlag für Berlin-Brandenburg, Potsdam, pp. 97-117.

Pyke, F. and Sengenberger, W. (eds) (1992), *Industrial districts and local economic regeneration*, ILO, Geneva.

Ratti, R. and Reichman, Sh. (eds) (1993), *Theory and Practice of Transborder Co-operation*, Helbing & Lichtenhahn, Basel.

Ribhegge, H. (1996), 'Euro Region Pro Europa Viadrina, Chancen und Schwierigkeiten einer grenzüberschreitenden Kooperation', in C. Montag and A. Sakson (eds), *Die deutsch-polnischen Beziehungen*, Brandenburgische Landeszentrale für politische Bildung, Potsdam, pp. 65-79.

Schabhüser, B. (1993), 'Grenzregionen in Europa, Zu ihrer derzeitigen Bedeutung in Raumforschung und Raumordnungspolitik', *Informationen zur Raumentwicklung*, 9-10, pp. 655-668.

Schröder, B. (1997), *Im Griff der rechten Szene. Ostdeutsche Städte in Angst*, Rowohlt, Reinbek.

Storper, M. (1995), 'The resurgence of regional economies, ten years later: The region as a nexus of untraded interdependencies', *European Urban and Regional Studies*, 2 (3), pp. 191-221.

Zidek, H. (1995), *Regionalanalyse für das produzierende Gewerbe in der Oderregion*, edited by IHK Frankfurt-Oder and European University Viadrina, European University Viadrina, Frankfurt (Oder).

PART III:
CONDITIONS AND POLICIES IN ACTIVE SPACE DEVELOPMENT

12 Managing the Openness of Border Regions in the Context of European Integration

ZDRAVKO MLINAR

The future of border regions is analysed in terms of the general long-term regularities of socio-spatial restructuring and in terms of the involvement of regional actors. Two dimensions of restructuring are elaborated: first, the shift from contiguous, area connectedness to network linkages and secondly, the extension from a region as a single unit to multilevel power-sharing. The legacy of the zero sum game – in terms of shifting rather than opening borders – influences the strategy of opening border regions from the point of view of distrust and confidence-building. Non-parity roles represent another limiting factor. Both are exemplified in the experience gained in the Croatian, Slovenian and Italian part of Istria and additionally in Euregions on the Czech – German border. In spite of the centralism of the newly formed states, local and/or regional authorities have a wide range of options for directly or indirectly influencing the intensity of cross-border flows.

Introduction

Discussions about border regions are usually concerned with the issue of how their role changes as state borders open up or are dismantled and various forms of cross-border co-operation expand. It can however be easily overlooked that this is only one of the ways that border regions integrate into Europe: that this is a case of interlinking of national peripheries on the basis of spatial contiguity and proximity.

Long-term regularities cannot be explained within a fixed territorial framework. Criticism of the "billiard ball" model of states can also be applied to regions. As differentiation penetrates into the substructures of states (Blau, 1977) and the subsystems autonomise, the framework of their integration widens beyond the border regions. Mainstream thinking seems to assume that connectedness between border regions will simply grow; but as the framework expands connectedness based on vicinity can also decline. The interlinking of peripheries on the basis of contiguity has to be examined together with changes in their relationships to the state centres and the interlinking of centres on a European scale. If the dynamics of regional cross-border interlinking overtake the interlinking of national centres, this is usually perceived as a threat to national unity and territorial integrity. If it just follows decisions made at the centre - like a policy of complete opening of the borders - it is unlikely that the region will activate its specific human and material resources optimally.

Thus, discovering the objective regularities in long term changes does not imply narrowing the scope for devising an autonomous and active role for particular regions; rather the contrary. In the present analysis an attempt is made to identify some of the objective and subjective dimensions of variation between regions in Europe. Although opening, removing, or even erecting borders is primarily in the hands of central authorities, there is still a great range of short and even some long-term measures available for the deliberate management of regional openness.

It is precisely *pre-determination* of the border regions, in the restrictive and affirmative senses, that is relaxed most by the state as borders open. At the same time, because of their previous constraints they face the great challenge of innovatively creating new cross-border combinations of resources and thereby ways of joining European integration processes (Mlinar, 1992). Thus no general European pattern can automatically be followed by all regions; this would have unintended consequences and even be counterproductive. Certain features, whether singly or cumulatively, call for distinctive strategies for regional opening, namely:

-   in some European regions historical burdens of hostility and conflict relations between adjacent countries still linger;
-   unequal size of countries or regional centres implies non-parity roles

and thus hinders opening[1];
- differences in economic development;
- pretensions by one side to cultural superiority over the other;
- differences in political system, or the effects of earlier differences such as those between West and East European states.

Figure 12.1 Italian, Slovenian and Croatian border areas

These peculiarities do not negate long term European integration trends or processes, but they certainly do require different modes of action which diverge from some general, uniform pattern, and show that there is not just

---

1 Italy's population of 57 million is almost 30 times greater than neighbouring Slovenia's which is a little under 2 million; Trieste, the Centre of the Friuli-Venetia Giulia region, has a population of 272,230 whereas Koper, the regional Centre of Slovenian Istria has 25,000.

the one but *several "roads to Europe"*. Particular attention is paid in this chapter to experiences in the regions lying along the Italian-Slovenian-Croatian borders, to Istria especially, and additionally to Euregions and cross-border co-operation along the Czech-German and the Czech-Austrian borders. Examination of the *horizontal (spatial)* and *vertical (hierarchical)* dimensions of restructuring shows that a region can behave as an object and as an active subject in these changes. This ties in with the concept of "active space" as defined by van Geenhuizen and Ratti (1998).

Border Regions in the Shift from Areal to Network Linkages

From predetermination and external determination to autonomous innovative programs

To glimpse the future of border regions their role has to be examined in relation to one of the fundamental socio-spatial transformations of the present-day. Namely, the shift from regional linkages based on *the contiguity* of relatively closed "territorial communities" to *selective and dispersed* interlinking of growing numbers of relatively autonomous actors to form trans-national networks. This process involves a transition from the situation in which the territorial framework was predetermined and accepted by the actors as self-evident, to a new situation in which technologically and politically increased accessibility allows the creation of an individual frame of reference within the broader area with greater selectivity.

The transition from the one to the other situation is the most radical precisely for border regions. With closed state borders, which lead to "billiard ball" type relations, these regions are primarily *determined from above*, through decisions made at the centre. The centre imposes the border regime and the specific content and mode of operation of such border services as the army and customs and thereby various restrictive measures governing movement and activity in the border region. It is similar for the usual fluctuations and instability along the border reflecting frequent shifts in exchange rates and the like. The importance of *external determination* comes not only from restrictive measures imposed by the centre but also from the centre's assistance to the border regions, again with a view to national interests, e.g. the assistance received by the Friuli-Venezia Giulia region, and especially Trieste and Gorizia, to compensate for the loss of their former hinterlands and to entrench or consolidate the border.

As noted above the study of *cross-border co-operation* and even the formation of cross-border regions or Euregions tend to focus on processes related to the opening of borders primarily in the context of *contiguous regions*. There is a certain normalisation process involved here: in place of the former one way links, turning the periphery "with its back to the border" (Strassoldo in Gottmann, 1980) and a predominance of radial communications with the centre, linkages are established in the opposite or tangential direction to the former. This both necessitates and enables *greater independence* and a quest for comparative advantages and programs that may be more successfully executed in collaboration with districts and regions across the state border. This entails a certain diversification of linkages, although primarily inside the border region. The *area of interlinking on the peripheries* of two states is thus expanded. It is *a priori* characteristic of state centres that they interlink only through a network.

The easing of political tensions between East and West after 1991 brought a resurge of initiatives for various cross-border linking projects, the formation of cross-border regions and of working groups, etc. With regard to spatial expansion in the sense of *contiguity*, the following developments can be identified:

- Initiatives to create the *cross-border region of Istria,* according to a general Euregion model, to institutionalise links and the autonomy of parts of Croatia, Slovenia and (symbolically) Italy (Malabotta-Richter, 1995). There are other similar proposals for all-embracing interlinking in the northern Adriatic which have various motives although all seek to transcend the more or less pronounced peripherality of the border regions of these three states.
- Greater experience has been gained with the Alp-Adria Working Group. This now involves more than just border region co-operation in the narrow sense, but it is still a case of *a contiguous area* (the first instance of interlinking through the iron curtain) which transcends the state-centric model of spatial organisation.
- Ultimately the expansion of the European Community and the present EU, with certain deviations such as Greece, could also be put in the category of *areal interlinking*. Although typically the scope and need for selective links which do not follow the contiguity pattern also grow along with territorial expansion.

This brings us to the second fundamental form, namely *network interlinking* of territorial and non-territorial actors, within which the changing role of border regions and local communities (towns) has to be examined[2]. In principle there is no predetermination of links either within or between regions on this model. *Selective establishment* of specific links and networks of actors for various purposes is of decisive importance here. The representatives of regions as a whole or of towns, districts, economic organisations, educational institutions, professional associations or even informal groups and individuals take the initiative to set up trans-national networks, whether as "founders" or by joining existing networks. In view of the process of autonomisation (diversification) it may be anticipated that this will render conceptions of cross-border co-operation, which to date have arisen primarily in the framework of regional interlinking on the basis of proximity or adjacency, *too narrow*. The implication of the second model, networking, is that "disembedding", in the sense of an actor's movement away from its immediate surroundings, occurs simultaneously with the increased dispersion. This means that bilateral links, and hence cross-border co-operation, will be diluted and *weakened relatively*, particularly in the long term. This weakening, however, will entail less one-way dependence and thus the strengthening of an actor's autonomy. Today there are already instances of linking such as between Croatian Istria and the Italian regions of Toscana and Veneto. At the town level there is a network of towns in the Alpine arc with its headquarters in Trento, which is a kind of *intermediate category* of areal and network linking. An instance of a trans-national town network is the "walled cities" of Europe network, to which Slovenia's Piran belongs. There are many more ramified and spatially wider networks of various economic, educational and cultural organisations, groups and individuals.

When this process of interlinking is examined over the European space as a whole, and of course on a global scale, numerous overlaps and intersections of areal and network units may be seen. Answers as to what is the *future of the border regions* should be sought within this perspective. The structural context indicated will beg for a response or even force a choice of options. Doing nothing will no longer mean just failing to win something, it will probably mean actually slipping backwards, a

---

2 In connection with this Cook (1995: 19) indicates '...a new approach to regional development, based on the central idea of the networked regional innovation architecture ...'. According to him networking involves *reciprocity, trust, learning, partnership* and *empowerment*.

deterioration. This presentation of the broad structural transformation links up here with the active role of the border or cross-border region, and the concept of *"active space"*.

The transition from areal to network interlinking, which entails increasing dispersion in space, implies the gradual *loss of those features* of border regions and their co-operation which previously stemmed from the *break in linkages* and *separation* in space. Features that express the intensification of unique combinations and links between people and resources in space will be strengthened.

In parallel with inner or sub-regional and sub-local differentiation and autonomisation which increase the number of actors in trans-national interlinking not mediated by the state, the usual conception of cross-border co-operation is also transcended. With respect to the simplified expectations of a *universal rise* in cross-border co-operation, it must be noted here:

- that cross-border interlinking will actually intensify in the period of the opening of state borders, and
- that in the long term there will be a relative decline in the significance of this process and border locations in general.

Initially there is a tendency to correct the earlier restricted and one-sided interlinking due to the presence of the state border. Then border regions will begin to normalise and be governed by the same general standards and models of interlinking in space as all other regions. Nevertheless for some decades the border regions can be expected to retain certain distinctive features, because of the possibility of:

- their more intensive combining of language-cultural features from both sides of the border which will persist for some time after state borders are formally dismantled, and
- the innovative incorporation of the border regions' specific cultural legacy in development programs to "capitalise" on the former differences, e.g. the establishment of colleges for translators.

From state periphery to trans-national network

In the past the state has either neglected its periphery or treated it as "sensitive" for territorial and state integrity. Thus, for example Italy gave considerable economic aid to its border regions and towns when they began to stagnate. Similar "national-defence" concerns may be seen in the location of various central state agencies or institutions, such as a school for tax inspectors directly beside the border near Gorizia in the territory of an ethnic minority, which sparked protests. Several international scientific research institutions and university faculties have been located on both the Italian and the Slovenian side of the border, i.e. in Gorizia and Nova Gorica, in Trieste, Koper, Portorož, etc. The motives and the explanations given for the siting of these educational and scientific institutions differ. In some cases they are quite pragmatic economic interests or ecological concerns, in others the need for national consolidation along the border is accentuated. Of course when it is a matter of entrenching national boundaries the focus is not on cross-border co-operation but rather on institutions that are turned towards the interior of the state. In respect to the very specialised international scientific centre, such as the International Centre for Theoretical Physics in Trieste, it is noteworthy that their frame of reference overreaches by far the space of cross-border co-operation between border regions. Their co-operation even extends beyond European space to global networks.

Various programs and initiatives for*"trans-national inter-university co-operation"* are also proliferating. These have either been introduced by the EU (Tempus, Copernicus, etc.) or by others such as Austria which initiated CEEPUS or the Central European Exchange Program for University Studies. This program promotes the mobility of students and professors and extends co-operation from that typical for *bilateral* inter-governmental agreements to a *multilateral* trans-national network of co-operating university units from a large number of countries, which diminishes the likelihood of non-parity roles. This model significantly surpasses the once typical centre-periphery relations because in CEEPUS, with the exception of administrative co-ordination which is located in Vienna, the *centre* consists of the proposers of concrete programs (over 200 "network carriers" and another 441 partner institutions at the beginning of 1995) so that the sites of the centre and the periphery are not pre-determined. The location depends on the origin of the initiative or the activity, and of course on objective conditions as these vary from one place or region to the other. This interlinking model points to new possibilities for a deperipheralisation of today's border regions. The new mode of activity also implies that the border region will have to count less and less on the state's classical role and more and more on its *own inventive combining* of different situational advantages and human resources to produce innovative developmental programs that can be offered more widely in Europe. This also confirms the view that knowledge of the objective long-term regularities of socio-spatial restructuring does not nullify the need to make concrete efforts. It is precisely awareness of them that gives actors an edge over others.

From the Region as an Individual Actor to Multilevel Power-sharing

Another general aspect of socio-spatial restructuring that has to be borne in mind in discussions of European integration and the future of border regions concerns the changes in the relevance, roles and interrelations of the different levels of territorial organisation of a society. Through some inertia the error is constantly made in the social science literature of dealing with only one or some of these levels; and in this case, the regional level. With integration in Europe the range of levels involved in resolving a particular problem is broadened. With the entry of the supranational level into the arena the importance of the autonomy of the region and of local and sub-local units grows, as well as that of non-territorial actors and especially individuals. The relationships between these levels also change[3]. Although the range of levels is widened, relations between them are dehierarchised and for example, various shortcuts between the highest and lowest levels are built (Mlinar, 1995).

The autonomy of regions cannot increase on the long-term if relations between the higher and lower levels of decision-making are seen as functioning like a *zero sum game*, in the sense that one side's gain is the other side's loss. After all, the goal is not decentralisation as an alternative to centralisation, but that of intensifying inter-level linkages and *power-sharing* to allow the zero sum logic to be transcended. A win-win logic is in prospect instead of win-lose. The wider the deviation from practice the more conflicts between levels intensify, especially between representatives of national versus border region interests. This is evident in the case of Croatian Istria (Mlinar, 1996). At the same time we have the example of projects that have been carried out successfully precisely because they involved an *optimal combination* of the resources and the interests of all the

---

3 In summary of several studies Barry Jones states that '... the EC's ability and willingness to deal with all levels of Belgian society, national, regional, local and private could very well point to a future pattern of increasingly decentralist, self-organising, and porous societies with policy networks at a variety of levels.' (Jones, in Keating, 1995: 293).

pertinent *decision-making levels.* Experience with one program related to the formation of a new university in Slovenia, the "Third" or "Primorska University", in the region along the Italian border is illustrative. It concerns the establishment of a College for Hotel Management and Tourism at Portorož in the Piran municipality where it is now a functional unit. Underlying the success of implementing this program was precisely the combining of numerous and complementary actors from *several relevant levels*, with a major role for individuals as initiators, engagement of private commercial companies from the tourism sector, the municipality, inter-municipal co-operation, which in fact substituted for the role of the region (regions do not yet formally function in Slovenia) and the role of the Slovenian centre, i.e. the expert council which must give approval, and support from European level programs, namely assistance through the Tempus program.

In a sense the role of the region, and of the state, was put in perspective in this case because it was only one of several levels and actors. Growing access and direct links between the actors in the global context will weaken the role of the region as an intermediate level of territorial organisation. It will lose its classical administrative role and its role hitherto as a relatively closed, isolated unit both inside the national space and in cross-border relations. Growing in importance are the roles of individuals and groups that are active in a region and manage to link together different elements from the broad European and world space in a way that takes into account the local and regional peculiarities and comparative advantages.

The establishment of the College at Portorož exemplifies the active combining of local, near-by Italian and EU Tempus experience and actors. It has even resulted in English being chosen as a language of instruction besides Slovenian and Italian. The European programs, Tempus in this case, obviously help expand the circle of participants and thereby significantly enrich the content and overcome the traditional non-parity in cross-border relations between the two neighbouring countries. By using programs that directly or indirectly involve cross-border co-operation, the EU thus serves as an important supranational complement in the multilevel decision-making and implementation of development programs. Instead of frontal clashes with the state centre the region can pursue its particular interests through affirmative programs of a multilevel nature. Programs like Phare and Interreg are aimed at promoting interregional and cross-border co-operation *together with* the states to which these regions belong rather than by *bypassing them*. It may be anticipated that in the future both regional and local representatives will increasingly take part in inter-state negotiations. Representatives of border areas in particular already occasionally participate in inter-state meetings as *co-participants in state delegations* or in other ways, although this is not suitably reflected in payment of the costs of such international activities by the municipalities and regions. There are plans to open an office of the Slovenian foreign affairs ministry in the border area.

Quite often representatives of border regions and local governments are best acquainted with specific situations and can make the most constructive contribution to settling disputes even at the inter-state level. Through their independent action (initiatives, statements) representatives of Istrian municipalities in Slovenia and Croatia have even intervened in the settlement of inter-state disputes over demarcation of the border and the border regime. The cross-border co-operation in the framework of Phare and Interreg programs, both in the planning and implementation of particular projects and in their financing, is an example of multilevel participation.

Legacy of the Zero Sum Game: Distrust and Real or Perceived Threat

From shifting to opening and dismantling borders

In many parts of Europe opening borders and expanding cross-border co-operation are hampered by the historical legacy of conduct that amounts to a zero sum game: as much as one party gains is lost by the other. This refers above all to territorial expansionism and the *moving of border*s rather than their opening and cross-border co-operation. This conduct came to an end in Europe in great measure with the end of the Second World War, although in the Balkans it is only just coming to an end in Bosnia-Hercegovina. The moving of borders leaves behind long lasting memories in people's minds when it is a matter of judging intentions on the other side of the border. This is the legacy of the past century or centuries of frontal *territorial expansionism and (sometimes) hostility*, when every movement on one side of the border aroused fear that it was directed against the neighbouring country. In fact, the transition "from confrontation to interpenetration" is characteristic of the present-day (Mlinar, 1997). This is the transition from territorial conflicts and advancement of national borders towards in some cases the dismantling of these very borders. As an expression of political might, at least in its open and visible form, territorial expansion has lost its legitimacy and its economic sense. In both cases this opens up different prospects for the future. It has already become practically impossible today for EU member countries to exercise their superiority forcibly to change borders at the expense of another country. Nevertheless, in some border regions in Europe the *legacy of the old pattern* (the zero sum game) is still highly pervasive. At times old territorial pretensions, or the fear of them, reappear in a new form, even in certain European programs; particularly in border regions which are loaded with a "rich" history of national-territorial shifts.

The Istria region has experienced dramatic shifts in both directions in the past century. The conflict between the Venetian Republic and the Austro-Hungarian Empire was followed by great population movements with immigration into the region from the Italian side and the emigration of Slovenes and Croats, and subsequently a reverse flow: the withdrawal to Italy of about 250,000 exiles and "opters" for Italian citizenship and domicile, which was later followed by immigration from other parts of the former-Yugoslavia. These movements were connected to actual or desired *shifts of the borders* or resulted from pressure from each succeeding regime to *assimilate* members of the neighbouring people.

Historical memory may persist as a residual category in the form of a *feeling of threat* from a bigger country or in the form of a sense of violation of "historical rights" to a particular territory. The main point is that in such situations it becomes very difficult to distinguish well-intentioned initiatives for cross-border co-operation from old one-sided expansionist pretensions in new clothes. Understandably, every action becomes ambiguous. At each step there is the question of *trust or distrust* and doubts about ulterior motives which might be hidden in the background. Although this distrust and feeling of threat is present primarily on the "weaker" side (Slovenia, Croatia) it is notable that there is also a certain *feeling of threat* on the Italian side, particularly in the border areas and in the city of Trieste. There is the fear that opening the borders will lead to some kind of "*invasion*" of Slavs, who are viewed with disdain.

In recent years there have been several clashes or adverse reactions in the border regions of Slovenia and Croatia (Istria) because of the ambiguity of particular steps taken by the Italian side. These can be summarised as follows:

- Italy's granting of *dual citizenship* irrespective of actual national origin.
- The award of Italian *pensions* to Slovenian citizens who worked for a

prescribed period of time on the one-time territory of the Italian state.
- The opening of Italian *educational* institutions to young people from the other side of the border.
- Assistance to individual members of the Italian minority undertaking independent *economic* activities.
- A *language policy* which incorporates a one way approach and the assumption that only one side learns the language of the other, and not *vice versa*.
- Similarly, the convening of the "Istrians World Congress" (in Pula, Croatia) at which an Italian orientation predominated in terms of the composition of participants, standpoints taken, language and symbols.

One-way *information flows*, in views of Italy's technological and economic advantage, ignorance of neighbouring languages, and feeling of cultural and other superiority, are slowing cross-border co-operation with Slovenia and Croatia at least at the formal, institutional level (see also Cappellin and Batey, 1993). An international dispute was triggered by the demand that Italian exiles and "opters" should be allowed to buy *real estate* in Slovenian Istria. Under such circumstances, intensifying certain forms of cross-border co-operation, or attempts, tends to increase distrust and trigger defensive reactions in the face of hypothetical intrusions or expansion from the other side of the border. Groups that have never relinquished their "historical claims" and continue the zero sum game constantly give grounds for such misgivings. In view of past experiences, and not least the collapse of former-Yugoslavia and the armed conflict over "Krajina", in Croatia fears and warnings were especially pronounced that the demand for the autonomy of Croatian Istria concealed *a drive for secession* or some new form of statehood (Corsican model).

The issue of distrust also came into play when the *loyalty* of border region residents who accepted the Italian offer of dual citizenship was questioned. Doubts and warnings of *disloyalty* were voiced most strongly in central Slovenia which in turn provoked a general adverse reaction amongst the Primorska populace, residents of the region along the Italian border. This was not just a classic centre-periphery clash. It sharpened differences and demarcated a particular socio-cultural identity. This identity is neither defined spatially by the area inside the borders nor by language. It characteristically appears within the *cross-border region* which differs culturally from the central part of the country. Consequently, probably for the first time in such pronounced form, *a differentiation that seemed to be transcended* has begun to revive between the western part of Slovenia, which historically was exposed predominantly to Italian (Romanic, Mediterranean) influences, and the continental part with its predominantly German influences. Somewhat unexpectedly a difference that at least publicly had been erased has reappeared as a challenge to the undisputed centrality and representativeness of the Slovenian capital[4].

How confidence can and cannot be built

Building mutual confidence obviously depends on how much domination tendencies and especially the vestiges of previous aspirations to territorial expansion are excluded from mutual relations. In the case of the regionalist movement in Croatian Istria something quite different has come about, namely *a rise in distrust and open conflicts* and estrangement between the national and regional authorities as well as between the neighbouring countries. The lag in Istria's development into an Euregion may be attributed to the following circumstances:

- The program for the formation of the new cross-border region was not a gradual but a radical one and moreover, came at a time when the question of the final constituting of the state and its territorial integrity was at the forefront in Croatia. At that time even ties that had already been established, such as between actors in Croatian and Slovenia Istria were disrupted. The pretentious program was devised with too little regard for the factual situation and actors.
- In view of the foregoing, a *cross-border territorial construct* emerged which was more like the traditional pattern of areal organisation of society than a new network pattern. Thus, it raised a threat to national sovereignty because of the aspiration for immediate territorial institutionalisation in a cross-border framework. Had there been a greater variety of specific and concrete cross-border co-operation programs and actors there would have been less of a clash with the striving to preserve national sovereignty or more simply, with centralised control of the whole territory of the country (Croatia). This is nothing unusual, just a very pronounced case. James Wesley Scott

4 For analogous tendencies by the Northern league in Italy, but in the direction of the North, see Bull (1996: 81).

(1989) like many others, reports persistent tensions between regionalism and national sovereignty in Europe generally. He notes that if co-operative programs prove useful in the economic sphere it is easier to institutionalise cross-border co-operation over a broad field of political activity.

The prospects for joining a broad common political structure like the European Union fundamentally alters the way opening borders and cross-border co-operation are viewed. This is the prospect of moving from a win-lose to a win-win logic. Not all that surpasses the borders of one's country, at a time when communications are not automatically severed at the border, is necessarily lost. The proposal to establish Istria as an Euregion, however, came at a time when Italy was a member of the EU, Slovenia only later became an associate member, and Croatia had not even been accepted as a member of the Council of Europe. All the more reason then why gradual intensification and diversification of forms of co-operation were so important. The opposite may be argued of course that the region could blaze a trail into Europe. In fact, with the help of Italian mediators, especially Toscana and Veneto, Croatian Istria did become a member of the Assembly of European Regions. This could be one way of joining European integration processes; but it is not a completely "neutral" path, because it is based on the assistance of regions from a neighbouring country with which Croatia is competing for control and influence in Istria as one of its regions whose loyalty is questioned.

As we have seen, historically the ethnic composition of the population of Istria has been altered substantially, leaving an impression of instability and consequently all the more distrust. Although the representatives of the Istrian movement in Croatia deny allegations of *separatist tendencies*, under the circumstances a certain doubt and hence distrust remains. An indiscriminate forcing of a *general model* of "European regions" can prove counterproductive in this situation. The legacy of Istria's great *ethnic fluidity* still creates the potential uncertainty that perhaps old territorial pretensions are being pursued under the aegis of European programs.

Non-Parity Roles and Seeping Sovereignty

The experiences described by Frantisek Zich (1993) in the first sociological study of Euregions along the Czech-German and Czech-Austrian borders confirm our conclusions on Istrian cross-border co-operation to a great extent[5]. In the Czech Republic cross-border co-operation also involves partners of marked *non-parity*. Given the historical-political circumstances the partners on the German side of the border are more influential, like the Italian side in the Istrian case, than those on the Czech side. Moreover, the initiative came *from the German side*; interest on the other side was tied to expectations of additional resources from the central government for regional development. It was not just a question of teething problems, because there was no appropriate normative groundwork for the functioning of an Euregion and their activities were illegal to some extent. More importantly, these activities overreached their competencies arising from international agreements so that in time they began to function like independent international subjects. 'The foreign - and primarily German - influence was strongly felt in the activities of regional associations and raised questions within the executive power about the purpose and mission of Euroregions.' (Zich, 1993: 9). Arguments began between the ministries and the representatives of the newly established Euregions regarding their role and authority or competencies. Exercise of the *principle of state integrity* was especially in question and in relation to that it was proposed to amend the statutes[6]. Interventions by the government slowed work on the Euregions and also dampened the interest and enthusiasm of the participants themselves.

In connection with the clash between the regions and the central authorities, the Czech premier expressed a critical view of Euregions:

> 'The hidden meaning of Euro-regions is different. We must carefully monitor and distinguish between friendship, partnership, co-operation, trans-frontier projects, and attempts *to erode the identity of our state...*' 'We want to go to Europe as the Czech Republic. I do not presume we want to go to Europe as Euregio Nisa, Labe (the Elbe) or Shumawa...' 'Some mayors believe they can be in Europe earlier in this way. I do not. I see it as *jeopardy to state identity.* ' (Klaus, cited by Zich, 1993, italics Z.M.).

---

5 A more general formulation of the problem stated in the LACE Guide (1995) is: 'In Central and Eastern Europe, approaches to cross-border co-operation have understandably emerged rather hesitantly since these borders were firmly closed. Major disparities and historic factors mean that policy needs to be very carefully and thoughtfully prepared ...' 'Those border areas with minorities will require a particularly careful approach.' (p. 4).

6 This is of course highly reminiscent of the dialogue and disputes between the Istrian representatives and the central government in Zagreb (see also Mlinar, 1996).

Thus, the central issue is the autonomy of the cross-border region and the extent to which it conflicts with the preservation of national sovereignty. As already noted, regional-national antagonisms are common everywhere but with the cross-border region they are intensified because of the potential threat to the national centre's control. The threat is all the greater the greater the openness of the borders and the possibility:

- that the border region will intensify tangential cross-border links and thus evade traditional hierarchical control within the borders of the national state, and
- that it will become subordinated to centres on the other side of the border.

It may be concluded from various vain attempts to form Euregions on borders where differences on major criteria have accumulated, that these attempts were premature, as Zich observes (1993: 17). Above all the time required for their implementation was not estimated realistically. It cannot be a "one-step action".

> 'Any establishment of regional bodies in order to coordinate, activate and encourage crossborder activities appears to be an end in itself under the circumstances where direct human economic and other contacts are yet to be established. There is a risk that the whole process will come to a sticky end of organizational formalism. *Direct* human, communities', institutions' etc. *contacts* seem to be *more effective than joint bodies*. Joint projects solving specific regional problems also seem to lead directly to the end. '
> (Zich, 1993: 17, italics Z.M.).

This experience confirms our initial statement that a *uniform* application of particular concepts like the Euregion under different concrete circumstances could bring integration processes in Europe to a halt. Uniformity may be indicative of the absence of an individual region's active and autonomous role, thereby giving rise to the prevalence of a general model at the expense of expression and respect for specific distinguishing features. Emphasising *territorial organisational* change more than concrete programs, and diversified links between different actors on both sides of the border, in the institutionalisation of cross-border co-operation will provoke a negative reaction by the central state agencies which are the ones losing control. The provocativeness of this approach comes from its *frontal, general redistribution* and loss, from the standpoint of national sovereignty, of political power. This is not the case when cross-border links are intensified on the basis of individual projects and between concrete actors, or at least it is less evident and provocative and does not amount to such a clear zero sum option for the two countries. Greater resistance by the national centre may also be anticipated because this approach gives priority to the *territorial* principle over the *functional*. This takes us back to the logic of exclusiveness and ideas about alternative territorial distributions of social power.

It is notable that *interest in and the initiative* for cross-border Euregions have come from the more powerful states (Italy, Germany) or groups behind the border which are strongly tied to them. This heightens apprehensions that a role of dominance in the border region, rather than equitable co-operation, could be established in this way. At the same time reservations and doubts begin to arise that the label of Euregions is just covering a new form of territorial expansionism or widening "spheres of interest" by a more powerful country in the style of old power politics.

Dimensions of the Region's Active Role

Regardless of the different legal status of regions in different states there is a theoretical and empirical basis for identifying some of the dimensions of the strategy of opening up border regions and incorporating them in European integration processes. In spite of their subordination to the central authorities, the opening or dismantling of borders gives border regions greater scope for directly or indirectly influencing these changes, whether alone or as co-participants. The first condition is awareness of these possibilities and the dilemmas set out below.

The border regime and restructuring: direct or indirect involvement of the regional or local authorities?

The region can exert influence on the openness of the border directly, or indirectly. In the former case it can be at least one of the participants in decisions at the national level. This role was established in the regulation of small border trading. An example is the initiative of municipalities in Croatian and Slovenian Istria for "open borders". There is even greater scope for indirect influence on the degree of openness through various development programs, such as restructuring the workforce by expanding tertiary education, scientific research, etc. This has an indirect and affirmative effect on growth of spatial mobility of labour, even across present-day borders, and trans-national communications. Further, regional or local authority may impact on public opinion and thus indirectly on increasing or decreasing cross-border co-operation, and so forth.

Territorial opening and/or functional interlinking?

Experience has shown that tendencies towards greater openness may favour the territorial aspect when it is a matter of overall reorganisation and institutionalisation of a new cross-border unit like a Euregion, or concrete projects to deal with specific problems. There is greater chance for success when concrete co-operation programs are implemented first, i.e. the functional approach, and territorial reorganisation is complementary, rather than the reverse.

Bilateral and/or multilateral cross-border linking?

Bilateral cross-border co-operation is most expedient and characteristic for contacts with the physically closest partners, but it is nevertheless substantively limiting. Non-parity relations between the partners come most easily to the fore in this form, whether it is economic, political or linguistic inequality. This unevenness may be mitigated by including the partners in multilateral programs, such as EU programs, which are based on general norms and a higher level of internationalisation which averts particular relations of domination and subordination.

Mobilising historical memory and/or accentuating common European perspectives?

As a rule, because it is a national periphery, the border region has not had suitable opportunity to express its identity which frequently includes elements of cultural similarity with neighbouring peoples. Thus, new aspirations to reconstruct this identity can, on the one hand, lead to strengthening cross-border links; on the other hand harking back to history can also be a form of entrenching the demand for compensation for historical wrongs such as lands lost on the other side of the border. Stressing *common perspectives* inherently broadens understanding for overcoming the mutual exclusiveness that was part of the logic of the zero sum game played in the past. It can contribute to the recognition that what passes beyond or is beyond the other side of the border is not lost but is also "ours", and so diminishes the difference between what is local and what is foreign.

Bastion of national integrity and/or an international hub?

In the past the border region, its towns and countryside were at a disadvantage because of greater transportation costs and a smaller market for example. In varying degrees in different countries, in line with national interests, it received government subsidies to avert instability and consolidate the situation at the edges of the state. The central role of the state is however weakening in the context of European integration processes, and the border and the "hostile encirclement" of the past are also fading. The state is relinquishing control functions except in so far as it is becoming the guardian of general European civilisational norms and a corrective factor in situations when deviation from these norms would lead to one-way intrusions across the border in the style of erstwhile models and the zero sum logic.

As the state withdraws a vacuum emerges which challenges the border region to construct development programs more independently on a cross-border, international scale. Its comparative advantages give it a more congenial position in the broad European space to assert itself in particular fields of activity as a new hub. The opening and dismantling of state borders will reduce or even dissipate certain complementarities between regions on either side of the border, for example, home services and supply of farm produce in the informal or grey economy. Room for such activities will shrink and sooner or later there will be changes either as a result of tighter control on the borders or uniformisation of the legal order throughout the European space, for example, as a result of the Schengen treaty.

Gaining and losing comparative advantages?

A European perspective for the border region means both the loss of certain functions and the acquisition of new ones. In both cases it will be a challenge to the region to not just follow some national pattern of action or even specific directives from the central government, but *to activate* to a greater degree its own potentials with regard to the *particular* border situation. Instead of relying on "geographic rent" as preordained, it will find itself having to define its own development program and activities whether as a response to the changes already underway or in anticipation of them. A feature of the changes will be that various activities concerning or associated with border control, e.g. loss of jobs in the customs service, forwarding agencies, etc., will end. There will have to be a reorientation towards other fields of employment which will exacerbate the problems stemming from the general restructuring of the economy and from de-industrialisation.

Integration with spatial concentration or a redistribution of functions, i.e. decentralisation?

Opening to the broader space, whether inside a region, a state, or in relation to a neighbouring state, or on a European scale, comes to a halt when it entails increasingly the asymmetry of roles and a concentration of functions and hence domination or greater domination by a bigger centre. Continuation of the process of integration will presuppose more and more a wide spatial redistribution of functions and social power. Here too the regions and their municipalities will have more success if they attract and establish completely new and distinctive programs or functions, such as new institutions and tertiary education programs, than by achieving a redistribution of the existing ones. To overcome the lag in intra- and interregional co-operation the typical approach so far must be abandoned: a) the preservation of autonomy even by means of autarky, with opposition to integration and b) achieve integration even with concentration, i.e. domination. The key to this is finding suitable locations for specific functions of regional, national or European significance along the periphery as well, and thereby promote *polycentrism* and diminish the significance of the differentiation into centre and periphery in general.

Opening as increasing access to and for others?

Opening up regions on a European scale, politically as well as in terms of transport and communications, entails greater accessibility to others as well as for others. Attention is being focussed on removing borders as barriers to increase freedom of movement. Nevertheless the fact remains that this implies a loss of control over flows, of people, goods, and information, as well as an insurge of influences from elsewhere. These uncontrolled flows and the increased vulnerability of the region to socially, physically and especially ecologically harmful intrusions from outside are a strategic issue in further development. Along with the aspirations for freedom of movement in the "new Europe", numerous European countries are taking special care to avert "undesirable" changes in the ethnic composition of border regions.

Theoretical notions about post-modern society, and visions of the "new Europe", although inconsistent and indirect, for example with regard to labour mobility, include a trend towards heterogenisation of the population. This is different to simple replacement of a hitherto relatively homogeneous group with another. Precisely the possibility of influxes of relatively homogeneous populations from elsewhere and particularly in serried form has generally triggered the strongest rejection by local residents. In the long term at least, the "danger" of intrusions by such homogeneous populations will be diminished by the growing diversification of the work force. Consequently, in the future mass immigrations (or migrations) need not be expected within the EU, rather the pattern will be of individualised and dispersed spatial mobility. The loss of control in the sense of the possibility of frontal territorial exclusion of "foreigners", and hence discrimination, will shift attention towards the regions' greater involvement in the formulation of *non-territorial mechanisms of regulation*, particularly general European restrictive norms and affirmative programs.

Conclusions

The theoretical findings and their possible action implications presented above, may be understood as challenges and orientations for future research into their operationalisation and embedding in the "grand theories" of social development. In the context of post-modernity even the notion of development is being questioned. Evolutionary notions of predetermined, lawful socio-spatial processes are being countered by arguments in favour of "open-ended transformation" and the unpredictability of change.

The study of regional development is only a particular case in point which can benefit from and contribute to the clarification of the epistemological context. The empirical evidence does not support the "end of regularity and predictability" thesis - so far. For example, the process of supranational, European integration and parallel sub-national (regional) autonomisation are clearly demonstrated even though their interdependence may not be so apparent. If there are cases of deviation they do not necessarily negate the regularity. They may merely prove the rule, and should be a reason for more elaborate research design. Additional explanatory variables representing structural dimensions and characteristics of the actors have to be introduced.

The failure of the initiative from Croatian Istria to establish the Istria Euregion with Croatian, Slovenian and Italian components, while several Euregions are already functioning, calls for further examination of the dynamics of the relationships between central and regional level government in the broader context of national-state formation, assertion of national identity and protection of national integrity. "Historical memory" of conflictual relationships between neighbours and differences in the size and power of countries and actors on both sides of the border, i.e. non-parity roles, are some of the factors affecting the management of openness. Thus, openness does not simply increase linearly, nor does the autonomy of regions (states) in determining the degree of openness.

The outlook for regional development will therefore involve declining independence as a condition for rising autonomy. Exclusion and discrimination on the basis of territorial provenience have become unacceptable at least within the European Union. Frontal territorial exclusiveness is unfeasible as a mode of border management. The region will have to play an active role in selectively combining the material and human resources of both the border area and the expanded area of connectedness.

Future research should unravel the prima facie paradox of long-term regularity and predictability on one side and autonomy of action and choice of border management options or scenarios on the other.

References

Blau, P. (1977), *Inequality and Heterogeneity*, Free Press, New York.
Bull, A. (1996), 'Regionalism in Italy', in P. Wagstaff (ed), 'Regionalism in Europe', *Europa*, vol. 1, no. 2-3, Intellect, European Studies Series.
Cappelin, R. and Batey, P.W.J. (eds) (1993), *Regional Networks, Border Regions and European Integration*, Pion, London.
Cooke P. (1995), *Planet Europa: Network Approaches to Regional Innovation and Technology Management*, Case Study, Elsevier, New York.
Geenhuizen, M. van, and Ratti R. (1998), 'Managing Openness in Transport and Regional Development. An Active Space Approach', in K. Button, P. Nijkamp and H. Priemus

(eds), *Transport Networks in Europe: Concepts, Analysis, and Policies*, Edward Elgar, Cheltenham, pp. 84-102.

Gottmann, J. (1980), *Center and Periphery: Spatial Variation in Politics*, Sage Publications, London.

Jones, B. (1995), 'Conclusion in Jones Barry', in M. Keating (ed), *The European Union and the Regions*, Clarendon Press, Oxford.

Lace Guide (1995), *Project of the Association of European Border Regions*, Gronau.

Malabotta Richter, M. (1995), *'Transfrontier Regional Cooperation: Theory and Practice'*, 16th Conference Europe of Regions, November 16-18, Maribor.

Mlinar, Z. (1992), 'European Integration and Socio-Spatial Restructing: Actual Changes and Theoretical Response', *International Journal of Sociology and Social Policy*, vol. 12, no. 8.

Mlinar, Z. (1995), 'Territorial Dehierarchization in the Emerging New Europe', in J. Langer and W. Pöllauer (eds), *Small States in the Emerging New Europe*, Verlag für Soziologie und Humanethologie, Eisenstadt, pp. 161-179.

Mlinar, Z. (1996), 'Regional Autonomy, National Integrity and Transborder Cooperation: The Case of Istria', *Regional Contact*, IX, no 11, Maribor.

Mlinar, Z. (1997), 'From Confrontation to Interpenetration: Territorial Conflicts in Retrospective and Perspective', in J. Rotblat and M. Konuma (eds), *Towards a Nuclear-Weapon-Free World*, World Scientific, London, pp. 608-619.

O'Dowd, L. (ed) (1996), *Borders, Nations and States*, Avebury, Aldershot.

Ratti, R. and Reichman S. (eds) (1993), *Theory and Practice of Transborder Cooperation*, Helbing & Lichtenhahn, Basel.

Scott, J. W. (1989), 'Transborder Cooperation, Regional Initiatives and Sovereignity Conflicts in Western Europe: The Case of the Upper Rhine Valley', *The Journal of Federalism* 19, Winter.

Strassoldo, R. (1983), 'Frontier Regions: Future Collaboration or Conflict?', in M. Anderson (ed), *Frontier Regions in Western Europe*, Frank Cass, London.

Zich F. (1993), *Euroregions along Czech-German and Czech-Austrian Borders*, Institut of Sociology, Prague.

13 Active Cross-Border Regions: Institutional Dynamics and Institution Building

JOACHIM BLATTER

How can border regions adjust to the challenges of globalisation and the transformation of national borders from barriers to contact zones? This contribution addresses this question from the perspective of political science and organisational theory. First, it is shown how continental integration processes and new production regimes create new paradigms which stimulate stronger economic and political co-operation across national borders. Second, in order to understand political institution building processes the interdependence of the cross-border arena with many other political arenas has to be recognised. Finally, some suggestions for designing institutions for active cross-border regions are provided. A complementary variety of different institutions provides the best opportunities for innovation, social learning and democratic responsiveness.

The Importance of Ideas and Paradigms

In times of rapid change and increasing complexity the "interpretative" or "communicative"[1] approach is gaining more and more importance in many social disciplines. At the same time, when it is acknowledged that factors

---

1 Sometimes it is also called the "constructivist", "kognitive" or "reflexive" turn in social science. What these labels have in common is that they express the conviction that individual behaviour is neither determined by an "objective" structural environment nor by "objective" interests. Cognitive and normative orientations shape actors' perceptions, preferences and identities - and make them open for changes through social learning or persuasion (Scharpf, 1997).

like ideas[2], paradigms, perceptions, problem definitions and solution models are important for attempts to explain social change, these factors are also increasingly used for recommendations for the management of change. Recent thinking in the field of regional development does not represent an exception to these trends.

The importance of the spill-over of the "idea of integration" from the continental to the cross-border arena cannot be overestimated as an explanation of the timing of cross-border activities. The processes of continental integration can be seen as the dominant factors stimulating cross-border activities on a sub-national level during the last ten years. In Europe, and in North America, there has been a kind of mushrooming in the development of cross-border region-building since the end of the 1980s, which has paralleled the discussions before the introduction of the Single Market in the European Community in 1992 and the Free Trade Agreements of North America. Older cross-border linkages were reinvigorated and for the first time received enough political and, in some cases, financial strength to fulfil some of their long proposed goals. Even more significant, many initiatives emerged on virtually every border region. Even in regions where cross-border co-operation was not an issue and where there was limited socio-economic interdependence, the idea of a common region was a booming discussion topic and changed perceptions at the end of the 1980s and the beginning of the 1990s (Blatter, 2000a, 2000b).

One example is the Lake Constance region where the "Bodenseerat" - a private association of regional leaders from the political, economic and scientific field - was founded in 1991. Stimulated by this event, the older cross-border institutions of the executive leaders of the German and Austrian Länder and the Swiss Cantons deepened their activity in scale and scope and produced a new, common "Leitbild", a comprehensive development proclamation, for the entire border-spanning region. The environmental groups, which have a long history of cross-border co-operation in the region, gave their co-operation a new structure called "Umweltrat Bodensee" (Environmental Council of the Lake Constance) and created a "Bodenseestiftung" (Lake Constance foundation) (Müller-Schnegg, 1994; Scherer and Müller, 1994). Another example is "Cascadia",

---

2 The word "idea" is increasingly used in political science and especially in the subdiscipline of International Relations. Ideas include world views, principled beliefs and causal beliefs and provide road maps in complex situations (see Goldstein and Keohane, 1993).

a cross-border region at the western side of the U.S.-Canadian border which includes the U.S. states of Washington and Oregon and the Canadian province of British Columbia[3]. An initiative from members of the Washington State legislature in 1989 resulted in the founding of the Pacific Northwest Economic Region (PNWER), a public-private association including five American states (Oregon, Washington, Idaho, Montana and Alaska) and two Canadian Provinces (British Columbia and Alberta). With the ratification of the co-operation agreements in the sub-national legislatures, the incorporation of the Governors and Premiers into the association, and the support and participation of many private companies, the PNWER is now a strong and innovative political actor designed to promote economic development and trade abroad and within the border-spanning region. The Cascadia Corridor Task Force, the Cascadia Economic Round Table, the Pacific Corridor Enterprise Council and the Pacific Northwest Economic Partnership are other cross-border initiatives in the economic realm which have emerged in recent years. Achieving the critical mass to compete successfully in the greater continental and global market is the dominant motivation in this discussion about "Cascadia" (Alper, 1996).

All this cross-border activity is clearly stimulated by the debate about economic integration on a continental level[4]. However, it is not only the widening of spatial horizons through continental and global integration processes that has changed perceptions along national borders, there are also other scientific, technical and socio-economic developments which allow and stimulate new spatial alignments and growing cross-border linkages. In particular new problem definitions have changed the character of many policy fields from interterritorial (re-)distributive and competitive to interterritorial co-operative.

In the field of natural resources we find significant changes in problem definition which, in turn, creates new perceptions of cross-territorial co-operation: the dominant perception of resources, especially water, was that of a special territory-bound resource with characteristics of a "mass product" with relatively simple user-structures. Disputes about the quantitative distribution between territorial units dominated the conflicts. During the last

---

3 As in all cross-border regions one finds a broad variety of territorial definitions of the region - indeed, the phenomenon of "la geometrie variable" is one of the most striking characteristics of present cross-border co-operation.

4 Comprehensive descriptions of the political activities in these two cross-border regions can be found in Blatter (2000a).

decades, however, the perception has changed: it is no longer a discussion only about resources focusing on quantitative aspects, but much more a discussion about the environment emphasising mainly qualitative aspects. The latter debate involves a more diverse user-structure than the former. The conflict between different user-groups, including environmentalists as advocates of a much broader sense of usage, is now the dominant division line. Territorial cleavages are losing importance, although they are still very important, at least in that their legacy strongly determines rights and institutions in the newer policy processes.

In economics, there has been a similar development: the production regimes are fundamentally changing in the transformation process from a resource-based to a service- and information-based economy. The former are characterised by relatively simple mass products; the latter by a very sophisticated and differentiated production chain. Now regional "active spaces", "innovative networks" and "milieus" are the key words (Camagni 1991; van Geenhuizen and Ratti, 1998). In border regions, this means a perceptual shift from competition, since often both sides of the border had and have the same natural resources and products, to synergy and a move to trans-border process integration. This process is taking place in the economic field but is accompanied by political activities which are necessary to provide what Ratti calls "support spaces" (Ratti, 1993). In the political realm, the concept of "region states" in a "borderless world" (Ohmae, 1993) is the stimulating idea resulting in regional co-operation efforts within and across nation states.

These developments have their value in attempts to explain the emergence of cross-border linkages; and they can be seen as changes which open opportunities for cross-border activity. However, it can not be taken for granted that these opportunities are acted upon, whereas at the same time there are increased risks for border regions with the further opening of the border. The transformation of borders from barriers to filters, to open borders (Ratti, 1993) is producing new challenges. Old gateway-functions are being lost and there is a danger that the border regions are degenerating into pure transportation corridors or into "dead space" between the centres (Ratti in this volume). New fields (policies, products, services) must be discovered in order to play their roles as "contact zones" (Ratti, 1993), "transfer hinges" and "innovation poles" (Blatter, 1996). The concept of "regional active spaces" may provide an adequate paradigm to manage the transformation in an active and creative way instead of passive adaptation. Regional "active space" is shown as a meso-economic force field with an intermediating role in two dimensions: the territorial and the economic dimension. In respect to the former, "active regions" provide the right mixture between necessary social cohesion of the local community and communicative openness towards global dynamics. With respect to the latter, "active regions" provide three types of relations: relations connecting the factors of production; relations connecting suppliers and customers; and relations connecting economic actors with public and governmental institutions (van Geenhuizen and Ratti, 1998).

A complementary aspect to this dimension of changes in problem definitions, perceptions, ideas and paradigms is the dimension of institutional settings and dynamics. Actors are embedded in institutional roles which influence - restrict or facilitate - the acceptance and promotion of new ideas. Ratti (1993) includes those aspects of institutions and governance in his concept. Nevertheless, he is mainly concerned with relations within the economic sphere and with modes of governance between economic and political actors. As we will see, the political sub-system itself is a complex entity with its own logic and dynamics. Therefore, public and political cross-border linkages and institution building processes can only be understood when it is taken into account that there are strong interdependencies within a complex field of political institutions and arenas. All this should be acknowledged in explaining political cross-border activity, as it will be shown in the next section, but it has also to be taken into account in organisational models for active and successful cross-border regions. This chapter will be concluded with some guidelines in this respect.

Political Cross-Border Activity and Institution Building

The dynamics resulting from institutional interdependencies are of major importance for an understanding of the process of political cross-border activity and institution building. Institutional settings and dynamics provide the framework within which new problem definitions, ideas and paradigms can find resonance, support or limitation. Institutional interdependencies are to be understood here as political relationships between the cross-border political arena and other political arenas.

"Political arenas" are defined here as political interaction systems across boundaries which separate different subsystems of modern political-administrative systems. *Figure 13.1* shows the relevant political arenas in the context of cross-border co-operation. One can distinguish between:

a. the sub-national, cross-border arena
b. the international/continental arena
c. the vertical intergovernmental arena
d. the intra-state horizontal arena
e. the intra-borderlands (or inter-local) arena
f. the inter-sectoral arena
g. the relationship between executive and legislative branch
h. the ideological/partisan competition
i. the public-private relationship.

The cross-border arena on a sub-national level (a) is the arena of interaction on which analyses are focused. The international or continental arena (b) and the intra-state arena (d) are other dimensions of horizontal political interrelation. (c) stands for the intergovernmental relation between the different layers (levels) of government. (e) points on the principal-agent problem between the (executives of the) cross-border association and the individual, local members. (f) characterises an example (the most important) of inter-sectoral relations as a result of the differentiation of the society and the administrative structures into different policy fields, interest groups and departments (it is the "third dimension" in this model). Other political arenas, like the relationship between executive and legislative branches (g), ideological competition between political parties (h), and the public-private relationship (i) are shown here as aspects of political life within a cross-border association, but can be seen in a wider perspective as aspects of political culture which are important for every level of the participating political units.

While the phenomenon of interrelated or interlocked political arenas is not new in political science[5], this schematic shows that the interrelationships between political arenas are much more complex than is expressed by the term "two-level-games" (Putnam, 1988) or "Mehrebenenverflechtung"

5 Prominent in International Relations is Putnam's (1988) term "two-level games" to stress the fact that state executives simultaneously have to negotiate in the international realm and in the domestic arena. Research on German federalism has uncovered that vertical intergovernmental relations have an impact on the relationship between the executive and the legislative branch. Whereas Scharpf, Reissert and Schnabel (1976) pointed to the restricting impact of intergovernmental linkages for parliaments, Benz (1992) stressed the fact that it might also be the other way around: governments controlled by a competitive parliament have restricted leeway in intergovernmental bargaining.

(Benz, 1992); and until now, there has been no systematic application of these theoretical approaches in borderlands theory. This is somewhat surprising because there is much evidence that political cross-border interactions and institution-building depend strongly on developments in other political arenas as the cross-border arena is still of minor importance in comparison with other arenas. Some approaches to cross-border policy analysis (Mumme, 1985; Ingram and Fiederlein, 1988) focus on the influence of other political arenas on cross-border policy issues, while scholars in the field of "International relations of sub-national units" have developed very sophisticated models of actors and motivations for cross-border and international activity (Duchacek, 1986, 1990; Soldatos, 1990, 1993; Hocking, 1993a). However, what is still missing is a comprehensive approach to "interrelations" between the different political arenas in the field of cross-border co-operation.

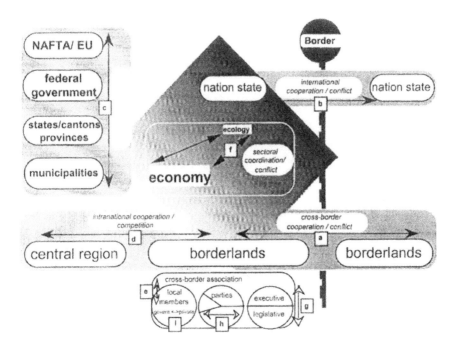

Figure 13.1 Political arenas in cross-border co-operation

Interrelations between the different political arenas can take very different forms: analytically, one can classify these interrelations on a continuum between one pole, with very loosely coupled relations, and another pole with very strong coupled relations. "Loosely coupled" means that the policy arenas have no direct linkage (i.e. there is no actor in both arenas), that they are not formally interdependent, and that the relationship between these arenas is characterised by a perceptional recognition of developments in the other arena.

Spill-overs of ideas - the idea of European integration, for example - and concepts, and also strategic adjustments, are examples of such loose interrelations. "Strong coupling" is characterised by formal - mutual or unilateral - dependence of the political arenas. Prominent examples are the right of the legislature to ratify international treaties that were negotiated by the executive branch, or the participation of sub-national governments in federal law making. In between these poles one finds relationships between the political arenas which are characterised by direct but not inevitable linkages - e.g. actors are in both arenas but they do not have to link the activities of these arenas. Here, one can think about unintended consequences in the border arena as a result of developments in other arenas, or about "coalition building strategies" which are characterised by the voluntary and active use of alliances in one arena with the aim to influence the outcome in another arena.

In such an analytical framework, interrelations with other political arenas can exist as restrictions or hurdles, but can also serve as incentives and facilitators of cross-border co-operation. Developments in different political arenas can correlate positively or negatively. This means that stronger integration or co-operation in one arena can result in more integration or co-operation, but also in less integration or in conflict in the other arena. Analytical scholars should start with such an open model, it is then a matter of empirical research to clarify the relationship between the different political arenas. Cross-border institution building can also be seen as the result of a policy-process. Therefore, the external influence can be classified with respect to which stage it occurs - the distinction between agenda setting, decision making and implementation is most important.

Having described the basic categories of this analytical framework for understanding political processes in emerging cross-border regions, the following sub-sections provide examples which show the influence of other political arenas in the process of cross-border political institution-building.

Influence of the international or continental arena

It has been shown that the continental arena was of major importance in initiating cross-border activities through the "spill-over" effect of the "idea of integration". In addition, in the stages of institutionalisation of cross-border linkages, the continental arena often provides a very important influence. That is the case in Europe, where the European Union created the INTERREG-program for cross-border regions parallel to the introduction of the Single Market requirements. INTERREG was a major incentive to start or to strengthen the cross-border linkages, but it also shaped the *form* of these cross-border institutions significantly because of formal requirements for financing and as a result of personal communications and recommendations from members of the European administration (for one concrete example of the latter, see Beck, 1997). Furthermore, this recommendation aspect was institutionalised by a further program which gives a particular Euregion, the one at the German-Dutch border, money to help other Euregions to get started (Scherer, Blatter and Hey, 1994: 15). In a different way, there is a similar phenomenon in North America. With the same motivations as in Europe for the INTERREG program - reducing the problems for border regions which are created with free trade and continental integration without harmonisation - the NAFTA agreement brought joint U.S. and Mexican funding and institutions to their common border region: in side-agreements to the NAFTA, the North American Commission on Environmental Co-operation (NACEC), the Border Environmental Co-operation Commission (BECC) and the North American Development Bank (NADBANK) were established in 1994. The tasks of NADBANK and BECC are border centred: they should facilitate the financing of environmental infrastructure and economic development projects on the United States-Mexico border. The NADBANK will act as the lead bank in the financing of border projects that have been environmentally certified by the BECC. Both institutions have advisory-boards and process approaches which provide for broad representation and participation of all levels of government, private investors and environmental groups (Hinojosa-Ojeda, 1994; Mumme, 1995).

In summary, the different paths of continental integration - a political union in Europe, and an expanded free trade regime in North America - have had strong impacts on institution-building in the border regions. The supranational union introduced border-spanning sub-national institutions with comprehensive tasks or policy goals - in some parts of Europe with marginal influence of the nation states. While the regime building process in North America created very comprehensive, but still nation-state-dominated institutions, with much narrower tasks, mainly assessing and financing of infrastructure. The indirect induction effect is very similar on both continents: the emergence and vitalisation of a broad variety of trans-national and sub-national cross-border networks with strong public-private co-operation.

Influence of the vertical intergovernmental arena

In all modern states international relations are the responsibility of the federal or national level of government (Michelmann and Soldatos, 1990; Hocking, 1993a: 48). Nevertheless, participation of sub-national units in international negotiations in fields where these sub-national units have po-licy responsibilities is very common, as are many cross-border contacts of these units. Nevertheless, such contacts had been limited to rather non-political issues and were very informal in nature (Duchacek, 1986: 241). General decentralisation trends during the 1980s provided incentives, capacities and opportunities for stronger international activities of the sub-national units, and as part of this development, a more systematic and strate-gic approach to cross-border co-operation evolved (Hocking, 1993b). New alignments between the different layers of government brought new incentives and opportunities almost everywhere.

Apart from these general trends - which all promote enhanced cross-border co-operation on a sub-national, regional level - there are many examples of direct influence between vertical intergovernmental relations and cross-border co-operation. A common phenomenon in cross-border negotiations is that the main dispute is not across the national border but between the different governments on one side of the border. A prominent example was the conflict in Canada over the Columbia River Treaty (Swainson, 1979); there are many more (Blatter, 1994a). A recent example was the blockade of the German-French agreement on local cross-border co-operation because of a dispute between the federal and the state gov-ernments in Germany as to who has the responsibility and authority to sign such an agreement. Background elements of this dispute were the new alignments of Länder participation in European affairs and the signing of a similar agreement at the German-Belgium border without federal participation. Often, the competition of different governmental levels influences the form of the cross-border linkage. For example, at the end of the 1980s the inclusion of Liechtenstein and Graubünden in the Water Protection Commission of Lake Constance was done in a more informal way in order to keep out the German federal government. The new Cantons are working together now with the older members of the commission as equal partners, but the international treaty has not been modified (Scherer and Mueller, 1994: 59).

It is not just the federal-state arena that provides incentives and restrictions for cross-border co-operation or shapes these linkages; there are similar elements in the state-local arena. The formation, and especially the established membership, of the first governmental linkage at Lake Constance was mainly a result of the rivalry between the state government and the chief executive of a border county (Bullinger, 1977). Cross-border co-operation on a state-provincial level or on a Länder-Canton level is often disturbed by the resistance of local municipalities to implement the cross-border policies. One example is a rejected proposal for a sewage treatment plant in the British Columbian capital of Victoria in the same time as provincial and state governments started to work together to protect the common watershed of the Georgia Basin/Puget Sound.

Influence of the intra-state horizontal arena

Territorial cleavages within a nation state or a sub-national unit have always been used to justify different kinds of cross-border activity. The most extreme examples are found in separatist movements such as in Quebec, but recently Central and Eastern Europe have also witnessed a resurgence of secessionist activities. This kind of activity and interpretation is seen in most border regions as a "19th-century" approach[6] but such an interpretation is still very relevant in many capitals, especially in countries with fragile national cohesion. The consequence is suspicion or non-co-operative attitudes; but even in regions where secession is not really an issue, the feelings of inequitable treatment within the state, the notion of "hinterland" and "western alienation" (in the Pacific Northwest Region), "periphery" or "borderlands" (at the U.S.-Mexican border), "im Abseits" and "auf dem Abstellgleis" (in North-Eastern Switzerland) are strong motivational aspects for building coalitions across the border (see also Mlinar in this volume).

6 Charles Kelly, publisher of "The New Pacific", The Economist, May 21, 1994

That is even more so with respect to institution building, when there are new institution-building processes at the centre of the states. For the Ministerpräsident of Baden-Württemberg, German unification and the move of the government from Bonn to Berlin are serious threats to his south-western state, and this was why he initiated a great deal of activity with respect to his French, Swiss and Austrian neighbours. This is one reason for a lot of activity at the borders of Baden-Württemberg; but ironically we find exactly the same process one level lower now within Baden-Württemberg. Due to the building of a new regional district around the state capital, Stuttgart, the border regions with no equivalent institution fear even stronger discrimination in comparison to the agglomeration of Stuttgart. What makes this tension between capital and periphery even stronger is the history of Baden-Württemberg, which is the only German state to have been merged from two distinct states after Second World War - but against the strong resistance of many people in the smaller state. Now this conflict is showing up again because the territory of the former state of Baden is the borderland of Baden-Württemberg, and the agglomeration of Stuttgart is at the core of the former state of Württemberg. All this produces a strong motivation for cross-border coalition building, culminating in a study by the Chamber of Commerce of Northern Baden where a new common European state of Baden and the French Elsass was recently proposed.

There is a growing list of cross-border coalition building in North America due to more or less visible struggles and competitions between such regional coalitions for money and investment. It started with northern coalitions between U.S. and Canadian actors against southern water demands and is currently especially vibrant in the various transportation and trade corridor coalitions competing for NAFTA related funds dedicated to north-south infrastructure expansion. Beyond these issue-related coalitions, there are general integration - more precisely: inclusion - processes in this north-south direction between Canadian and United States sub-national units: the participation of Canadian provincial actors in American regional associations like the Western Legislative Conference (British Columbia and Alberta) and the Eastern Legislative Conference (Quebec). All these developments contribute to a significant change in territorial interest building processes. The traditional process of interest integration within the national arenas as a dominant first step followed by international negotiations will very often be replaced by a much more interwoven policy process based on a patchwork of different and often changing alliances of territorial units within and across national boundaries.

Influence of the inter-local arena

One major restriction for stronger cross-border institutions results from the desire of the sub-national participants not to give up any autonomy. Therefore, cross-border voluntary associations concentrate their activity on "positive-sum-games", but even here the co-operation is threatened because of local rivalry. Often, the most serious restriction in the formation of a cross-border alliance or association lies not in the nature of the border or in the differences across the border. Differences, competition and envy between the units on *one* side of the border are the dominant conflicts. In contrast to cross-border relations these units within a national sub-region of the trans-boundary region share a long history of interrelations and interactions. Contrary to some well-meaning advocates of increased social relations, the result is often not just a simple improvement of understanding, identification and brotherhood; more often it is a mixture of positive and negative feeling between groups. In the area of politics the competitive aspect can be strong, especially when there is an urban-rural cleavage within the participation region which is often mirrored in different party preferences. One example where such rivalries can be observed is Südbaden, where Freiburg, the dominant city, elected a Social Democrat as mayor and the surrounding counties are ruled by Christian conservatives. A similar situation can be found in Western Canada, where Alberta, proud of "rural values" and lead by a right-wing Premier is involved in a lot of battles with British Columbia, dominated by boom-town Vancouver and ruled by Social Democrats. In contrast to the cross-border arena, political actors of different parties compete against each other within a national territory. Very often co-operation across the border is easier than co-operation within the country.

Influence of the inter-sectoral arena

In most cross-border regions, the interactions have, until recently, focused on relatively narrow issues and on technical solutions. Technical experts in specialised departments found solutions within their responsibilities. The main disputes arose in benefit and cost sharing aspects across the border (for example, Swainson, 1979; Blatter, 1994b). Nowadays most cross-border issues have spill-overs in different sectoral policy arenas or various fields of departmental responsibilities. This makes the issue and the interactions much more complex. The field of water policy illustrates the developments and impacts on institution building very well. In almost all cross-border regions one finds old user-oriented international institutions acting in a relatively narrowly focused technical manner: commissions for stream flow regulation, shipping, water withdrawal and electrical power usage. As the problem definitions have changed during the last decades new protection oriented policies emerged and new institutions were built or new tasks were integrated into older institutions. There is an interesting difference between North America and Europe: In North America, the new tasks were more or less successfully integrated into the older institutions[7].

In contrast, the Europeans constituted new institutions to protect their shared water resources. Examples are the International Commission for the Protection of Lake Constance in 1960 and the International Commission for the Protection of the River Rhine against Pollution in 1963. Recently, with the NADBANK and the BECC, there are new institutions for one North American border but the tasks between these and the older institution are not defined as clearly as in Europe along the user-protection division line.

The new institutions can be seen as cross-border administrative focal points of younger interest coalitions in their attempt to introduce new measures in a policy field strongly occupied by users and their administrative regulatory bodies. Under the pressure of these new interest coalitions, older conflict lines within the users, and between the private users and the regulatory commissions lose significance. Hence, these older cross-border actors build own advocacy-coalitions[8] against the environmental advocacy-coalitions (see Blatter, 1994a). However, open conflict between the different advocacy-coalitions and their cross-border commissions is not the general feature. The younger cross-border institutions first concentrate their activities in fields where they can act on their own. Public sewage treatment plants are examples in the case of water protection commissions. After consolidating their power-base they get into fields with cross-sectoral impact. Most often on the cross-border or international level the protection oriented commissions make only

7 More successfully into the International Joint Commission (IJC) at the U.S.-Canadian border and less successfully into the International Boundary and Water Commission (IBWC) at the U.S.-Mexican border (Mumme, 1985: 632).

8 Sabatier (1991: 151-2) defines that an '... advocacy-coalition consists of actors from many public and private organisations at all levels of government who share a set of basic beliefs (policy goals plus causal and other perceptions) and who seek to manipulate the rules of various governmental institutions to achieve those goals over time.'

recommendations. The implementation of environmental policy goals in other policy fields (like agriculture or transport for example) - resulting in unavoidable conflicts - takes place within the national arenas and seldom on an international or cross-border level (Scherer and Mueller, 1994; Blatter, 1994b); an exception can be found in Blatter (1994a).

Such inter-sectoral conflicts are even more obvious in the process of creating cross-border regions on a sub-national level. Here the conflict can no longer be described as inter-sectoral interest conflict, as a much deeper cleavage shows up: contrary values and visions of what the new cross-border region is for. On one side there are the "free-traders", arguing for joint activities to get the critical mass for global competition and pushing mainly transport infrastructure developments. On the other side there are the "bioregionalists", defining the cross-border region on a watershed basis and favouring protection measures and growth management (Alper, 1996). Both sides agree that the cross-border region is a political arena of growing importance but they have totally different views on what this new region should look like and are fighting about the definition and the institutions of this emerging policy arena. Two illustrative examples: after the foundation of the economy oriented "Bodenseerat" at Lake Constance, environmental groups established a foundation called "Bodenseestiftung" to capture this name because they feared the Bodenseerat could have the same idea. The Pacific Northwest witnessed a harsh dispute between "bioregionalists" and "free-traders" over the usage of the name "Cascadia Institute" (Alper, 1996).

In summary, sectoral differentiation of the society and the administration has facilitated the co-operation across territorial boundaries; but at the same time, when interterritorial "epistemic communities"[9] strengthen their ties, the inter-sectoral gap between these communities gets wider. Not by accident, at the same time the very comprehensive phrase of "sustainable development", which includes economic, social and environmental goals, is receiving wider recognition. Whether this is the beginning of a new inter-sectoral integration movement, or has to be described as a "red herring" for the real conflicts, is still an open question.

---

9 Haas defines epistemic communities as '...a specific community of experts sharing a belief in a common set of cause-and-effect relationships as well as common values to which policies governing these relationships will be applied.' (Haas, 1989: 384).

Influence of the relationship between executive and legislative branch

The relationship between executive and legislative branches is very different in the political systems of western democracies. In parliamentary systems such as those found in Germany or in Canada, the governmental executive is in and of the legislature. In this "fused power model" (Rutan, 1981) almost no distinction can be found between the executive branch of government and the majority in the legislature. Where the executive is directly elected by the peoples as in the United States, in the Swiss Cantons or in the Southern German municipalities, there are two distinct, separate units, each with their own identities. In cross-border regions with different political systems, this may produce serious limitations in institutionalising linkages which go beyond intergovernmental relations. This was the case in the Pacific Northwest during the period from the beginning of the 1970s until the end of the 1980s, when attempts made by the Washington State legislature to establish formal linkages with the legislature of British Columbia failed, because the Premier of BC made it clear that such linkages would not fit into the BC political system (Rutan, 1981: 74). Yet because of the leadership and participation of legislatures in building the Pacific Northwest Economic Region, we now know that such differences are not absolute hurdles for more cross-border integration but - combined with competition between political parties - can be used to block the formation of linkages which erode the exclusivity of governmental connections.

Influence of the ideological arena through competition between political parties

In an example contrary to the last point, partisan competition, combined with a differentiated political system, can provide the incentives and opportunities to complete the web of cross-border linkages. In the Lake Constance area, all the governments are dominated by conservative parties, and neither the "Bodenseekonferenz" nor the "Bodenseerat" provides a platform for the opposition parties. That was one reason why a Social Democrat, as a member of the parliament of Baden-Württemberg, started to establish a meeting of members of the parliaments around the lake. The conservatives countered with a meeting of the presidents of the parliaments and with statements that stressed their opinion that the "Bodenseerat" is the "quasi-parliamentarian" voice of the Lake Constance region. Recently the Green Party started an initiative to install a "real parliamentarian body" for the common Euro-region: they want to create a cross-border association with delegates from the legislatures of the German counties, the Austrian Land of Vorarlberg and the Swiss Cantons.

Influence of the public-private relationships

One of the striking characteristics of recent cross-border activities is a strong private-sector influence. The Chambers of Commerce play a prominent or even dominant role in the Swiss, French and German "Regio" associations in the Upper Rhine Valley. The "Bodenseerat" was founded by 19 politicians, 5 scholars and 19 business members (Sund et al., 1992). The Pacific Northwest Economic Region has its own Private Council. The working groups have a private sector chair and a public sector co-chair. The private sector is expected to set the direction of the Working Groups.

In environmentally oriented cross-border co-operation, the institutional integration of NGOs has not gone as far. In most border regions the environmental groups build their own cross-border networks and institutions. They are integrated into the governmental cross-border activities through personal contact, hearings, workshops and similar procedural forms, but in commissions, boards and working groups are almost always exclusively bureaucrats and scientists. A lack of integration between administrative cross-border linkages and environmental NGOs is very characteristic in European cross-border regions (Scherer and Blatter, 1994: 13, 56). On this point, the BECC at the U.S.-Mexican border also provides the exception: the ten member Board of Directors is comprised of five members from each country, with two representing the federal environmental ministries, two the national section of the International Boundary and Water Commission, and the remaining six members drawn two apiece from border states, border municipalities and members of the public (Mumme, 1995a: 11). It seems that integration processes in the cross-border and the public-private dimension are going hand in hand, but with significant differences between the private groups. Until now, in most border regions we find a strong imbalance in the institutionalisation of cross-border regions with the consequence of institutionally marginalised social and environmental interests.

What does this institutional complexity mean for the discussion about regional "active spaces"? First, one can conclude that a differentiated political system provides much openness and many access points for new ideas and concepts. The dynamics of multidimensional institutional interrelations can be seen as a force towards creativity and stimulation of social learning. In border regions, where the internal political system is less differentiated and the cross-border contacts are monopolised by the nation state, such a creative institutional setting is missing. On the other hand, this complexity means a tremendous challenge for a positive management of openness. An important conclusion from this section is that such attempts for a better management have to start with a recognition of this complex institutional environment.

Institutional Design for Active Cross-Border Regions

Institutional questions are not the only relevant aspects for successful cross-border co-operation[10], the way of organisation and management of trans-border relations is also an important factor of influence. For successful cross-border co-operation it is important to differentiate co-operation structures in a special dimension: topics which are different in their policy-character should be treated by separate institutions. In the literature of political science you find various typologies of policies. Benz recently (1994) presented a typology which distinguishes four types of policies: regulation, production, distribution, re-distribution[11]. These policies differ significantly in respect to the probability of conflict and voluntary co-operation. Productive and distributive tasks with the expectation of positive results for both sides are easier to manage than problems with regulatory or re-distributive character. Examples of the former are transportation facilities where positive effects are expected on either side of the border - even more so if there exists a governmental fund dedicated for such investments. An example of the latter is the regulation of trans-border water usage, which represents a "zero-sum-game". If the regulation enlarges user options on

10 For comprehensive recommendations for better practice of cross-border co-operation see: Scherer and Blatter, 1994.

11 Benz (1994) defines production and regulation as tasks of co-ordination; distribution and re-distribution are characterised by their focus on gains and losses. Only the first kind of tasks, production, contains a constellation with congruent goals among the participants, the main problem being to avoid the "free-rider-problem". Distribution is used here as defined by Benz close to the original definition of Lowi (1972) as a "positive-sum-game" where local actors can get money from central governments through co-operation. In the border regions this kind of policy was established throught the EU INTERREG programs.

one side of the border and reduces them on the other side, the policy gets a re-distribution character. In the case of constructing transportation facilities it is obvious, that in "productive tasks" distributional questions also often arise: who profits how much from the project and who pays for which parts? The dominant perception of the problem or project is crucial: is the common profit in the foreground or does the question of distribution dominate?[12]

In consequence, conflicting problems should be separated organisationally from co-operation projects where consensus is easily reached. Separation is necessary since one needs different institutions and mechanisms to solve these unequal problems in an effective manner. "Productive chances" can be exploited best with flexible networks, but in order to manage very conflicting themes, strong and formal institutions with clear and explicit rules and norms are necessary. This means the institutionalisation of cross-border co-operation should be based on two different, often antagonistic, organisational concepts which should be optimised separately in respect to their specific functions and should work together in a synergetic way.

The first organisational concept is characterised by the words networks and "la géométrie variable". Networks and "la géométrie variable" are the key words for adequate organisational forms to realise the chances of productive co-operation projects. Networks are characterised by a loose coupling between the actors. In networks one finds rather informal contacts; moreover the actors work together on a basis of equality, partnership, and trust. Informal networks open the opportunity to integrate actors from different subsystems and therefore, provide one solution to the above delineated problem of multiple institutional interdependence. Since few formal rules exist, networks can be composed according to specific issues. Furthermore, they can be activated and changed very pragmatically. The autonomy of its members is not very restricted. This is crucial for cross-border co-operation due to the unwillingness of the involved states to give up sovereignty - but it also means that only such policies or projects can be implemented when all participants are interested in a change. The rationale for networks does not imply that one does not need formal agreements,

12 Because the dominant perception is the crucial point, one strategy for better cross-border co-operation is to stress the common interests and values, but it is not always possible to change the perceived character of a task from a "zero-sum-game" to a "positive-sum-game".

contracts and organisations in these cases, but they should be part of the final stages of the co-operation process. Co-operation in some regions - for example the Upper Rhine Region - was dominated by relatively formal and rigid organisational structures. Hence, they often lacked creative ideas and motivated actors (Blatter, 1994b).

"La géométrie variable" is used here as an opposing concept to clearly defined and separated sovereign territorial units, such as modern towns and states of the past nineteenth and twentieth centuries (Scherer and Blatter, 1994). The territorial impact of problems varies widely between the different topics and can change over time as can be seen in the case of public transport. Therefore, today you find a broad variety of co-operation organisations and mechanisms. They have different but sometimes overlapping territorial or sectoral areas of responsibilities and have individual membership arrangements. In spite of this rationale for variable and flexible networks, which goes along with the present trend in organisational theory and management, this is not enough to solve complicated and conflicting cross-border problems in a sufficient way. You also need rather formal institutions in the sense of "international regimes"[13] (Krasner, 1983; Rittberger, 1993). Here, not immediate mutual benefit but commonly accepted, clear and explicit principles, norms, rules and procedures are the base for co-operative habits. These institutions should be constituted as strongly or "hierarchically" as possible. This means that they should be established by an international treaty on a high governmental level and that their national participants should have the obligation to follow their decisions and to implement their programs; moreover, it means that a conflict solving mechanism should be integrated in the international treaty which is not based on unanimity but on a neutral judge. Even if not used, this mechanism will help as a "shadow of hierarchy" (Scharpf, 1997) to make complicated negotiations continue[14]. These recommendations are dedicated to the long established border commissions at a national or state level. Such a helpful "shadow of hierarchy" is not only reached through a stronger international institution, another option is international law and the national

13 "Regimes" can be defined as a set of implicit or explicit principles, norms, rules and decision-making procedures around which actors' expectations converge in a given arena of international relations (Krasner 1983: 2).

14 Alternatively, such a "hierarchical" institution can play a mediating role in cross-border negotiations between decentral units, as it was the case with the International Joint Commission in the Skagit River Treaty case (Alper and Monahan, 1986).

courts. The opening of national court systems for lawsuits by foreigners functions as incentive for co-operation across national boundaries. Widely publicised examples are the agreements between the Dutch municipality of Rotterdam and several chemical firms in Germany, France and Switzerland. These agreements under private law committed the companies to reduce their pollution, Rotterdam on the other hand dropped its liability claims and did not file a lawsuit (Bernauer and Moser, 1996).

There are also good arguments for more formal cross-border institutions at a local or regional level. The best possibility to enhance a common identity and orientation, to inform the public and to legitimate cross-border politics is to establish a parliamentarian platform with representatives from the local or regional parliaments as in the Euregion at the German-Dutch border in Gronau. Overall, the questions of legitimacy and participation should not be underestimated. In regions, where this has not been taken into account, there is permanent trouble and the acceptance of cross-border co-operation suffers.

A broad and differentiated variety of co-operation forms and mechanisms is a compelling necessity to handle cross-border problems and projects in an adequate way. This is still right, and must be accepted and even promoted although the problems of complex structures such as confusion and redundancy[15] have to be acknowledged. At Lake Constance, for example, there is a broad variety of institutions in different policy sectors, with different administrative levels involved, with participants from different subsystems of the political-administrative system and working on different legal bases. These institutions can be described as:

- Conference of the national ministers for environment.
- German-Austrian commission for spatial planning.
- German-Swiss commission for spatial planning.
- International commission for navigation and shipping on Lake Constance.
- International commission for protection of Lake Constance against pollution.
- International fishery commission.

---

15 Since many cross-border regions show a multiplicity of cross-border institutions there are often complaints about double work and ineffectiveness. Grabher (1994) has shown how important "slack" or "redundancy" is for successful regional development.

- Conference of the heads of government of the Länder and Cantons around Lake Constance (IBK).
- Permanent office of the IBK for public information and co-ordination.
- Bodenseetreffen der Parlamentspräsidenten: regular meetings of the presidents of the sub-national parliaments around Lake Constance.
- Regular meetings of members of parliament from around Lake Constance.
- Advisory council for the implementation of the EU INTERREG program "Alpenrhein-Bodensee-Hochrhein".
- Private association with high-ranking members from politics, business and universities.
- Association of the environmental groups around the lake.
- Foundation for environmental projects in the Lake Constance region.
- Association of the municipal water utilities in the Lake Constance region and Upper Rhine Valley, and
- Conference of the municipalities of Konstanz, Kreuzlingen and Tägerwilen.

Structural complexity, not one comprehensive institution for cross-border co-operation is the key for successful cross-border co-operation. Nevertheless, on the concrete operational level handy projects are much more likely to get implemented than comprehensive, big programs. In cross-border co-operation the complexity increases immediately because of the involvement of actors at least in the horizontal and vertical dimension (and often also in other dimensions, see *Figure 13.1*). Therefore, to get something done, small encapsulated projects are needed. If such pilot projects are successfully implemented, they will provide an impulse to continue. An example, which shows this point in an impressive way, is the comparative analysis of cross-border co-operation on public transport in the Regio Basiliensis (Upper Rhine region) and the Lake Constance region (see Pötsch, 1994; Schnell, 1994). While in the Lake Constance region a transborder train was established in no more than twenty month from the first contacts, in the Regio Basiliensis more than ten years were necessary to produce first results. The opening of the first line of the proposed Regio-S-Bahn in June 1997 was only possible once the comprehensive, trilateral approach was dismantled into single bilateral projects. The slow pace is basically due to organisational problems in the Regio-S-Bahn: a too complex institutional structure and too many actors involved.

To reduce complexity in cross-border co-operation to a workable size means that the focus of co-operation often has to be very narrow. There is a great danger that attempts to bridge territorial barriers are accompanied by the widening of cleavages between sectors (between environmental and economic goals and actors for example, for a detailed discussion of this problem, see Blatter, 2000b). As Ron Mader (1998), a profound observer of the US-Mexican border activities puts it: 'I see only business people talking with business people, scientists with scientists and park directors with park directors'. In the same way, the network-approach has to be complemented by formal institution-building, the pragmatic project-approach should be embedded in an open and comprehensive discussion on the future of the common region resulting in a development program that provides orientation, in the sense of priorities and restrictions, and not only a list of wishes as it is often the case. Here also "complementary" is the prescription that makes sure that cross-border regions are "active spaces" for the difficult way to a more sustainable global village.

Conclusion

In a world characterised by simultaneous trends toward globalisation and regionalisation national boundaries are loosing significance. Border regions have to adjust to these trends. Ideas and paradigms serve as stimulators and focal points for collective action within a cross-border region. Nevertheless, to be "active spaces" there is a need for governance and institutions in these regions. As a first step towards realistic and fruitful recommendations for institution building in cross-border regions one has to understand the institution building processes in complex institutional environments. Political institution building processes can only be explained through a closer look at interrelationships of the cross-border arena with other political arenas. This was done in this chapter with examples from Europe and North America. A first conclusion was that a differentiated political system provides much openness and many access points for new ideas and concepts. The dynamics of multidimensional institutional interrelations can be seen as a force towards creativity and stimulation of social learning. In border regions where the internal political system is less differentiated and the cross-border contacts are monopolised by the nation state such a creative institutional setting is missing. On the other hand, this complexity means a tremendous challenge for a positive management of openness. Based on these insights the last section offered some recommendations for institution building in cross-border regions. Since neither informal and decentral networks nor central and formalised regimes alone can cope with all problems and tasks, a broad and differentiated variety of co-operation forms and mechanisms is a compelling necessity. A pragmatic project-approach has to be accompanied by an open discussion about the goals and directions of the common region. The basic feature of modern societies, complementary differentiation, is the only realistic prescription for border regions to cope with the latest turn of the ongoing process of modernisation.

Not as obvious is how to avoid the negative consequences of those processes. It seems that the most serious price one has to pay for better interterritorial co-operation is the deepening of the cleavages between sectoral units and communities. Sectoral segmentation and departmentalisation is not only the basic characterisation of institution building on an international level, it is also a basic feature in many cross-border regions. Further research is necessary to get a better overview of how border regions deal with the problem of co-ordination and integration of the various cross-border communities and institutions, and of which strategies are the most successful.

Acknowledgement

I would like to thank Prof. Rod Dobell, Prof. Norris Clement and Justin Longo as well as Prof. Remigio Ratti and Prof. Marina van Geenhuizen for their help. All remaining mistakes are of course my responsibility. A further thankyou belongs to the Studienstiftung des Deutschen Volkes, Bonn, and the Gottfried Daimler und Karl Benz Stiftung, Ladenburg, for their financial support.

References

Alper, D. and Monahan, R. L. (1986), 'Regional trans-boundary negotiations leading to the Skagit river treaty: Analysis and future application', *Canadian Public Policy*, vol. XII, 1, pp. 163-174.

Alper, D. K. (1996), 'The idea of Cascadia: Emergent Trans-border Regionalism in the Pacific Northwest-Western Canada', *Journal of Borderland Scholars*, vol. 11, 2, pp.1-22.

Beck, J. (1997), *Netzwerke in der transnationalen Regionalpolitik. Rahmenbedingungen, Funktionsweisen, Folgen,* Nomos, Baden-Baden.

Benz, A. (1992), 'Mehrebenen-Verflechtung: Verhandlungsprozesse in verbundenen Entscheidungsarenen', in: A. Benz, F. W. Scharpf and R. Zintl (eds), *Horizontale*

Politikverflechtung: zur Theorie von *Verhandlungssysteme,* Campus Verlag, Frankfurt/Main, New York, pp. 147-205.

Benz, A. (1994), *Kooperative Verwaltung. Funktionen, Voraussetzungen und Folgen,* Nomos Verlag, Baden-Baden.

Bernauer, Th. and Moser, P. (1996), 'Reducing Pollution of the River Rhine: The Influence of International Cooperation', *Journal of Environment & Development,* vol. 5, 4, pp. 389-415.

Blatter, J. (1994a), *Erfolgsbedingungen grenzüberschreitender Zusammenarbeit im Umweltschutz. Das Beispiel Gewässerschutz am Bodensee,* EURES-discussion paper 37, EURES, Freiburg i.Brsg.

Blatter, J. (1994b), *Erfolgsbedingungen grenzüberschreitender Zusammenarbeit im Umweltschutz. Das Beispiel Gewässer- und Auenschutz am Oberrhein,* EURES-discussion paper 43, EURES, Freiburg i.Brsg.

Blatter, J. (1996), 'Cross-border Co-operation - Development and Organization', in J. Skrbec (ed), *Sbornik prispevku z II. Mezinarodni konference Vyvoj a rizeni Ceske ekonomiky v obdovi transformace,* University of Liberec, Liberec (CR), pp. 18-31.

Blatter, J. (2000a), *Entgrenzung der Staatenwelt? Politische Institutionenbildung in grenzüberschreitenden Regionen in Europa und Nordamerika.* Nomos, Baden-Baden.

Blatter, J. (2000b), 'Cross-border Cooperation and Sustainable Development in Europe and North America', in P. Ganster (ed), *Border Regions in Transition,* San Diego State University Press, San Diego (forthcoming).

Bullinger, D. (1977), *Grenzüberschreitende Zusammenarbeit in der Regionalpolitik. Theoretische Ansätze und ihre Bedeutung für das Bodenseegebiet,* (Diploma thesis), Universität Konstanz, Konstanz.

Camagni, R. (ed.) (1991), *Innovation Networks: Spatial Perspectives,* Belhaven Press, London.

Duchacek, I. D. (1986), *The Territorial Dimension of Politics. Within, Among, and Across Nations,* Westview Press, Bolder and London.

Duchacek, I. D. (1990), 'Perforated Sovereignties: Towards a Typology of New Actors in International Relations', in H.J. Michelmann and P. Soldatos (eds), *Federalism and International Relations. The Role of Sub-national Units,* Clarendon Press, Oxford. pp. 1-32.

Geenhuizen, M. van, and Ratti, R. (1998), 'Managing Openness in Transport and Regional Development: An Active Space Approach', in K. Button, P. Nijkamp and H. Priemus (eds), *Transport Networks in Europe. Concepts, Analysis and Policies,* Edward Elgar, Cheltenham, pp. 84-102.

Goldstein, J. and Keohane, R. O. (eds) (1993), *Ideas and Foreign Policy: Beliefs, institutions and political change,* Cornell University Press, Ithaca and London.

Grabher, G. (1994), *Lob der Verschwendung: Redundanz in der Regionalentwicklung - ein sozioökonomisches Plädoyer,* Edition Sigma, Berlin.

Haas, P. M. (1989), 'Do regimes matter? Epistemic Communities and Mediterranean Pollution Control', *International Organization,* 43, 3, Summer 1989, pp. 377-403.

Hinojosa-Ojeda, R. (1994), 'The North American Development Bank. Forging New Directions in Regional Integration Policy', *Journal of the American Planning Association,* vol. 60, 3, Summer 1994, pp. 301-304.

Hocking, B. (1993a), *Localizing Foreign Policy: Non-central Governments and Multilayered Diplomacy,* The MacMillan Press, London and New York.

Hocking, B. (1993b), 'Managing Foreign Relations in Federal States: Linking Central and Non-Central International Interests', in B. Hocking (ed.), *Foreign Relations and Federal States*, Leicester University Press, London and New York, pp. 68-89.

Ingram, H. and Fiederlein, S. L. (1988), 'Traversing boundaries: a public policy approach to the analysis of foreign policy', *Western Political Quarterly*, vol. 41, 4, pp. 725-745.

Krasner, S. D. (ed) (1983), *International Regimes*, Cornell University Press, Ithaca and London.

Lowi, T. (1972) 'Four Systems of Policy, Politics and Choice', *Public Administration Review*, vol. 32, pp. 298-310.

Mader, R. Internet address: http://www.geocities.com/Silicon Valley/84668/index.html

Michelmann, H.J. and Soldatos, P. (eds) (1990), *Federalism and International Relations. The Role of Sub-national Units*, Clarendon Press, Oxford.

Mueller-Schnegg, H. (1994) *Grenzueberschreitende Zusammenarbeit in der Bodenseeregion. Bestandsaufnahme und Einschaetzung der Verflechtungen politisch-administrativer und organisierter privater Gruppierungen*, Rosch-Buch, Hallstadt.

Mumme, S.P. (1985), 'State Influence in Foreign Policy Making: Water Related Environmental Disputes Along the United States - Mexico Border', *Western Political Quarterly*, vol. 38, 4, pp. 620-640.

Mumme, S.P. (1995), The North American Commission for Environmental Cooperation and the United States-Mexican Border Region: The Case of Air and Water, in *Transboundary Resources Report*, Vol. 9, No. 2, Summer 1995. Albuquerque, New Mexico.

Ohmae, K. (1993), 'The Rise of the Region State', *Foreign Affairs*, vol. 72, pp. 78-87.

Pötsch, P. (1994) *Erfolgsbedingungen grenzüberschreitender Zusammenarbeit im Umweltschutz. Das Beispiel ÖPNV in der Oberrheinregion*. EURES-discussion paper 35. Freiburg: EURES.

Putnam, R.D. (1988), 'Diplomacy and domestic politics: the logic of two-level games', *International Organization*, vol. 42, 3, pp. 427-460.

Ratti, R. (1993), 'Strategies to Overcome Barriers: From Theorie to Practice', in R. Ratti and S. Reichman (eds), *Theory and Practice of Transborder Co-operation*, Helbig & Lichtenhahn, Basel, pp. 241-268.

Rittberger, V. (ed) (1993), *Regime Theory and International Relations*, Clarendon Press, Oxford.

Rutan, G. F. (1981), 'Legislative Interaction of a Canadian Province and an American State - Thoughts upon Sub-National Cross-Border Relations', *American Review of Canadian Studies*, vol. XI, 2, pp. 67-79.

Sabatier, P. A. (1991), 'Toward better theories of the policy process', *Political Science & Politics*, June 1991, pp. 147-156.

Scharpf, F.W. (1997), *Games Real Actors Play. Actor-Centred Institutionalism in Policy Research*, Westview Press, Boulder CO.

Scharpf, F.W.; Reissert, B. and Schnabel, F. (1976), *Politikverflechtung: Theorie und Empirie des kooperativen Föderalismus in der Bundesrepublik*, Campus, Kronberg/Ts.

Scherer, R. and Blatter, J. (1994), *Preconditions for successful cross-border co-operation on environmental issues. Research results and recommendations for a better practice*, EURES-discussion paper 46, EURES, Freiburg i. Brsg.
[http://www.unisg.ch/~siasr/people/bla.htm].

Scherer, R. and Müller, H. (1994), *Erfolgsbedingungen grenzüberschreitender Zusammenarbeit im Umweltschutz. Das Beispiel Bodenseeregion,* EURES-discussion paper 34, EURES, Freiburg i. Brsg.

Scherer, R.; Blatter, J. and Hey, Ch. (1994), *Preconditions for successful cross-border Cooperation on environmental issues. Historical, theoretical and analytical starting points,* EURES-discussion paper 45, EURES, Freiburg i. Brsg. [http://www.unisg.ch/~siasr/people /bla.htm].

Schnell, K.D. (1994), *Erfolgsbedingungen grenzüberschreitender Zusammenarbeit im Umweltschutz. Das Beispiel OePNV in der Bodenseeregion,* EURES-discussion paper 36, EURES, Freiburg i. Brsg.

Smith, P.J. (1993), 'Policy Phases, Sub-national Foreign Relations and Constituent Diplomacy in the United States and Canada: City, Provincial and State Global Activity in British Columbia and Washington', B. Hocking (ed), *Foreign Relations and Federal States,* Leicester University Press, London and New York, pp. 211-235.

Soldatos, P. (1990), 'An Explanatory Framework for the Study of Federated States as Foreign-policy Actors', in H.J. Michelmann and P. Soldatos (eds), *Federalism and International Relations. The Role of Sub-national Units,* Clarendon Press, Oxford, pp. 34-53.

Soldatos, P. (1993), 'Cascading Sub-national Paradiplomacy in an Interdependent and Transnational World', in D.M. Brown and E.H. Fry (eds), *States and Provinces in the International Economy,* Institute of Governmental Studies Press, Berkeley CA, pp. 65-92.

Sund, H., Maus, R. and Ritscherle, W. (eds) (1992), *Vom Bodenseeforum zum Bodenseerat,* Universitätsverlag Konstanz, Konstanz.

Swainson, N.A. (1979), *Conflict over the Columbia. The Canadian Background to an Historic Treaty,* McGill-Queen's University Press, Montreal.

14 Knowledge as a Crucial Resource in Policy Making for Mainport Rotterdam

MARINA VAN GEENHUIZEN AND PETER NIJKAMP

Access to information and synthesis of advanced knowledge form critical success conditions for a knowledge society. This also holds for regional development and infrastructure. Rotterdam exemplifies a region that benefits from a longstanding openness as a seaport serving large parts of Europe. Nowadays, Rotterdam is facing increased competition with seaports serving the same areas. An answer to this situation is to expand the port area, in such a way that new demands for land for port activity can immediately be satisfied. This policy is based on the old logic of mass transport and mass manufacturing. Critics prefer a structural shift to innovative activities focusing on added value, a policy line that requires an improved use of the learning capability in the region. This chapter shows that new policies including a radical turn in development may face resistance from traditional forces.

Knowledge as a Crucial Resource

The past decades have shown an unprecedented increase in the openness of regions towards the global world and, on a lower geographical scale, towards regions behind political borders. This development rests on the disappearance or change of character of many borders. In addition, information and communication technologies allow information to be exchanged almost everywhere where computers can be connected, and physical transport is less and less hindered by obstacles such as mountains and sea straits. An increased openness, however, does not automatically mean a higher living standard in terms of incomes, and environmental and social well-being (van Geenhuizen and Ratti, 1998). For example, regions may turn into transit areas and become seriously exposed to negative externalities of transport, or turn into a living-place for numerous immigrants who remain isolated in terms of employment and culture, from the greater society surrounding them. Openness also means exposure to competition from other regions and cities. As a result of the increased factor mobility and the weakening of national protective measures, sources of competitiveness are increasingly limited to regional endogenous strength: knowledge being one of the remaining localised resources.

When attracting new activities and investments, regions can follow two basic strategies, i.e. competing with low costs and competing with high value-added (Reich, 1997). The latter strategy is connected to higher incomes per capita than the former. Of course, in reality there is a mix with different accents. Competing with high value-added means competing with creativity, the best and latest information, the highest standards of production, and an easy access to resources and companies around the world (cf. Andersson, 1991; Kanter, 1995). Accordingly, this strategy is based upon the production and use of knowledge as an essential economic resource. Thus, the first reason for drawing attention to knowledge as the outcome of learning processes is the better performance of knowledge-based economic growth. A second reason is the structural change, as an autonomous development, of the economic base of many city regions in advanced economies. There has been a transformation of the economic base from commodity-based activities in the production sector to knowledge-based activities in the broader service-sector; and there has been a move from mass production to more flexible modes of production, the latter requiring greater variation in knowledge and knowledge applications. A third reason for paying attention to knowledge is that its use can offer potential solutions for environmental problems, in terms of application of new technology and new organisational formats. A fourth and distinct reason for paying attention to knowledge and underlying learning is the nature of policy making itself. There is an increased uncertainty in regional policy making, caused by factors such as policy outcomes in related policy fields, support of stakeholders involved and macro-economic developments. Policy making becomes a learning activity in itself by including methods of policy design that take uncertainty into account (e.g. Friend and Hickling, 1997). Knowledge and the underlying learning processes, therefore, need to be treated by regional and urban policymakers as an essential source of economic power and competitiveness, as well as a source of sustainable development (Knight, 1995; Lambooy, 1997; van Geenhuizen and Nijkamp, 1998; van Geenhuizen and Ratti, 1998). Accordingly, it is a major challenge to formulate policies and strategies for enhancing and valorising knowledge cultures and transforming knowledge into economic development, which at the same time have a focus on sustainable growth.

In this chapter we draw on the "active space" approach to regional development (see, Ratti this volume) by elaborating the concept of learning capability. In the empirical part, we discuss Rotterdam in the Netherlands as an example of a region in a highly competitive environment. Further, we evaluate some policy responses to this competition according to the paradigm of "active space" development.

Regional Active Space Development and Learning Capability

The key process in regional "active space" development is creative learning. Creative learning enables pro-active behaviour among companies and policy makers when responding to internal and external changes, in terms of objectives, norms and particularly system-oriented strategic rules. The capability for creative learning rests on a variety of regional attributes, i.e. human capital in terms of skills, experiences, learning attitude, etc.; transactional relationships between manufacturers, suppliers and customers; informal contact networks; synergy effects between various knowledge actors such as universities and companies; and the output of serendipity and research and development. The learning capability of a region, economic sector, or the greater society, can be defined as the capability to create and attract new knowledge and to make use of this knowledge in a socially and economically efficient way. A full use of the learning capability in regional development requires the following essential activities and actions (Knight, 1995; Jin and Stough, 1998; van Geenhuizen and Nijkamp, 1998, 1999, 2000):

- *Building trust and reciprocity.* Trust between regional actors is a *conditio sine qua non* for learning as a bottom-up and consensus-based process. It has to do with the confidence that no party involved will exploit the vulnerabilities of the other. Reciprocity can be defined as a mutual understanding between parties that a given action will be returned in kind. Trust and reciprocity are important in maintaining effective information flow and co-operation within the networks involved.
- *Recognition of an urgent problem.* A sense of need for policy intervention is needed to find sufficient support for systematic and coherent policy

efforts to improve learning and to have it high ranked on the policy agenda.

- *Management of (public) stocks of knowledge.* Human resource management presupposes a keen knowledge management. This includes *inter alia* keeping the skills of the resident population up-to-date ("education permanente") and establishing and up-dating information systems that serve as support in policy decisions (DSS).

- *Advancement of knowledge creation and flow through networks and nodes.* Learning and new knowledge creation and transfer typically take place in networks of creators, users and intermediaries. Networking is important for advancing synergy (serendipity) between different actors and disciplines (e.g. Charles and Howells, 1992). Furthermore, networking is necessary to improve the integration of knowledge actors in the local community and to connect them with global knowledge actors. Networks need to be sufficiently transparent for outsiders to make use of them, and sufficiently flexible and open to accommodate an effective information flow and co-operation between the actors involved.

- *Transformation of knowledge.* There is a difference between scientific knowledge and knowledge understood by entrepreneurs, policy makers and most community members, and there is a difference in vocabulary and framework between scientific disciplines (e.g. Kamann, 1993). In order to smoothen interaction and make new knowledge more useful, knowledge is transformed in various ways such as by adapting the vocabulary and by increasing transparency.

- *Transmission of knowledge.* This includes formal education such as that provided by universities, higher educational institutes, and company schools. It also includes education in local history, economy and culture, and elaboration of regional crafts using informal channels. Clearly, access to knowledge disseminating institutions is a necessary condition.

On a higher level there is the self-organising power in the region. This ability becomes evident in reflection on the rules and norms that direct economic and social behaviour, using institutional memory and intelligence. In a situation of well-developed self-organising power, institutional monitoring and self-diagnosis are an embedded feature in society, along with continuous improvement which might be termed "learning by learning" (Camagni, 1991; Braczyk et al., 1998). It is particularly the quality of the networks, in the sense of incorporating enforceable social habits and routines, which underlies this learning (Storper, 1993).

The fact that the above range of activities is nowhere fully developed, can be ascribed to the comprehensive nature and complexity of learning and complexity inherent in knowledge policies. A number of factors can be advanced to explain this situation (Morgan, 1997; van Geenhuizen and Nijkamp, 1998), such as multiple actor and multiple role situation, multifaceted nature, changing setting of knowledge creation, measurement problems, multilayer policy (management) framework, a missing "problem owner" and absence of immediate policy results. These factors will be explained in the remaining section.

Many different actors are involved in learning and knowledge creation and use, such as universities and higher educational institutes, research institutes, consultant firms and think tanks, manufacturing and services firms, transfer institutes, brokers in network contacts and other intermediaries, and governments at different levels. These actors may have diverse or conflicting aims, such as improving the competitive edge, for firms, and creation of high-tech jobs, for local governments. In addition, particular actors simultaneously perform different roles. For example, universities are gradually moving to a multiple role organisation, including commercialisation of knowledge alongside the traditional tasks of research and education. In such a multiple actor and multiple role situation it is rather difficult (time-consuming) to gain sufficient support for particular policy decisions. A related cause of complexity is the multifaceted nature of learning capability, which includes aspects of science dynamics and serendipity, micro-economic behaviour of firms, sociology of clubs and informal networks, and economics of public finance, etc. Policy makers often have a monodisciplinary background, a situation that makes it difficult to cope with learning capability.

In the past few years, we have seen a gradual change in the setting of learning and knowledge creation and use. There is a shift from hierarchical, disciplinary and division of labour-based knowledge production to a mode in which research problems are set across disciplinary boundaries, with a strong focus on application and with new benchmark criteria such as flexibility and response time (Gibbons et al., 1994). At the same time, the number of actors involved is increasing aside from universities and established research centres with a growing emphasis on teams (consortia) working on a temporary basis. As a consequence, there is a trend for knowledge creation to become more volatile within the fast changing network configurations.

Aside from intrinsic complexity, learning and knowledge creation and use are also difficult to measure (OECD, 1996). For example, knowledge takes many different forms, such as embodied and disembodied knowledge, codified knowledge and tacit knowledge. The latter particularly seems important in innovation (e.g. von Hippel, 1994) but is difficult to map. It is also almost impossible to keep knowledge accounts, because input into knowledge creation and use is hard to measure and the impact of a "unit of knowledge" on economic performance is difficult to determine, assumed that one can determine a value for a unit of knowledge. There is no straightforward production function and no straightforward input-output relationship. Due to the absence of a uniform knowledge market, there is a lack of price information required to combine individual transactions into broader aggregates. Thus, knowledge cannot be measured using standard indicators and statistics unlike traditional goods and services.

Complexity also follows from the policy or management framework of regional learning and knowledge creation because it is essentially multi-layered. For example, a local government sets important local conditions for learning and innovation, such as making premises and accommodation available for high tech companies and making housing available for particular income groups. At the same time, public and private actors at higher spatial scale levels also influence learning capability on the local level to a considerable degree. For example, multinationals can decide to open or close down local laboratories and national governments can decide to increase or cut down university research budgets.

A further complicating factor in policy making is the fact that, despite the many actors involved and despite perhaps the gravity of the situation, there is often no clear "problem owner" for the task of improving learning capability. This means there is no clearly defined actor to push the issue of learning into the policy arena in a systematic and coherent way. In addition, learning policies only yield results in the medium term, not in the short term. Thus, when seeking support for learning policies, there tends to be competition from those socio-economic measures that yield immediate and clearly visible results, like job creation schemes and physical infrastructure improvement.

It can be concluded that learning and knowledge creation and use are difficult to understand due to actor complexity and systemic complexity and due to measurement problems. At the same time policy making faces a number of inherent difficulties. This situation hampers efforts when attempting to make solid policy decisions.

Rotterdam's Future: a Pressing Policy Question

Introduction

The "grandfather" of economics, Adam Smith, has already sung the advantages of locations in delta areas.

> He claims '... but those of a city, situated near either the sea-coast or the banks of a navigable river, are not necessarily confined to derive them from the country in the neighbourhood. They have a much wider range, and may draw them from the most remote corners of the world, either in exchange for the manufactured produce of their own industry or by performing the office of carriers between distant countries, and exchanging the produce of one for that of another. A city might in this manner grow up to great wealth and splendour.' (Smith, 1776: 406).

His hymn seems to apply to Rotterdam surprisingly well. Rotterdam is the largest port in a range of seaports along the European Channel and North Sea coasts, that starts in the South with Le Havre (France) and ends in the North with Hamburg (Germany). Rotterdam connects the North Sea to the river Rhine, the major waterway system in Western Europe which runs through Germany and France (tributaries) to Switzerland and has a length of approximately 1,250 km. Rotterdam is also connected to the Black Sea via the Rhine-Main-Danube waterway, but unlike the Rhine, this connection suffers from particular physical drawbacks such as a large number of locks. Looking back into history, it becomes clear that Rotterdam exemplifies a region of longstanding openness. The Treaty of Mannheim paved the way for the free use of the river Rhine in 1868 and the direct (open) waterway connecting Rotterdam to the North Sea was opened in 1872.

Seaport activity in Europe is currently undergoing change influenced by two factors. One, the completion of the single European market and the opening of Eastern European markets influence Rotterdam's market share. Here uncertainty is related to the pace of integration of Eastern Europe in the European and global economy and the ability of Rotterdam to capture growth in throughput of goods for Eastern Europe in competition with ports in Germany. Two, sea and land transport are becoming increasingly incorporated in complex networks of chains starting with raw material production and ending with delivery to customers, and in which speed and flexibility are crucial. Global sourcing and selling and just-in-time logistics require suppliers to ship smaller quantities more frequently and quickly over long distances, this also means that air cargo will play an increasingly important role in overseas transport (Kasarda, 1996). The globalisation of markets and new systems of internal and external logistics have led to a new use of networks and nodes in new spatial configurations (Priemus et al., 1995; Ratti, 1995). In addition, global shippers and carriers are becoming increasingly powerful players due to ongoing merging and acquisition (Janelle and Beuthe, 1997; van Klink and de Langen, 1999). These developments indicate that existing hinterlands are increasingly difficult to retain (van Klink and van Winden, 1999).

Throughput in the port of Rotterdam consists mainly of bulk goods. In terms of tonnage, approximately 75 per cent is dry and liquid bulk (*Table 14.1*). The main category of bulk is crude oil, with 32 per cent of all throughput. Port activities in the region of Rotterdam contribute 2.4 per cent to GNP in The Netherlands. When including multiplier effects, using a multiplier of 4.5, this amounts to approximately 10 per cent. Value added from port activities has slightly increased over the past five years, whereas employment within the port region has faced a structural decline. This decline amounts to 12 per cent in the past five years, and is the result of an ongoing mechanisation and automation drive, and an ongoing containerisation.

Table 14.1 Economic indicators of Rotterdam port activities

Indicator	1994	1998
Total throughput (million metric tons)	293,9	314,8
Of which:		
- Bulk (dry)	89,0	89,7
- Bulk (liquid / gas)	135,6	143,7
- General cargo	69,3	81,3
Employment (port-related activities) *	67,649	60,374
Direct gross value added of seaport complex		
(DFL billion) **	11,1	14,5
Idem, as share of Gross National Product	2.0%	2.4%

* In the region, including stevedores, storage and distribution, shipping, port-related manufacturing (e.g. petrochemical industry).
** Factor costs (current prices).

Source: Municipal Port Authority Rotterdam Databank.

Rotterdam has by far the leading position in the Le Havre- Hamburg range of ports with a good 40 per cent of all throughput (*Table 14.2*). Throughput in Rotterdam is however growing somewhat slower than in the second and third largest ports, Antwerp and Hamburg, i.e. by 7 per cent versus around 10 per cent. Le Havre and Wilhelmshaven are growing relatively rapidly but these had a smaller base level in 1994 compared with Rotterdam.

Table 14.2 Throughput in ports in the Le Havre-Hamburg range

Seaport	Total throughput in 1998		Change in 1994–1998 (%)
	Absolute *	Share (%)	
Rotterdam	314,8	40.2	+ 7.1
Antwerp	119,8	15.3	+ 9.4
Hamburg	75,8	9.7	+ 11.0
Le Havre	66,4	8.5	+ 22.0
Amsterdam **	55,8	7.1	+ 16.0
Wilhelmshaven	43,8	5.6	+ 25.5
Dunkirk	39,2	5.0	+ 5.7
Bremen	34,4	4.4	+ 11.3
Zeebrugge	33,3	4.3	+ 1.2
Totals	*783,3*	*100*	

* In million metric tons.
** North Sea Canal Region

Source: Municipal Port Authority Rotterdam Databank.

When considering the fast growing container sector, competition seems to be fierce (*Table 14.3*). Rotterdam holds the largest share in the Le Havre - Hamburg range (31 per cent) but three ports, including a relatively large one, are growing in significantly stronger manner than Rotterdam, i.e. Le Havre (51 per cent), Antwerp (48 per cent) and Felixtowe (43 per cent). The growth of the latter port, in the United Kingdom, indicates the development of a hub function which also partly serves the European mainland. Moreover, four southern European ports have increased in importance rapidly, i.e. Gioia Tauro in southern Italy (since 1995), Genua (106 per cent), Barcelona (81 per cent) and Algeciras (81 per cent). This pattern confirms that hinterlands are becoming increasingly difficult to demarcate, a situation of uncertainty that seriously impacts on policy for future port development.

Table 14.3 Throughput of containers in European container ports

Seaport	Throughput in 1998 Absolute *	Share (%)	Change in 1994 – 1998 (%)
Le Havre –			
Hamburg			
Rotterdam	6,011	31.2	+ 32.4
Hamburg	3,547	18.4	+ 30.0
Antwerp	3,266	17.0	+ 47.9
Felixtowe	2,500	13.0	+ 43.1
Bremen	1,826	9.5	+ 21.5
Le Havre	1,319	6.9	+ 51.1
Zeebrugge	776	4.0	+ 27.4
Totals	*19,245*	*100*	
Other Ports			
Gioia Tauro	2,126	-	Since 1995 in operation
Algeciras	1,812	-	+ 80.5
Genua	1,266	-	+ 105.9 **
Barcelona	1,095	-	+ 81.0
Valencia	1,005	-	+ 50.0

* Number x 1000 TEUs (Twenty Feet-Equivalent-Units).
** 1995-1998.

Source: Municipal Port Authority Rotterdam Databank.

Future policy for Rotterdam

Given the uncertainty of hinterland development the question arises as to what policy is it wise to use to improve the competitive position of Rotterdam, e.g. a policy to build additional port facilities, or a policy to use existing facilities better and move to more innovative (transport-related) activities. In 1993 the Municipal Port Authority of Rotterdam (MPA) made a case for extending the port with a "Second" Maasvlakte of 1,450 hectare gross area (1,000 hectare net) to be constructed in the open North Sea (MPA, 1993). In The Netherlands a decision regarding such a huge infrastructure work is considered to be of national importance and must therefore, be taken at the national level. The decision making process was still running in 1999. Opponents of the project point to the alternative option of growth of port activities at a number of sites within the old port boundaries, and to the potential for developing a network model of ports in which Rotterdam co-operates with existing smaller ports in the region, such as Vlissingen (van Klink, 1998). Moreover, some opponents question the ongoing growth of port activities as being largely traditional in that it is concerned with transit and storage. It has already been mentioned above that employment from port activities is falling in the region (van de Berg and van Klink, 1996; van Boven and Machielse, 1996). Growth in transport employment and added value is manifesting increasingly at higher spatial scales. Apart from the petro-chemical industry, most port activities have a focus on transit and storage leading to the generation of employment and value added in a much larger area than the Rotterdam region. Thus, from a perspective of the port as an economic "engine" for the Rotterdam region only there seem to be no pressing reasons for port expansion.

The above situation calls for new policy approaches to the economic development of Rotterdam which break with traditional paths to a certain degree. This is indicated in *Table 14.4* under "active space" development. In this development, the port will not necessarily be expanded. Knowledge will be used to improve flexibility and productivity per unit space and to manage transport chains in an intelligent way within larger networks. This trajectory clearly includes a turn in which the focus of policy content moves away from volumes to added value and the focus of policy design moves away from fixed goals based on trend analysis to learning policy with a strong adaptation to uncertainty. In contrast, in a regional "passive space" development new physical infrastructures accommodate increasingly larger flows and storage in order to benefit from scale economies. In this case the trajectory of the past is continued under the influence of established lobbies and networks. Coastal facilities will be increased but land-side transport may lag behind causing the risk of congestion and other negative externalities. The latter development is the one indicated by Grabher (1993) based upon established network ties that potentially blind the actors involved.

In the next section we will explore actual policy making for the regional economy of Rotterdam and show that the policy contains elements of both "passive" and "active space" development.

Table 14.4 Seaport development policy in terms of active space logic

"Passive space" development	"Active space" development
Content	**Content**
- Competition by additional supply of space and efficient transhipment	- Competition using knowledge and flexibility, and an increased productivity per unit space
- Economies of scale	- Economies of scope and speed
- Oriented on volumes (bulk)	- Oriented on value added
- Labour concern in terms of dealing with surplus of workers	- Labour as an asset, in terms of quality and flexibility
- Seaport activity as the main regional source of economic growth	- Seaport activity as a part of large scale value added logistic networks
- Modest concern for environmental and social sustainability	- Large concern for environmental and social sustainability
- Focus on hardware, i.e. physical infrastructure and premises	- Focus on software and orgware connected to a more efficient (innovative) use of physical infrastructure
Process	**Process**
- Dominant role of established lobbies and inward looking (blinding) networks	- Open orientation and alertness to change and uncertainty, incl. problem diagnosis and monitoring
- Trend analysis and forecasting, conservative scenario analysis	- Innovative scenario analysis and experiments
- Fixed goals and top-down approaches	- Adaptive (flexible) planning and participatory approaches

Progress and Constraints

The following elements of the economic policy for Rotterdam will be discussed: policy responses to the modest level of innovation in the economy and two different types of scenario approaches to future economic development.

Arriving at a diagnosis of a socio-economic problem fits an "active space" development in that it indicates a move to higher levels of learning. It serves to teach policy makers about crucial shortcomings in terms of seriousness and comprehensiveness of the situation, and in terms of causality and need for adaptation to changes. The problem in the Rotterdam economy is a relatively low level of innovation, in terms of research and development, innovation output and new firms development (van Boven and Machielse, 1996; van Geenhuizen and Nijkamp, 1998). In a causal analysis it became clear that the tissue of local networks underlying learning and innovation is sometimes weakly developed in Rotterdam (Bureau Bartels, 1996). There is a situation of hesitation for local co-operation, in some cases due to lack of trust and in other cases due to lack of mutual interest. Most knowledge institutes in Rotterdam are national and global in orientation, leading to little use of local learning opportunities and use in the local economy. A further bottleneck is a lack of coherence and co-ordination and, as a result, a lack of transparency in knowledge transfer. Many different actors and initiatives are involved, leading to a disjointed policy. Given such a diagnosis the next step is to find some solutions to the problems and to monitor developments, but improving trust and creating fertile local networks is a long-term effort without immediate results. One achievement in increasing co-ordination in transport-related research, however, stands out. This is discussed below.

In 1996 two managing (intermediary) organisations were established, one with a *disciplinary* approach (transport technology) and an emphasis on links between research institutes and knowledge users in the business world (CTT, 1996), and the other with a *regional* focus on knowledge transfer and use, in particular education (KMR, 1996). The former organisation was set up to initiate and co-ordinate research projects concerning innovation in transport activity, including land-side transport. The projects are partly financed by the Dutch government following the government's recognition that the competitive position of Rotterdam could be improved by a better use of knowledge (CROW, 1994; MEZ, 1995). So far, transport research in The Netherlands has been done in a fragmented way with a limited practical use of results. It has often been one-sided with a strong emphasis on technology and transport economics, and without a fundamental approach.

Nowadays, various organisational improvements are becoming evident, when compared to past approaches; first, the present research projects are increasingly based upon *public-private partnerships*, meaning joint financing by the national government (two-third) and companies (one-third). This model implies greater commitment from companies (demand driven) and also greater intent to use the newly developed knowledge. Secondly, present research projects have surpassed problems of individual companies and single transport modes, to arrive at more *integrative* frameworks. Additionally there also seems to be a change in attitudes with respect to the actors involved, i.e. a shift from reluctance to participate towards a constructive atmosphere of co-operation between universities, research institutes and transport and logistics companies. A recent development is the spatial concentration of the first mentioned intermediary (CTT), the PhD school of transport (TRAIL) and a test (simulation) laboratory at a location close to Delft University of Technology. Such a spatial concentration can be seen as favourable because it enables researchers to benefit from close personal interaction. More importantly perhaps, CTT (under the new name of CONNEKT) has recently increased its tasks by strengthening its links with the Ministry of Transport and Waterworks by incorporating a connected research bureau. Accordingly, there has been a shift from an intermediary position to a more focal position. No doubt, co-ordination and coherence have increased, but it remains a challenge for this strong actor to serve independent and basically innovative transport research and not to fall back to serving traditional research.

In the remaining section we will discuss the use of scenario analysis as a tool in policy design. In the case of Rotterdam, scenarios have been developed for port development only but also for the regional economy in a broader sense. With regard to port development the scenarios served to indicate land shortage for port development under different circumstances (*Table 14.5*). Remarkably, the Central Planning Bureau of the Netherlands (CPB) produced lower estimates of land shortage compared to the Municipal Port Authority (MPA). The difference can be explained by a lower estimation of the Rotterdam share in total throughput of all ports in Northwest Europe, and a higher estimation of increase in productivity per unit land and re-use of vacant land in the port of Rotterdam.

Table 14.5 Shortage of land in Rotterdam in 2020 in three scenarios

	Global Change	European Co-ordination	Divided Europe
Exploring 2020 (MPA 1998)	1, 260 ha	750 ha	0
CPB Work Document (CPB 1997)	610 ha	370 ha	0

Source: PMR 1999.

We should note that both scenarios are relatively conservative because they do not basically question the future role of the port. There is no alternative scenario in which there is a surplus of port area on the old land that can be used for intelligent and innovative (transport-related) activity. The driving forces identified are all external, such as world economic development, international economic relations and demographic developments, and reflect moderate situations. There are no "bold" driving forces like fierce dematerialization or, from a different perspective, dominating Mediterranean ports.

In contrast, the scenario analysis for the broader regional economy contains a number of "bold" elements (*Table14.6*). Such an approach to the future conforms to the specific function of the scenario analysis chosen by the municipality here, namely to learn about future uncertainty and to be able to pick signals of change. Two driving forces have been recognised, i.e. the geopolitical and trade development leading to a further globalisation of the economy or, conversely, to a retreat into trade blocks and small units, and policy forces leading to a slow and sticky process of decision-making or, conversely, to fast and efficient decision-making (Machielse and Verkennis, 1996). It is clear that two scenarios encompass turns in the future of Rotterdam in that the port has lost its important position, i.e. Rotterdam Disconnected and Rotterdam Talented. Such a development is based upon economic fragmentation and isolation. Only in the latter scenario is there compensation in terms of strong innovative activity. The latter two scenarios give policy makers the opportunity to creative thinking placed in new frames of reference.

At the same time, it needs to be realised that when it comes to a translation of rather innovative scenario thinking into operational policy in practice, there is a danger of falling back on fixed paradigms, old success stories and well-known networks. The reason is that policy making organisations have not learned to deal with scenario results that break with current trends and produce open ends (van Geenhuizen et al., 1998). This experience in Rotterdam illustrates that the entire municipal organisation needs to offer conditions that facilitate learning in a creative way, not only those sections or departments with the specific task to develop innovative approaches.

Discussion

There is no doubt that regional economies can only increase in competitiveness by improving the use of their learning capability. The seaport of Rotterdam in The Netherlands is subject to competition in terms of amounts of throughput, particularly in the container sector.

Table 14.6 Exploratory scenarios for the Rotterdam economy (2015)

	Rotterdam Global	Rotterdam Connected	Rotterdam Disconnected	Rotterdam Talented
Role of Rotterdam port	Global player, largely based upon knowledge	Important player reacting on developments elsewhere	Small role in highly fragmented world	Small role of port
Basis of competitive advantage	Better learning capability Scale, scope and speed	Traditional factors Scale, speed and service	Not relevant	Full use of learning capability
Overall orientation (values)	Competition and innovation Primacy of the market	Traditional Innovation hampered by need for broad policy support Solidarity and justice	Self-organisation and social cohesion Informal economy	Self-realisation Creativity Rewards for talent Culture Experiments
Network position in the world	Central position Strong co-operation	Central position Dependence on multinationals	Not central Firms have a local orientation	Not central Intern. firms connect with regional value-added networks
Relation port and urban economy	Highly connected	Disconnected Local orientation of urban economy	Not relevant	Not relevant
Government focus	Orgware Knowledge Competence and networks	Hardware Land and infrastructure	Not relevant No separation informal-formal economy	Software Human skills as major resource

Source: Adapted from Machielse and Verkennis (1996).

Given the fact that both employment and added value in transport and port-related industry are increasingly created outside the region of Rotterdam, an innovative answer, meaning a partial turn in development path, needs to be found for this situation.

The attention in this chapter has focused on promising changes in policy making in this context, i.e. the creation of a powerful co-ordinating actor in knowledge networks and the use of scenario thinking for the future economic development of Rotterdam. An important finding is that if there is solid progress in one particular respect, conservative forces may prevent further basic innovation and cause traditional developments to continue. This may be due to the role of established power networks and lobbies, but also, simply, to the fact that most actors in policy making have not learned to deal with uncertainty, i.e. with learning approaches. Here lies an important field for further research. In terms of an "active space" development it includes identifying those conditions required to come to an acceptance of uncertainty as a given fact in policy making and a broad acceptance of policy making tools that match with uncertainty. There is also a need, in a wider sense, for finding ways to design institutions that serve a learning attitude in all segments of society.

References

Andersson, A.E. (1991), 'Creation, Innovation and Diffusion of Knowledge - General and Specific Economic Impacts', *Sistemi Urbani*, 3, pp. 5-28.

Braczyk, H-J, Cooke, P. and Heidenreich, M. (eds) (1998), *Regional Innovation Systems*, UCL Press, London.

Berg, L. van der, and Klink, A. van (1996), 'Beyond the limits of a Mainport' (in Dutch), *Economisch-Statistische Berichten*, 28-2-1996, pp. 180-184.

Boven, J. van, and Machielse, K. (1996), 'Dealing with uncertainty in policy' (in Dutch), in F. Boekema, D.J.F. Kamann and W. de Graaff (eds), *Local Economic Policy. New ideas in spatial research in the Netherlands*, Geo Pers, Groningen, pp. 91-109.

Bureau Bartels (1996), *Improving Trust: A Strategy for Knowledge and Innovation for the region of Rotterdam* (in Dutch), Bureau Bartels, Utrecht.

Camagni, R. (ed) (1991), *Innovation networks: spatial perspectives*, Belhaven Press, London.

Charles, D. and Howells, J. (1992), *Technology Transfer in Europe. Public and Private Networks*, Belhaven Press, London.

CTT (Foundation Centre for Transport Technology) *Annual Report* (various years) (in Dutch), Foundation CTT, Rotterdam.

CROW (1994), *Towards a new research program on traffic, transport and infrastructure* (in Dutch), Foundation CROW, Ede.

Friend, J. and Hickling, A. (1997), *Planning Under Pressure. The Strategic Choice Approach,* Butterworth-Heinemann, Oxford.

Geenhuizen, M. van, Nijkamp, P. and Rijckenberg, H. (1997), 'Universities and Knowledge-Based Economic Growth: The Case of Delft', *GeoJournal,* 41 (4), pp. 369-377.

Geenhuizen, M. van, and Nijkamp, P. (1998), 'Improving the Knowledge Capability of Cities: The Case of Mainport Rotterdam', *International Journal of Technology Management,* 15 (6/7), pp. 691-709.

Geenhuizen, M. van, and Nijkamp, P. (1999), 'Regional Policy beyond 2000: Learning as Device', *European Spatial Research and Policy,* vol. 6(2), pp. 7-20.

Geenhuizen, M. van, and Nijkamp, P. (2000), The Learning Capability of Regions: Patterns and Policies, in F. Boekema, K. Morgan, S. Bakkers and R. Rutten (eds), *Knowledge, Innovation and Economic Growth. The Theory and Practice of Learning Regions,* Edward Elgar, Cheltenham, pp. 38-56.

Geenhuizen, M. van, and R. Ratti (1998), 'Managing Openness in Transport and regional Development: An Active Space Approach', in K. Button, P. Nijkamp and H. Priemus (eds), *Transport Networks in Europe, Concepts, Analysis and Policies,* Edward Elgar, London, pp. 84-102.

Geenhuizen, M. van, Zuylen, H. van, and Nijkamp, P. (1998), *Limits to Predictability in Traffic and Transport,* Ministry of Traffic and Water Management, Advisory Council, The Hague.

Gibbons, M., Limoges, C., Nowotny, H., Schwartzmann, S., Trow, M., and Scott. P. (1994), *The New Production of Knowledge: The Dynamics of Science and Research in Contemporary Societies,* Sage Publishers London.

Grabher, G. (ed), *The embedded firm. On the socioeconomics of industrial networks,* Routledge, London.

Hippel, E. von (1994), 'Sticky information and the locus of problem solving: implications for innovation', *Management Science,* 1994, pp. 429-439.

Jannele, D. and Beuthe, M. (1997), 'Globalization and Research Issues in Transportation', *Transport Geography,* 5 (3), pp. 199-206.

Jin, D.J. and Stough, R.R. (1998), 'Learning and learning capability in the Fordist and post Fordist age: an integrative framework', *Environment and Planning A,* 30, pp. 1255-1278.

Kamann, D.J.F. (1993), 'Bottlenecks, Barriers and Networks of Actors', in R. Ratti and S. Reichman (eds), *Theory and Practice of Transborder Cooperation,* Helbing & Lichtenhahn, Basel, pp. 65-101.

Kanter, R.M. (1995), *World Class. Thriving Locally in the Gobal Economy,* Simon & Schuster, New York.

Kasarda, J. (1996), 'Transportation Infrastructure for Competitive Success', *Transportation Quarterly,* 50 (1), pp. 35-50.

Klink, H.A. van (1998), 'The port network as a new stage in port development: the case of Rotterdam', *Environment and Planning A,* 30, pp. 143-160.

Klink, H.A. van, and Langen, P. de (1999), *Scale and Scope in Mainport Rotterdam* (in Dutch), ETECA/Erasmus University, Rotterdam.

Klink, H.A. van, and Winden, W. van (1999), 'Knowledge as a new factor in hinterland competition: the case of Rotterdam', *Tijdschrift Vervoerswetenschap,* 1999, 1, pp. 41-50.

KMR (Foundation Kennisinfrastructuur Mainport Rotterdam) (1996), *Annual Report 1995* (in Dutch), KMR, Rotterdam.

Knight, R.V. (1995), 'Knowledge-based Development: Policy and Planning Implications for Cities', *Urban Studies*, 32, pp. 225-260.

Lambooy, J.G. (1997), 'Knowledge production, organisation and agglomeration economies', *GeoJournal*, 41 (4), pp. 293-300.

Machielse, K., and Verkennis, A. (1996), 'Four Times Rotterdam' (in Dutch), *Economisch Statistische Berichten*, 4-12-1996, pp. 988-991.

MEZ (Department of Economic Affairs) (1995), *Knowledge on the Move* (in Dutch), MEZ, The Hague.

Morgan, K. (1997), 'The Learning Region: Institutions, Innovation and Regional Renewal', *Regional Studies*, 31.5, pp. 491-503.

Municipal Port Authority Databank. Rotterdam.

MPA (Municipal Port Authority) (1993), *Havenplan 2010*, Rotterdam.

MPA (Municipal Port Authority) (1998), *2020, Integrated future scans for port and industry* (in Dutch), Rotterdam.

MVW (Department of Traffic and Transport) (1996), *An International Comparison of Infrastructures*, SDU Publishers, The Hague.

OECD (1996), *Science, Technology and Industry Outlook 1996*, OECD, Paris.

Priemus, H., Konings, J.W. and Kreuzberger, E. (1995), *Nodes in Freight Transport: Typology and Dynamics* (in Dutch), Delftse Universitaire Pers, Delft.

PMR (Project Mainportontwikkeling Rotterdam) (1999), *PMR on the right way* (in Dutch), PMR, The Hague/Rotterdam.

Ratti, R. (1995), 'Dissolution of Borders and European Logistic Networks', in D. Banister, R. Capello and P. Nijkamp (eds), *European Transport and Communication Networks*, John Wiley, Chichester.

Reich, R. (1997), *Boston 2040, Planning for the Future*, Lecture in Faneuil Hall Boston, May 24, 1997.

Smith, A. (1776), *An Inquiry into the Nature and the Causes of the Wealth of Nations*, Liberty Classics, Indianapolis (English translation).

Storper, M. (1993), 'Regional worlds of production: Learning and innovation in the technology districts of France, Italy and the USA', *Regional Studies*, 27, pp. 433-455.

15 Co-operation between Local Governments: Can a Holding Company be a Solution in Public-Private Partnerships?

REMIGIO RATTI AND CHRISTIAN VITTA

The mismatch between the supply size of high-order services and the size of municipalities, often called spillovers, gives rise to a need for municipalities to co-operate. In Switzerland, canton laws on municipal organisation provide the directives concerning co-operation among local public entities. Inter-municipal collaboration can take on different forms: co-operation based on private law, inter-municipal conventions, associations of municipalities, and in the extreme case of complete integration, the merger. All of these forms of partnerships have advantages, but can also cause disadvantages and difficulties when applied. One possible alternative is to adopt the mode of a Holding Company for Public Shareholdings (HCP), which should favour the public-private partnership.

Introduction

Public finance is currently facing various changing under the influence of the emergence of new types of relationships between public actors and between public actors and private actors (Ruegg et al., 1994; Buchanan and Musgrave, 1999; Connolly and Munro, 1999).

In this chapter we discuss an innovative solution in the context of public-private partnership in the co-operation between local governments: the Holding Company for Public Shareholding (HCP). This solution is the result of empirical research involving tourism-related services in the Locarno region (also called Locarnese in Italian), conducted at the Institute for Economic Research (I.R.E.) in Lugano (Ticino, Switzerland). The presentation of the HCP project is connected to the following main aspects:

- At the functional level, providing services to the community takes form today in an emerging approach based on demand (market space). The ensuing adjustment of supply (production space) requires consideration be given to the characteristics of a given infrastructure, and above all to a project's feasibility. This calls for a network strategy that increases flexibility and makes "geometrically variable" solutions possible (Ratti, 1992). In this context, an HCP should represent an innovative way to create a support space.
- In regional economic theory, the problem is to succeed in providing a response in line with a region's changes in internal and external conditions (Ratti, 1997). The possibility to develop the regional potential of a collective depends on this. It is necessary to define a new "territoriality" that solves current problems of complexity and fragmentation of horizontal and vertical institutional relations. A new governance going beyond traditional institutional and juridical schemes is necessary (Ratti, 1995). An HCP should represent the ideal instrument for providing new responses to internal and external challenges to the tourism industry in the Locarno region.

The Locarno Region

The study concerning the creation of a holding company was conceived with the view of applying it to a particular area, the Locarno region of southern Switzerland, in the Canton of Ticino (Dafflon and Della Santa, 1996). To fully comprehend the project, it is necessary to have background information on the context in which the HCP project was developed.

The Locarno Region: an area with a focus on tourism

Geographically, the Locarno region has the advantage of a convenient location that is easily reached both from the north and the south; in particular the metropolitan areas of Zurich and Milan can be easily and quickly reached. In addition, the Locarno region has a mild climate and an attractive landscape, situated on the shores of Lake Verbano (Torricelli, 1997). These characteristics make the region one of the most important areas for tourism: it represents 15 per cent of the canton's population, while it accounts for 42 per cent of its tourism.

The area consists of eleven municipalities connected by a strong interest in tourism and the surrounding mountain valleys lying to the north. The current institutional structure does not provide for an organised form of co-ordination among these municipalities. In the past, the development of tourism facilities and services was essentially based on the initiative of individual municipalities. This absence of co-ordination led in some cases to a dispersion of resources, for example, the creation of the same infrastructures in communities a few kilometres apart.

The planned HCP should first of all guarantee this necessary co-ordination and represents a structure able to respond, from an organisational point of view, to current challenges in the tourism industry. A model of challenge and response of the "open region" helps us better understand the role of this new organisational structure in a regional context (*Figure 15.1*).

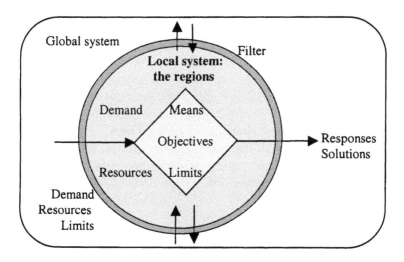

Figure 15.1 A model of challenges and responses of the open region

The Locarno region represents a system that aims to meet external challenges in tourism, in particular, increasingly greater competition from other areas in the tourism industry, i.e. other Ticino areas and nearby Italian areas that draw tourists, such as Varese and Como (*Table 15.1*).

Table 15.1 Total overnight stays, including overnight stays in hotels

Area	Total overnight stays *	Of which in hotels (%)	Overnight hotel stays 1995	Change in overnight hotel stays 1994/95
Locarno Region and Valle Maggia	3,322849	45.27	1,263878	- 4.30 %
Varese-Como-Ossola (VCO Province)	2,305414	58.32	1,344583	+ 10.20 %

* CH 1993; VCO 1995

Source: Torricelli (1997).

The response is based on external resources, such as public financial aid from the canton and the federal government, and on internally available means, for example, what the area offers visitors, financial resources made available by the region's municipalities. The response to challenges can be represented both in terms of new objectives and new operative instruments. An important response is the HCP project, a new tool for managing tourism-related projects. This new structure, in addition to having consequences for the internal structure of the area, also generates changes in external relationships. We are faced with a new institutional actor that strengthens the region's ability to negotiate with other areas (Vitta, 1997b).

The Locarno Region's tourism services

There are important services available to support tourism activity in the Locarno region. In this context various existing (planned) companies could become part of the "Holding Company Group".

Companies in existence or in the launching phase (Vitta, 1997b)

- Riservazioni turistiche SA: a company operating locally, in the Ascona-Losone area. Its objective is to reduce the number of intermediaries between the provider and the consumer of tourism products, and as a consequence reduce costs. This objective can be reached enabled by new computer and telecommunications technologies. The future goal is to expand this company's activity to the entire Locarno region.

- Lago Maggiore Tours SA: a company that focuses on the organisation of excursions, conventions and events in the area. It is currently experiencing major financial difficulties; reorganisation of its management structure and activities is necessary.
- Cardada Impianti Turistici SA: a company that is to lead to the re-launching of another company, Floc SA, which currently manages lift and cable transport infrastructures. This service is concentrated in the summer months. The re-launching project has involved considerable financial backing by public entities. The business plan indicates financial problems for the new company in the event it does not succeed in increasing the number of passengers from 100,000 to 170,000.
- Casino' Kursaal Locarno SA: a company that manages a small casino, a restaurant and a dancing establishment. After slot machines were allowed by law (1993), Kursaal has managed to earn sizeable profits, which it distributes into the region to favour culture and tourism. Part of its activity is at risk because of a new federal law on casinos.
- Fondazione Casino' di Locarno: in pursuit of its objectives of promoting culture, Kursaal di Locarno SA created a foundation that manages a theatre and promotes culture.
- Grossalp SA: a plan to restructure the Grossalp ski lift system, which is located in Bosco Gurin, has been developed.
- Porto regionale di Locarno SA: a recently established company that should construct and manage a regional port for small- to medium-sized watercraft in Locarno. Plans include the creation of 400 boat landing-stages. Financing should be guaranteed by public entities and private parties in response to a stock issue.

Plans to establish new companies (Vitta, 1997b)

- Holding Service SA and Progetti SA: these two companies directly depend on the realisation of the HCP project. The first will act to provide logistic support, while Progetti SA should facilitate the realisation of new initiatives (see next section for more details).

- Area di servizio SA: this project calls for establishing infrastructures aimed at providing information on accommodation, culture and recreation for visitors arriving in the Locarno region. A restaurant and a petrol station are to be built next to the information service point.
- Aeroporto di Locarno SA: a company that is to be part of a plan for the partial privatisation of the Locarno airport. This new company should manage airport taxes, the purchase and sale of fuel and passenger accommodation.
- Golf SA: a company that is to plan and build a golf course in the area. The plan is to be promoted by an association in Losone.
- Casino di Locarno SA and Casino di Locarno Operating SA: these two companies are to handle the realisation and management of a full-scale casino, this alternative can not be provided by Kursaal. Problems connected to its establishment can be traced to proposals presented in the plan for a new federal law on casinos that call for limitations on the number of licenses for casinos granted in Switzerland and for high taxation. The Locarno project is threatened by these uncertain factors.
- Piscine regionali (regional swimming pools): a regional plan that calls for building new bathing facilities, such as outdoor and indoor pools, and possibly pools for massages, water sports, and other activities.

The HCP Plan: the Holding Company for Public Shareholding

The HCP is designed to facilitate co-operation among municipalities and other public entities in the Locarno region (Rossi et al., 1997). The aim is to develop, establish and co-manage with private parties, through the various publicly and privately financed companies, initiatives the importance of which goes beyond the municipal level (infrastructures and services) of public interest, mainly in connection with tourism, recreation, culture and sports. The qualifying elements of the HCP can be summarised as follows:

- A structure allowing co-operation among municipalities and with other public entities in the area, i.e. tourism boards and associations of municipalities of the Locarno and Vallemaggia regions.
- A structure to facilitate combined investments with private parties.
- An instrument to raise the level of and re-launch tourism, which is a strategic industry for the Locarno region.

- An instrument for the realisation of recreational, cultural and sports products that extends beyond the individual community level to benefit residents of the entire Locarno region.

HCP's share capital and its holdings

The HCP is a financial holding company. As such, it does not produce goods and services, nor does it realise infrastructures. Its activity consists of its board selecting companies in which it intends to hold a share and establishing the amounts of share capital it holds. The HCP then acts as shareholder, sole, majority, "leading", or minority. The HCP participates in the shareholders' meeting, it chooses members of the board of directors and the auditors, and, in particular, it chooses its representatives on these bodies if the size of its share provides for this possibility. The HCP can thus play a considerable role regarding the direction and co-ordination of regional development, in particular the development of tourism. Finally, as part of its role, the managers of the HCP will have to decide where to invest profits.

As its name indicates, Holding Company for Public Shareholdings, the share capital of the HCP, which is used to acquire shares in other companies, will have to be wholly owned by public entities. Although there is no juridical obligation to exclude private parties from owning a share in an HCP's share capital, as long as these shares are held in a minority, it appears opportune to follow the principle that shareholdings are to be held exclusively by public entities, in particular the municipalities. Indeed, the aim of having municipalities as HCP shareholders, which must be able to defend their actions to citizens-taxpayers, is to guarantee that resources invested in the HCP are clearly used in the public interest. The presence of private shareholders, who work towards other legitimate aims, would therefore add an element of confusion. The reconciliation between public and private interests takes place in the companies in which the HCP holds shares. At this level, the companies in which shares are held, HCP representatives will come together to develop and manage infrastructures and services together with other shareholders, both private and public.

HCP operative instruments: Holding Service SA and Progetti SA

Upon its establishment, the HCP is to set up two service companies, Progetti SA and Holding Service SA, in which it will hold 100 per cent ownership. These companies will allow the HCP to operate effectively and to meet efficiently specific needs of its holdings. This arrangement can be explained as follows:

- *Progetti SA*: when a group of promoters, including the HCP, intends to plan a new project or develop an existing one, it can confer a mandate to Progetti SA. Planning costs will be covered by the different partners as a proportion of the interest they show in that project. If the project is carried out, the project commissioners, if they are not the same parties who realise it, will sell the project to the company that realises it. If, instead, insurmountable obstacles arise, the project costs will need to be covered by the commissioners without any loss for Progetti SA. The losses, tied to the normal entrepreneurial risk of any project in the preparatory phase, will therefore be equally distributed among the commissioners, including the HCP. Progetti SA favours and raises the level of relationships among regional actors, in particular, among public and private parties. The activities of this company should favour those privileged relationships between private and public which form an integral part of a support space.
- *Holding Service SA*: this company meets the need for rationalisation and efficiency. It will be able to carry out the administrative, accounting, controlling and marketing needs of companies belonging to the holding company group on their behalf, and for payment. The concentration of these activities provides an opportunity for greater communication of information and for a transparent vision of the situation, problems and opportunities connected to the projects in which the holding company participates. Holding Service SA may also offer analogous services to other companies not belonging to the holding company group.

Operational Aspects of the Planned HCP Project

To make this project operational, it is necessary to define who will hold the capital of the HCP, what the proportion of this holding is to be, and in which projects the available capital will be invested.

The geographical reference area and shares of HCP capital

The concept of geographical reference area needs to be clarified for future public shareholders of the holding company. The demarcation is established as a function of direct and indirect benefits generated by the companies owned by the HCP group. The particular activity of the companies held by the holding company generates positive external effects causing a problem in terms of equivalence (Jossa and Semo, 1990; Jacquemin and Mignolet, 1994): in fact, the circle of those who enjoy a service does not correspond to those who contribute towards covering the costs of the service. The HCP allows these projects to be grouped and co-ordinated, and it ensures that set up and operating costs are divided among the shareholders of the holding company. The distribution of these costs is not arbitrary, but is carried out using a specially designed allocation system that takes into account which actor benefits the most from the services offered.

We have been able to subdivide the Locarno region into two concentric circles: the first consists of the area's eleven municipalities in which the largest demographic, economic, tourism and financial potential is concentrated. Most of the existing or planned projects that might interest the HCP relate to this area. The second circle comprises the peripheral municipalities of the area, mainly located in the valleys lying outside of the first circle.

A third circle could be added, in the event the HCP's activity being extended beyond the Locarno region; this final circle would consist of the rest of the Ticino canton as well as the bordering area of Italy, i.e. the provinces of Verbano-Cusio-Ossola and Varese (Torricelli, 1997). This area and the Locarno region are in a co-operative-competitive relationship, and in this regard, the creation of projects in which the HCP participates cannot be excluded a priori.

Initially the capital of the HCP will be acquired directly from the eleven municipalities that make up the first circle, i.e. the shares of capital of the different municipalities will be determined using the allocation system. The second circle of municipalities will form a group to acquire together a share of capital that is proportionate to their financial means. In effect, these municipalities are in a "regime of financial compensation," which means they receive subsidies from the wealthy municipalities in order to be able to cover their current operating expenses. The allocation system that allows the division of share capital inside the first circle of municipalities is based upon the variables (Rossi and Baroni, 1997):

- tax income tied to tourism, coming from individuals,
- tax income tied to tourism, coming from corporate entities, and
- municipality population.

The first two variables are used to estimate the economic advantages due to tourism for each municipality, while the last variable considers the advantage for residents in terms of additional opportunities for recreation, sports and culture. It is possible to obtain the allocation system that takes into account both economic advantages as well as advantages to local residents, by combining the three variables (*Table 15.2*).

Table 15.2 Allocation of HCP share capital

Municipality	Share capital (%) shareholding	Share of capital (in SFr)
Locarno	28.23	654,000
Ascona	18.30	424,000
Minusio	12.56	291,000
Losone	9.15	212,000
Muralto	8.24	191,000
Gordola	5.05	117,000
Brissago	4.32	100,000
Brione s/Minusio	1.21	28,000
Tenero-Contra	3.50	81,000
Ronco s/Ascona	2.85	66,000
Orselina	2.29	53,000
All Municipalities (11)	*95.68*	*2,217,000*
Other shareholdings	4.32	100,000
TOTAL	*100*	*2,317,000*

Source: Rossi and Baroni (1997).

The eleven municipalities of the first cycle together hold SFr 2,217,000 (95.68 per cent of HCP's share capital). The item "other shareholdings" groups together the municipalities of the second circle and public entities that could be interested in the project. Their share is not calculated on the basis of the allocation system but as a function of their financial means.

Companies held by the HCP

The HCP is a financial holding company. As indicated above, it does not produce goods and services, nor does it realise infrastructures. Its activity consists of selecting companies in which it intends to hold a share and establishing the amounts of share capital preferred.

For a company to be partly owned by the Holding Company, it must respect certain selection criteria listed in the following order of importance (Vitta, 1997a):

1. Public interest in the project[1].
2. The project's supra-municipal dimension.
3. Private interest (in addition to public interest), so that a combination of public and private investors is possible.
4. Financial self-sustainability of the project.
5. Interest in the project that also includes tourism interest.

The application of these criteria must be flexible: we can have, for example, a company with a strong public interest, but this does not guarantee that its annual costs are covered. If these financial losses are not regular and do not threaten the global equilibrium of the HCP group, it is acceptable for the HCP to hold a share in this company. It is interesting to note how the third criterion calls for a partnership between public and private actors that can favour the establishment of privileged relations between these actors, which can contribute to strengthening the trajectory of the development and re-launching of the tourist industry in the entire Locarno region.

Based on data available to us, on the indications obtained from interviews with those directly involved, and on a financial analysis of each company, we have been able to carry out a selection of projects that could make up the initial framework of the HCP (see *Figure 15.2*). In the future the following foreseeable companies could also become part of the HCP: Piscine regionali (regional swimming pools), Golf SA and Grossalp SA.

As a result of the financial contribution guaranteed by Kursaal SA, the HCP will be able, from the start, to accumulate capital that will allow it to acquire new shareholdings in the future without having to ask public entities for fresh capital injections.

---

1 The criterion regarding public interest is evaluated as a function of the following criteria: natural or legal monopoly, positive external economy, negative external economy, market inability to ensure supply (Derycke and Gilbert, 1988; Löwenthal, 1993; Dafflon, 1994). The supply of public or merits goods has not been taken into consideration because none of the companies analysed offers goods or services of this type, except for the foundation, which is concerned with culture.

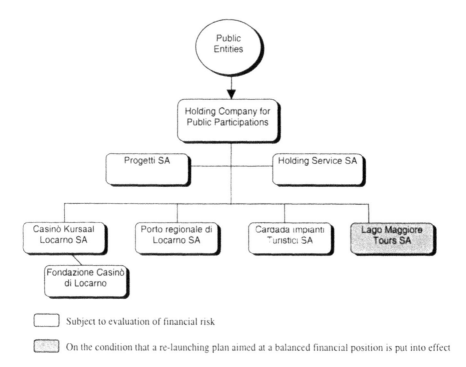

Figure 15.2   Initial framework of the Holding Company for Public Shareholdings (HCP)

Source: Rossi et al. (1997).

Conclusion

At the end of this study we can conclude that the project for a Holding Company for Public Shareholdings (HCP) is important, innovative and feasible. The HCP is an *important* project because it is an instrument that enables raising tourism to a higher level, tourism being a strategic sector for the Locarno region and currently in need of restructuring. The Locarno region is in competition with other Ticino areas and in particular with the nearby Italian areas. The HCP should provide the area with a dynamism that enables the strengthening of tourism and its re-launch in the entire area, and improves the position of the Locarno region at the canton and trans-border level. The possibility that the HCP is opened to nearby areas, including trans-border areas, can not be excluded; this would thus allow collaboration among regions that are faced with similar problems and that find themselves in a collaborative-competitive position.

The HCP is an *original* and *innovative* model of partnership among public entities (Della Santa, 1996; Pfister et al., 1999). It is an instrument particularly suitable for co-operation aimed at providing tourism-related services and for which a partnership with the private sector is necessary. The holding company is controlled by public entities and is a structure that respects democratic principles: the decision to acquire HCP capital and the designation of members of the board of directors are the competence of municipal governments.

The HCP is a *feasible* project. From the economic point of view, the HCP can guarantee its own future if it is able to maintain profitable companies among its shareholdings, which together results in a balanced financial position. In regional theory, the HCP is an innovative creation of a support space in the Locarno region. This structure aims to influence relations among public and private actors. Direct economic advantages, i.e. new resources to finance new investment, should arise from this collaboration, and indirect advantages deriving from new privileged relations among the various actors should contribute to improving the overall conditions of the entire region. Establishing the HCP is also aimed at increasing the ability of the region to respond to external challenges in tourism. The HCP is based on the search for a new "territoriality" characterised by a "variable geometric design", which is the expression of a higher level of an "active regional space". The presentation of the projects caused some negative reactions on the part of actors who had remained passive all along, but who felt threatened by the creation of HCP, which might take away some of their power. Arguing on the basis of diminished democracy (particularly, as they expect part of their authority or competence within the town Council might be questioned), there is an attempt on the part of these "circles" to throw discredit upon the project.

To conclude, we maintain that the structures and mechanisms of the "institutional state" have a limiting effect on, and sometimes even stifle, the innovative responses that a region may provide to solve the problems. Promoting new innovative projects which demand and depend upon a stronger co-operation between municipalities is far from simple, especially in a region which consists of small- and medium-sized municipalities with a whole history of strong rivalries behind them. It does take time before one becomes fully aware of the existing problems and before one accepts new rules of the game that confer a new role and a new jurisdiction on the public authorities.

References

Buchanan, J. M. and Musgrave, R.A. (1999), *Public Finance and Public Choice*, The MIT Press, Cambridge MA.

Connolly, S. and Munro A. (1999), *Economics of the Public Sector*, Prentice Hall, London.

Dafflon, B. (1994), *La gestion des finances publiques locales*, Economica, Genève.

Dafflon, B. and Della Santa, M. (1996), *Holding per le partecipazioni pubbliche SA: studio della proposta dal profilo della gestione delle finanze pubbliche*, Université de Fribourg, Friburgo.

Della Santa, M. (1996), *Dalla collaborazione alla fusione. Analisi degli aspetti economici, istituzionali e sociologici del comune*, Vico Morcote.

Derycke , P.H. and Gilbert, G. (1988), *Economie publique locale*, Economica, Paris.

Jacquemin, J.C. and Mignolet, M. (1994), *Finances publiques régionales et fédéralisme fiscal*, Presses universitaires de Namur, Belgique.

Jossa, E. and Semo, I. (1990), *Les finances locales. La décentralisation inachevée?*, La Documentation française, Paris.

Löwenthal, P. (1993), *Economie et finances publiques. Principes et Pratiques*, De Boeck Université, Bruxelles.

Pfister, B., Crivelli, R. and Rey, M. (1999), *Finances et territoires*, Presses Polytechniques et Universitaires Romandes, Lausanne

Ratti, R. (1992), *Innovation technologique et développement régional*, IRE/Méta-Editions, Bellinzona.

Ratti, R. (1995), *Leggere la Svizzera. Saggio politico economico sul divenire del modello elvetico*, ISPI/Casagrande Editore, Milano/Lugano.

Ratti, R. (1997), *L'espace régional actif: une réponse paradigmatique des régionalistes au débat local-global*. Revue d'Economie Régionale et Urbaine (RERU), n. 4/1997, pp. 525-544.

Rossi, M. and Baroni, D. (1997), *Quote di partecipazione dei comuni al capitale sociale della HPP*, Istituto di Ricerche Economiche, Lugano.

Rossi, M., Vitta, C., and Nosetti, O. (1997), *Holding per le Partecipazioni Pubbliche SA. Rapporto finale dello studio di fattibilità*, Istituto di Ricerche Economiche, Lugano.

Ruegg, J., Decoutère, S., and Mettan, N. (1994), *Le partenariat public-privé*, Presses Polytechniques et Universitaires Romandes, Lausanne.

Torricelli, G. P. (1997), *La posizione concorrenziale del Locarnese nel contesto regionale allargato*, Istituto di Ricerche Economiche, Lugano.

Vitta, C. (1997a), *Caratteristiche e valutazione delle società indicate dai promotori come interessanti per la HPP*, Istituto di Ricerche Economiche, Lugano.

Vitta, C. (1997b), *Valutazione del progetto da parte degli attori*, Istituto di Ricerche Economiche, Lugano.

16 New Technology-Based Firms and the Activation of the Styrian Economy

MICHAEL STEINER AND THOMAS JUD

The chapter is based on an empirical inquiry into the problems and the situation of new technology based firms (NTBF) in the Austrian province of Styria. The change within this region is interpreted in the context of an "active space" approach. The presence and growing number of NTBFs is regarded as a strong supporting element for the region in its attempt to become more of an "active space" and in its change from a closed to an open border region. The main problems of Styria's NTBFs are classified, and endogenous and exogenous factors of success and failures are given. The regional distribution and the technological fields of the firms are identified. Policy responses to the problems within a regional technology policy concept are outlined and corresponding instruments described.

Introduction

What improves the socio-economic performance of a region is still an open question. Yet the "regional active space" approach and the interpretation of regions as a "force field" point to important conditions for activation of regional potentials (Ratti, 1997; van Geenhuizen and Ratti, 1998). One of them is the creative ability available in a regional territorial system which enables regional actors to produce a coherent set of initiatives in response to challenges coming from within the system and from outside. The activity degree of the region depends partly on the managerial ability and entrepreunerial potential available.

In this chapter it will be argued that the presence of a growing number of new technology-based firms is a strong supporting element for a region in its attempt to become a more "active space". These firms play an important role both for the development of regions and for the dynamics of market processes, especially for the realisation of research intensive innovations. In contrast to already established firms, their innovative activities are mainly oriented towards their clients, towards existing product programmes, towards specific demand potentials and towards the desire for modifications (Mowery and Rosenberg, 1979). For small and new technology-based firms, innovations are "every day business" - the generation of innovations is their core-business, innovation oriented thinking and acting is taken for granted (Gundrum and Walter, 1995). Despite their minor contribution to employment creation, such firms play an essential role in the dynamics of regional economies.

Focusing on the Austrian province of Styria, the situation and the development of new technology-based firms (NTBFs), including their strengths and weaknesses, will be discussed. This region has suffered for decades from more or less closed borders to the East, Hungary, and the South, former Yugoslavia, and was also handicapped by the existence of an "old industrial area" in its northern part with a dominance of large, mostly nationalised firms; both facts were impediments to entrepreneurial activity and the development towards an "active space". Yet the period since the 1990s has brought about some relevant changes: with the opening of the borders towards its eastern and southern neighbours and through European integration Styria has changed from a closed border region on the outskirts of the West European market societies to a core area of the European Union where it offers the advantages of a stable economy with access to the internal market and a simultaneous close proximity to the South-Eastern regions of Europe. This can also be interpreted in terms of a change from a closed to an open border region (Ratti and Reichman, 1993), and it represents a strong challenge for Styrian firms as it calls for increased innovation, flexible production, networking activities and co-operation with R and D institutions. NTBFs play an important role in this process. The following tasks make up their contribution to economic change and regional development (Kulicke, 1990; Koschatzky and Kulicke, 1993):

- NTBFs accelerate the transfer of knowledge from science to the economy. Very often the new entrepreneurs were employees of universities, research institutions or research intensive firms before they left to found their own firm. It is exactly this spin-off behaviour that brings about the transfer of knowledge, which is so essential for economic development.

- NTBFs ease economic structural change by developing new products for new markets. Less developed areas especially have problems in keeping pace with other regions in their economic performance. NTBFs often help to generate competitive economic structures by creating new markets and stimulating established firms to increase productivity innovative activity.

- NTBFs are complementary to established enterprises in the sense that they help to build up competitive economic networks of firms. Closely connected to the structural impact is the cluster supporting power of NTBFs. The competitiveness of the leading firms of a region is very often highly sensitive to the performance of their related and supporting industries in the same area. NTBFs originating as spin-offs are mostly motivated by innovative ideas closely connected to the economic tradition, to products and services already offered, of the region. NTBFs develop innovative products or services used directly or indirectly in the production process of the leading firms, thereby generating positive effects for the whole cluster.

- NTBFs create qualified employment. The overall impact of NTBFs on employment is normally not considerable, yet in some rather "innovative regions" where NTBFs are concentrated it can add up to significant dimensions. Even more important is the qualitative aspect: often an essential part of the qualified workforce is employed in NTBFs, thereby also generating learning effects with connected firms. NTBFs may therefore be regarded as the dynamic part of learning organisations.

In the following we give a short description of the Styrian economy and its recent changes. We focus on NTBFs by first describing the available data and information sources. This is followed by an explanation of endogenous and exogenous factors influencing NTBFs' success and failure, and an analysis of empirical results of the situation in Styria. Finally, the present activities and future possibilities to support these firms are discussed.

Styria: from a Closed Border Region towards an Active Space

Styria is one of the nine provinces of the Federal Republic of Austria with a population of around 1,2 million, an area of 16,000 km² (about 19 per cent of the total area of Austria), a share in total employment of about 14 per cent, and a share in the Austrian GNP of around 12 per cent. Accordingly Styria is ranked 4th, 2nd, 4th and 4th respectively compared to the other 8 Austrian provinces. Its GRP per head is markedly below the Austrian average which puts it at the 8th place in a ranking of the provinces. Styria's regional economy is strongly dominated by the manufacturing sector which has a share of 16 per cent in total employment of Austria's manufacturing sector, which covers 18 per cent of Styria's total employment and which produces 33 per cent of the Styrian GRP. Thereby it is well above the Austrian averages of 11 per cent, 15 per cent and 23 per cent respectively. This dominant position of Styria's manufacturing sector can be traced back to the historical situation of two economically most important areas within the province: Upper Styria and the capital of Styria, Graz, with its surrounding areas.

In Upper Styria especially an industrial tradition based on the melting of iron ore and the availability of coal, caused inflexible economic structures to emerge. Businesses have problems in renewing their product-line and production techniques and in adapting to changing economic conditions. Economic problems brought about by this situation are indicated by low growth rates for industrial production, unemployment rates above the Austrian average and considerable migration rates jeopardising the economic competitiveness of the whole region (Tichy et al., 1982; Steiner, 1984; Geldner, 1989). Existing problems are aggravated by insufficient economic stimuli being generated by the central region, Graz and its surrounding areas, of Styria. Firstly, the capital of Styria in some respects faces problems quite similar to Upper Styria. Though there is a high concentration of innovative and technology-based firms in the central region, most existing businesses operate in the technology branches facing stagnating markets, e.g. textiles, paper, leather manufacturing, etc. Secondly, Graz has not taken on the role of a service centre for Styria. The size and number of firms in Graz are too small to supply their services proactively to the rest of Styria. Thirdly, Graz has not been able to take full advantage of the liberalisation of the eastern countries until now, partly because of its poorly developed transport infrastructure (Gruber et al., 1996; Jud and Piber, 1996). In addition the southern area of Styria has suffered historically from a lack of resources and low accessibility. Its economic problems have been aggravated by being a border region located near former Eastern European countries thereby being on the outskirts of developed western market economies.

Numerous studies have been carried out during the last ten years analysing the economic situation of Styria in different contexts to elaborate appropriate strategies and instruments for fostering Styria's long term competitiveness. Strongly simplified, the results suggested:

- to improve technology transfer in all of its possible forms[1] to generate innovative stimuli for the regional economy; innovation and collaboration between productive entities in Styria should be promoted, e.g. between firms, research, transfer, advisory institutions, and
- to restructure nationalised industry to make it more flexible and adaptable.

Based on these proposals, big nationalised firms have been reorganised and broken up into smaller parts, new applied research institutions have been established, technology parks have been founded, a technology transfer and advisory service infrastructure has been built up and regional development and financial support programs in general have become more innovation orientated. All together these policies seem to have fuelled the economic development of Styria bringing about a greater propensity for innovation and technological development within the region. Innovation activities have increased (see e.g. Steiner et al., 1996), employment in the private sector has risen and Styrians show a more entrepreneurial spirit, indicated by the increasing number of firms being created (see e.g. Steiner et al., 1993; Gruber et al., 1998). NTBFs represented a major factor in supporting the renewal of the Styrian economy; below we give a more detailed interpretation of this specifically "active" type of firm in the Styrian context.

A Twofold Empirical Approach to Explore the Situation of NTBFs in Styria

Useful data about specific groups of economic actors like NTBFs is traditionally very scarce in Styria just as in other European countries (see e.g. Kulicke et al., 1993). Information about foundation rates, the

---

1 Transfer of technical, organisational, managerial know how etc. by educational measures, incentives for establishing firms, establishment of research institutions, supply of advisory services, etc.

development of foundation projects and detailed economic data about NTBFs is not available. The reasons for this situation are the following: on the one hand, there is no tradition of collecting data about new technology-based firms because their effects on regional economic development have been analysed for only a short time; on the other hand, it is extremely difficult to differentiate NTBFs from normal firms using standardised economic data; to single them out one needs comprehensive information about their research activities, their economic activities and their financial transactions. Limited by these data restrictions the following twofold approach was chosen:

- Based on international research results an analytic framework was constructed identifying major factors of success and failure for NTBFs and relevant relations between these factors. This framework was used to elaborate a semi-structured questionnaire to carry out qualitative expert interviews with representatives of institutions engaged in advising, informing, teaching and giving financial support to young innovative firms, and to interpret the ascertained data accurately to reveal the specific problems faced by Styrian NTBFs.
- A database containing addresses of Styrian NTBFs was analysed to determine their regional distribution within Styria. The Styrian Institute for Promoting Economic Development (Wirtschaftsförderungsinstitut - WIFI Steiermark) supplied this information. The data was established by experts, working at the institute, with a profound knowledge of the activities and the performance of individual firms gathered through their day-to-day work as advisers and promoters of small firms in Styria.

Endogenous and Exogenous Factors of Success and Failure

In the following, the analytic framework, derived from a wide range of different theoretical and empirical works on NTBFs and their performance (see e.g. Sweeney, 1987; Prakke, 1989; Koschatzky and Kulicke, 1993; Kulicke et al., 1993; OECD, 1993; Oakey, 1994) is explained in more detail. At the highest level of abstraction of the framework (*Figure 16.1*) a differentiation is made between "Elements of the foundation process" (endogenous factors) and the "Economic environment" of the NTBFs (exogenous factors).

Analytic framework: elements of the foundation process

The elements of the foundation process can be classified into two groups. Firstly, there are six core elements which can be determined directly by the founder. Secondly there are two market-related elements (competitors, demand) which can be strongly influenced by the strategies and activities of the founder. Together they exert a strong influence on the economic success or failure of an NTBF during the process of foundation and market penetration. The six core elements are summarised in the remaining section.

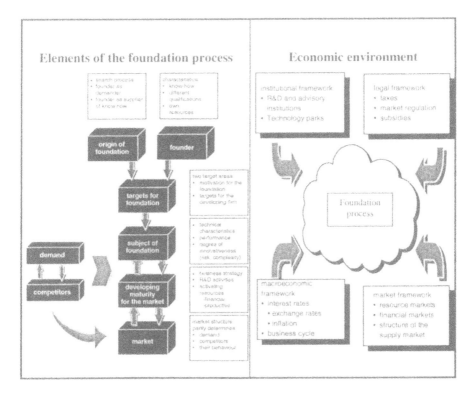

Figure 16.1 Analytic framework

Origin of foundation There are three starting points for founding a firm:

- During an explicit *search process* the new entrepreneur finds a market opening, which offers great profit opportunities; in this case he/she is highly informed about potential markets and competitors, which can be seen as an advantage both during the development phase, i.e. foundation, developing maturity, and the market-introduction phase.

- Formerly the new entrepreneur was a *demander*, or user, of a particular product; thereby he/she developed an idea for improving or substituting it, a situation in which the founder is often poorly informed about the market and may have deficits with respect to technical know-how.

- The founder developed his/her idea as a *supplier of a product*; in this case his/her information about relevant markets and technical knowledge is generally good.

Founder Qualifications, contacts with potential customers and own resources are the major assets of the founder. Many of the new entrepreneurs are technicians, who lack management qualifications, which raises the possibility of weak market performance for the young enterprise.

Targets for foundation The success of an NTBF is highly sensitive to the kind of motivation responsible for the foundation of the new firm. If it is profit expectation, the founder can be assumed to be highly informed. If it is the wish to be independent, the situation is indifferent. If it is frustration with the old job, the chances of success are rather low. In addition, there is a close relation between the kind of motivation and the explicit or implicit targets laid down for the new enterprise. For example, a founder wishing to become independent will not be eager to fill financial gaps with equity capital from external sources, e.g. venture capitalists, investment banks, etc., which can become a problem, especially during the marketing phase and later growth phases of the firm.

Product base The characteristics of the product which the new entrepreneur plans to bring to the market determine the competitors, the customers and the resource markets. They are also responsible for the scientific and technological know-how necessary and the financial capital used to produce and market the innovation. Therefore these characteristics play an essential role in dividing winners from losers. To get all the information necessary and to draw adequate conclusions is a complex process that founders are mostly not able to handle alone: at this point the availability of technological infrastructure gains importance, i.e. information, transfer and advisory services, etc.

Developing maturity for the market In this phase of the foundation process, the new firm and the characteristics of the product materialise in co-ordination with the market conditions. The product has to be designed, the manufacturing process has to be planned and the manufacturing start-up has to be prepared. The entrepreneur has to determine the legal status, the location of the firm, and he/she has to elaborate a business plan. Long term decisions have to be taken and external resources have to be mobilised, e.g. qualified personnel, additional know-how of R and D institutions, financial and physical capital, information about markets, information about service infrastructure, etc. Mistakes made in this phase can have dramatic effects on the economic success of the young firm.

Market (customers and competitors) It is the behaviour of competitors and customers finally which will determine the economic performance of an NTBF. A highly qualified and well-informed, i.e. in terms of business and technical know how and information about markets and available service- and R and D-infrastructure, management which develops sound targets and strategies during the foundation process will have the best chances of survival and growth in an uncertain and competitive economic environment.

Analytic framework: economic environment

The economic environment contains all the elements that influence the success and failure of an NTBF, but cannot be directly influenced by the founder. Therefore these elements have to be treated as parameters within this decision-making processes. The relevant components of the economic environment, their interconnections and their effects on NTBFs are very complex and cannot be treated fully within this chapter; but two aspects of great importance, especially for Styrian NTBFs, should not go unmentioned, i.e. advisory services and external financial capital. One of the prerequisites for the successful founding of technology-based firms is sufficient availability of information, e.g. customer needs, relevant competitors, resource markets, sources for external, financial capital, market openings etc., and qualifications to interpret them adequately.

Since the founding of technology-based firms is hindered by a number of factors, e.g. external effects, budget restraints, availability of information, small regional markets etc., a market for specialised advisory services on a regional level[2] is unlikely to develop. This in turn is an additional obstacle for potential founders. To escape this vicious circle, public initiatives are required to initiate the development of an adequate technology transfer and advisory service infrastructure. Additionally, such initiatives would offer the opportunity for promotional institutions to supply comprehensive project orientated development programs for NTBFs which could co-ordinate types of support, such as information services, advisory services, financial assistance, etc., with the different needs during the various phases of the foundation and market penetration processes.

One of the major problems faced by NTBFs is their scarce budget. Their own financial resources are often extremely limited and their access to external sources is hindered by their high, often difficult to assess, risk of failure. Banks, which supply credit at current interest rates, are very cautious about NTBFs because they are not able to participate in their, often above average, profits but have to share in their risks. Venture capitalists participate in the profits of NTBFs and are thus more prepared to take risks, but they prefer fast-growing already established firms, high investment sums and technologically simple innovation projects. They thus have a bias against small, newly founded technology-based firms. Principally there are two ways policy makers can choose to overcome these financial restrictions for NTBFs; they can offer direct financial aid or they can use indirect measures to reduce risks of investment for potential financiers and to activate private venture, e.g. Business Angel Capital.

The Situation in Styria: a Problem-Oriented Summary of Empirical Results[3]

The problems of Styrian NTBFs revealed by data collected within the framework described above can be divided into two classes and three sub-classes respectively (*Figure 16.2*). The problems in each class are summarised in the remaining section.

---

2 The market should be available at a regional level to take full advantage of the spatial proximity between suppliers and demanders.

3 For a more detailed analysis see Steiner et al., 1996.

Internal problems	External problems
Information problems	Demand related problems
Qualification related problems	Competition related problems
Financial problems	Financial problems

Figure 16.2 Classification of NTBF specific problems

Information problems Styrian NTBF founders do not adapt explicit search processes before starting the process of foundation. Therefore they often have information deficits. They are loosely informed about the products of their competitors and the needs of their potential customers and they are mostly not aware of their deficits. No technology, commercial or competition watch activities are introduced to guide the developing phase and the market introduction and no external resources, e.g. advisory and promotions agencies, R and D institutions, etc., are used to offer appropriate help in coping with the problems at stake. In this respect Styrian NTBFs are quite similar to their counterparts in other European countries. For example, a German study revealed that no more than 20 per cent of the founders of NTBFs search for market opportunities before starting their business (see Kulicke, 1990: 6). The OECD found that small and medium-sized enterprises (SMEs) in general, and innovative ones among them in particular, face numerous constraints in gathering and analysing relevant information and establishing "watch activities": lack of time, high costs of obtaining information, and lack of skilled specialists staff are just a few examples for such constraints (see OECD, 1993: 37).

Qualification related problems Styrian NTBFs have a strong bias toward the technical aspects of their products. They use overly expensive components in construction and leave out design aspects almost completely, thereby ignoring the needs of potential customers. They mostly lack management qualifications and are not aware of the complex processes which have to be directed and controlled in dealing with legal, technical, economic and communicational aspects. In addition, they do not develop explicit targets and strategies for their business, which has negative effects on their market performance and their ability to cope with pressing competition. Existing public and semi-public institutions in Styria engaged in supporting NTBFs have already recognised these problems and have set up adequate development programs. They supply financial support for advisory services demanded by NTBFs, thereby initiating the development of a new market segment for consultants. Though most of the international work on innovative SMEs and NTBFs places the entrepreneur and his or her characteristics on the top of the list of factors influencing success and failure, not much systematic empirical evidence is offered on deficits in their qualifications. Many authors, however, offer interesting indications, Kulicke (1993: 8) for example finds that most of the founders of NTBFs have technological skills but may lack experience and know-how in business related fields. As suggested by Young (1996), Binks (1996) and Philpott (1994) one of the main obstacles for SMEs to get funded by banks is their lack of ability to present financial information in the way required by investors.

Financial problems (internal and external) New entrepreneurs mostly do not have enough equity on their disposal to finance the foundation of a technology-based firm, so they are dependent on external financial capital; but very often they are not willing to share management with an external partner offering equity which in turn leads banks to be restrictive in supplying funds because they refuse highly geared and therefore risky lending (high ratio of credit to equity capital). In addition, the available venture capital in Styria and in the whole of Austria is very scarce and therefore not really risk-orientated. In short, the financial situation of NTBFs in Styria can be said to be precarious. Public institutions in Styria offer equity capital independently, but currently there are no initiatives to activate private capital from banks, venture capitalists or business angels. This situation is disturbing because, apart from its positive effects, public equity capital for NTBFs has serious drawbacks, which are for example:

-   The need for funds especially during the last phases of the foundation or innovation process is extremely high, often between 60-80 per cent of total costs (see e.g. Brooks, 1994), leading to a demand for capital which cannot be matched by public funds alone. This situation results in an overly restrictive distribution of pubic equity capital, often below the critical mass.
-   Funding of the last phases of the innovation or foundation process is very often equal to financing private profits with public money.

- Pay-back formulas for public equity capital very often cause serious liquidity problems for NTBFs, especially in their growth phases.

NTBFs in other European countries are facing similar problems. Though the supply of venture capital in Europe has boomed during the last years much of the funds available are used to back mergers and acquisitions or sustainable growth of SMEs. Early stage financing is still badly developed (see EVCA, 1995). Currently, policy decision makers at the national and international level are eager to improve the situation and to close the financial gap faced by technology-based firms during the early stages of their development: Business Angel networks have been established to match capital seeking firms with informal investors, guarantee schemes have been launched to reduce risks for equity investors, grants have been offered to reduce assessment cost for venture capitalists, etc; but things are moving slowly and it will take time to boost the European equity capital market for high risk ventures.

Demand related problems For many high-tech products the Styrian market is much too small. Hence innovative entrepreneurs have to export their products to "the rest of the world" or have to look for a location for their firm outside Styria. NTBFs do not like to co-operate with other enterprises, especially competitors, and consequently, they have no appropriate distribution system at their disposal and thus prefer locations with short distance to their markets. The effect of all this, is the specialisation of Styrian NTBFs in a few technological fields, as described in the next section. Those firms which choose to stay in Styria face another problem with demand. The potential customers of NTBFs are mostly other firms using the innovative products as inputs in their production processes. To buy from an NTBF can be very risky for them, because of the uncertain future of the young firms. If an NTBF goes bankrupt, its customers have to dispense with all after sales services and future up-dates of the product. In the face of technological lock-in effects, this event can be very costly. Thus Styrian firms very often prefer the products of expensive but established suppliers to the products of NTBFs.

At an international level much research has been done into the implications of external scale economies created by high-tech networks, innovative milieus, regional clustering, etc. for the location decision of NTBFs. The influence of factors like spatial proximity to customers and product markets has rarely been singled out. Therefore no statement can be made about the behaviour of NTBFs in other European countries and regions in comparison to Styria. There is some evidence, however concerning the behaviour of customers of NTBFs in other countries. A German study for instance revealed that 50 per cent of a sample's NTBFs were faced with cautious customers who decided not to buy the new product but to wait and see what will happen to the new firm in the near future. Thus to sell their product meant to devote considerable time and resources to persuading potential customers (see Kulicke et al., 1993: 91f). These results are quite similar to the situation found in Styria.

Competition related problems Once NTBFs have found market openings for their new product, competitive pressures are not too strong; but if they offer close substitutes to products of established enterprises, several different instruments are used to exert pressure on the new suppliers, e.g. price-cuts, imitation strategies, manipulating distribution systems, marketing abilities, etc. In a situation of hard competition the management qualifications of the new entrepreneur and the knowledge of adequate business strategies are essential for the NTBFs' survival. Again it is the German study mentioned above which delivers comparable results. All NTBFs in the sample observed supplied products for market niches without experiencing any considerable competition. One third of the sample did not know any competitors.

Summing-up this section one can conclude that a majority of the deficits and problems faced by Styrian NTBFs are generic in the sense that there are considerable similarities between technology-based firms in Styria and their counterparts in other European countries (e.g. During and Oakey, 1998; Oakey and During, 1998).

Regional Distribution of NTBFs in Styria

Two elements were used to determine the regional distribution of NTBFs within Styria. Firstly, the previously mentioned database made available by WIFI Steiermark which contains addresses of Styrian NTBFs and their technological specialisation. The classification of technological fields used by the WIFI is the following: energy and environment, engineering, new materials, electronics and biotechnology. The WIFI Steiermark is an organisation financed by the Styrian Chamber of Commerce. It consists of several departments offering different support services for Styrian firms. One of them is engaged in advising and supporting small firms in general and NTBFs in particular. Its staff keeps continuous contact with mangers and entrepreneurs of many businesses thereby obtaining a rich insight into their strength and weaknesses, their activities and their orientation towards technology and innovation. Using their experiences the experts at the department have built up the database mentioned above. It is dynamic in the sense that all newly founded firms are added as soon as they contact the institute but it does not contain the entire population of NTBFs. It is just a sample of firms but the most comprehensive sample available in Styria. Secondly, a regional classification of Styria introduced by Palme (1989) which identifies four different geographic areas according to their a) economic performance as indicated by labour costs in relation to net production value, b) their accessibility as indicated by size of firms[4], c) their resource endowment as indicated by the number of white collar workers in industry in relation to total employment in industry and wage rate, and d) the dominating economic sectors as indicated by employment in capital intensive or labour intensive branches in relation to total employment[5]. Accordingly, the Styrian areas can be characterised as follows:

- *Central area:* a high accessibility, a good resource endowment, a balanced mix of economic branches and a good economic performance.
- *Old industrial area:* a predominance of capital intensive branches, a poor resource endowment and a relatively poor economic performance.
- *Semi-industrialised area:* a mix of branches similar to the central region producing low-value products, a low accessibility, a weak resource endowment but an economic performance nearly similar to the central region indicated by low wage rate and high net production value.
- *Rural area*: a predominance of labour intensive branches, a poor resource endowment, a low accessibility and a poor economic performance.

4 The relationship between size of firms and accessibility may be surprising for the reader; but it becomes obvious if the underlying assumption of Palme is considered. He assumes that firms choose highly accessible locations endowed with an excellent transport infrastructure to take advantage of economies of scale. If he is right the size of firms actually would be a sound indicator for accessibility.

5 Economic data at the level of Styrian districts were used by Palme to determine indicator values.

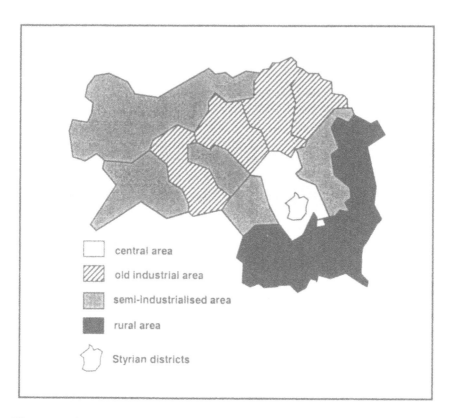

Figure 16.3 Regional classification of Styria

Source: Adapted from Palme (1989).

Based on the address data from 1994, the regional distribution of NTBFs could be determined. In more detail the following was revealed (See *Table 16.1, Table 16.2* and *Table 16.3*). First, the types of regions that are the most attractive for NTBFs and differences in the location preferences of NTBFs compared with other Styrian firms in the manufacturing sector, could be identified. Secondly, the relative importance of particular technological fields for the different Styrian regions and for Styria as a whole came to light in terms of NTBFs located in the respective regions.

The results can be interpreted in a twofold way. From *Table 16.1* it is evident that the central area and the old industrial area have a dominant position as a location for NTBFs in terms of their share in the number of

Styrian NTBFs and concentration of NTBFs in these areas indicated by their share of NTBFs being well above their share of all manufacturing firms with more than 20 employees. These findings suggest that the degree of economic development of an area, the spatial proximity of universities, research and transfer institutions located in Graz, e.g. Technical University Graz, Uni-Graz, Joanneum Research, and Upper Styria, e.g. Technology Transfer Centre, Technical University-Leoben, and the spatial vicinity of the developed service sector in Graz attract NTBFs from other areas and/or increase the number of spin-offs within an area. Additionally, they corroborate the view held by some authors (see e.g. Koschatzky and Kulicke, 1993) that the promotion of technology-based firms is an inadequate instrument for regional policy to balance regional disparities.

Table 16.1 Regional distribution of NTBFs compared to the regional distribution of all manufacturing firms in Styria*

Central		Old industrial		Semi-industrialised		Rural	
NTBFs	*All firms*	*NTBFs*	*All firms*	*NTBFs*	*All firms*	*NTBFs*	*All firms*
56%	35%	19%	15%	13%	26%	12%	24%

*more than 20 employees

Table 16.2 Regional distribution of NTBFs by technological fields

	Central	Old industrial	Semi-industrialised	Rural
Biotechnology	7 (64%)	1 (9%)	2 (18%)	1 (9%)
Electronics	250 (76%)	38 (12%)	20 (6%)	19 (6%)
New Materials	17 (18%)	46 (49%)	15 (16%)	16 (17%)
Engineering	110 (43%)	49 (19%)	49 (19%)	49 (19%)
Energy and environment	50 (54%)	12 (13%)	19 (21%)	11 (12%)

Further, *Table 16.2* indicates that the regional distribution of NTBFs by technological fields is closely related to the technological focus of research institution located in respective areas[6]. This relationship is shown in *Table 16.3*, which is an extension of *Table 16.2*.

Table 16.3   Relation between the technological focus of NTBFs and Styrian areas *

Technological focus	Research institutions		Regional distribution of NTBFs (%)			
	Ca	Oia	Ca	Oia	Sia	Ra
Energy and Environment	Technical University Graz, Joanneum Research		54	13	21	12
Engineering	Technical University Graz		43	19	19	19
New materials		CD-Laboratories, Technical University Leoben	18	49	16	17
Electronics	Technical University Graz, Joanneum Research		76	12	6	6
Biotechnology			64	9	18	9

* Ca: Central area; Oia: Old industrial area; Sia: Semi-industrialised area; Ra: Rural area.

The relationship can be interpreted in the context of two factors. Firstly, the availability of research institutions attracts NTBFs because it makes their access to information and know-how, urgently needed in the high-tech

6 A comprehensive description of Styrian research institutions and their technological focus is given in Steiner et al., 1996.

sector, much easier and reduces transaction costs for the small firms. After introducing their new products into the market, most NTBFs face bottlenecks concerning their financial and personnel resources. Increasingly resources have to be allocated to business activities at the cost of reducing R and D activities which in turn leads to a higher demand for technology transfer services from universities and research institutions. Secondly, the existence of research institutions within a particular area will influence positively the number of spin-offs.

The Policy Response - Present Activities and Future Tasks

Styria's economy has been based on manufacturing industry for the greater part of the 20th century; in its southern part (the "Grenzland") it was dominated by the agricultural sector. In the 1970s and 1980s this economic base created severe problems both in the agriculture and industry dominated parts. Since then, Styria's economy has been marked by external and internal change and the area is slowly developing into a technological region. The intensification of new firm foundation and the emergence of a high technology sector with NTBFs is part of this change.

Equally changing are its economic policy institutions and their instruments. The increasing ineffectiveness of the traditional regional policy, the progress in regional economic research and the availability of foreign examples for new solutions to existing problems are about to cause a shift from the old model to a new philosophy of regional development in Styria. This can be seen in the organisational structure of regional policy institutions, in their target orientation and in the application of their instruments. Within the last decade a vast network of technology-based infrastructure has been developed, e.g. industry and technology parks have been created, and transfer agencies and research institutions have been reorganised and extended. In 1991 the public administration of regional economic policy was partly privatised and formed as an independent institution outside the main administrative operators under private law. Its main focus is on granting various types of financial assistance, providing information and advisory services, founding of new firms, establishing and running technology parks, initiating border-crossing initiatives and organising fairs, seminars and workshops to improve the technology transfer. This focus is often supported by national institutions such as the Innovation and Technology Fund, the Innovation Agency and the Chamber of Commerce, the latter with a strong regional and local basis. All of these, mostly Styrian based, promotions and advisory agencies are strongly oriented towards the problems of NTBFs. The individual promotional initiatives attempt, through a phased or project oriented structuring of their supply of promotional services, to respond to the different needs of NTBFs during the foundation process. Over and above purely financial assistance, information and qualification services are offered which are extended through the contact arrangement activities *(Table 16.4)*.

Table 16.4 Problems of NTBFs and corresponding support measures

Information problems	Information meetings, procurement of co-operation partners, information about existing promotion schemes in Austria, Tech-Trend-Monitoring, market tests
Qualification related problems	Advisory services at low costs
Financial problems	Financial promotion services, e.g. subsidies for advisory services, seed capital, subsidies for R and D activities, etc.
Demand related problems	Procurement of contacts to customers, advisory services for customers, e.g. advantages of new technologies, how to use them, etc.
Competition related problems	Provision of patent data information

Styrian NTBFs are also well represented among the total of Austria's NTBFs in the framework of the seed financing programme of Austria's Innovation Agency: in the period from 1989 to 1993 their number was overproportional, and the relative amount of money received by Styrian firms was even stronger, as *Figure 16.4* and *Figure 16.5* show[7].

---

7 *Figure 16.5* has to be interpreted with care. The Austrian Innovation Agency only supports newly founded high-tech firms in their first development stage and the total funding offered is quite limited. It is therefore just a small number of firms which can take advantage of the support scheme. Thus the fluctuations shown by the curve can be traced back to an "effect of small numbers".

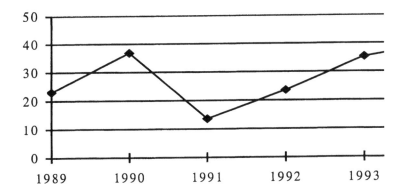

Figure 16.4   Percentage of Styrian NTBFs supported by the seed financing programme

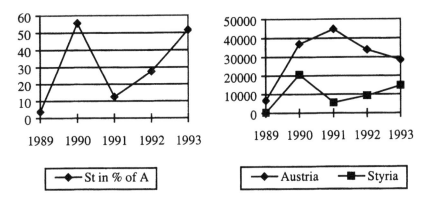

Figure 16.5 Total financial support by the seed financing programme (in 1,000 AS)

In the face of the specific problems of NTBFs described above (see *Table 16.4*) and their more familiar problems closely related to innovation, e.g. external effects, deficient protection against imitators, high risk, network externalities, etc., the established initiatives are characterised by a number of weaknesses, which create a need for strategies of renewal.

In 1993 the regional government of Styria together with other regional actors decided to develop a comprehensive "regional technology policy concept" to propel innovation and economic development in Styria. Numerous empirical investigations of the economic and technological situation of Styria were conducted to reveal the strengths and weaknesses of the region. Based on the empirical results main strategies and several different measures and initiatives were proposed, discussed with political decision-makers, representatives of promotion agencies, transfer institutions, Styrian firms etc. and reconciled with their suggestions to design an effective set of innovation policies which fit the needs of Styrian businesses and at the same time take into account the various political restrictions. Besides many other aspects the "technology policy concept for Styria", finalised in 1995 and implemented in the present years, includes recommendations about three strategies to improve existing promotion schemes directed at NTBFs.

Improve existing schemes
- Educational and advisory institutions should offer services specialised to the needs of the NTBF, e.g. courses in project management, business plan construction, etc. to deal with qualification related problems.
- (Semi) public institutions could offer technology, competition and commercial watch services to cope with informational problems. This would help new entrepreneurs to form rational expectations about competitors, customers and the structure and performance of their future market.

Integrate existing schemes
- Different promotional institutions should co-ordinate their activities much more to develop consistent promotion schemes orientated towards the needs of NTBFs during the process of foundation. They should start with educational and advisory services, continue with informational services and financial subsidies and end with the provision of venture capital to finance market penetration and growth of the new firms.

Supplement existing schemes
- It is the risk of failure which is responsible for the demand related and financial problems of NTBFs. If they go bankrupt, customers have to dispense with after sales services and with future updates of the product and financiers have to forego their investments.

- Promotion agencies should be well informed of the risks of the NTBFs assisted (managed) by them. If these agencies were to actively signal the prospect of success of NTBFs to customers and financiers, their contract-risk could be considerably lowered.
- The institutional network established to co-ordinate promotional institutions should also include banks and other financial intermediaries to mobilise adequate amounts of capital and loosen the financial restrictions of NTBFs.
- The ability and willingness of NTBFs to co-operate with other firms, e.g. jointly using distribution systems and resources, performing R and D activities, etc., should be supported, by financially assisting co-operative innovation projects of different firms, which are characterised by different comparative advantages.

Summary and Final Conclusion

This chapter is based on an empirical inquiry into the problems and the situation of new technology-based firms in Styria. The main puzzle was to obtain appropriate data to describe and analyse the situation of NTBFs in Styria. To cope with this deficit a twofold approach was used: firstly, an analytic framework was introduced identifying major determinants of success and failure for NTBFs and serving as a tool for obtaining appropriate information about their situation and problems in Styria via expert interviews and for an accurate interpretation of this information. Secondly, address data made available by WIFI Steiermark was used to determinate the regional distribution of NTBFs within Styria. The main problems of Styrian NTBFs revealed by expert interviews can be classified as follows:

- Information problems: very often NTBFs are loosely informed about their competitors and the needs of their customers; in addition they are mostly not aware of their deficits.
- Qualification-related problems: most of the young entrepreneurs have a bias towards the technical aspects of their products and lack management qualifications.
- Financial problems: though dependent on external funding many entrepreneurs are not willing to share management with external

partners supplying equity; moreover, the supply of equity is too limited, especially in the seed and start-up stages of the new business.

-   Demand-related problems: for many new high-tech firms the Styrian market is too small, so they move to locations near their main customers; in addition some potential customers are very cautious and will not buy products from young firms without any reputation as suppliers.
-   Competition-related problems: competitive pressure is not too strong for most of the firms because they serve market niches, but competition can get fierce if they supply close substitutes to established firms.

Compared to some of the results generated by different researchers for other countries and regions, the problems identified are to a great part generic and not specific to Styrian NTBFs.

Looking at the regional distribution of NTBFs one can see the dominant position of Styria's central area as a location for them, followed by the industrialised area in Upper Styria, the semi-industrialised area and the rural area. In addition the technological specialisation of NTBFs and Styrian sub-regions in terms of existing research centres focusing on particular technological fields seems to be closely related. These findings indicate that, firstly, the stage of economic development is an important factor for new entrepreneurs in determining the location of their businesses and, secondly, the availability of research institutions attracts NTBFs or raises the number of spin-offs. In recent years much has been done to support the creation and development of technology-based firms by public or semi-public promotion agencies. The technology policy concept for Styria, however, proposed three strategies to further improve the situation. Existing support schemes for Styrian businesses should be: 1) directed more closely at the problems of NTBFs, 2) co-ordinated with each other to fit the needs of NTBFs during the process of creating a new business, and 3) supplemented by new services improving NTBFs' relationship to customers, their collaborative behaviour and their access to private funding.

All these are small but important steps for creating an entrepreneurial spirit and an "active space". There are many dimensions of it, yet NTBFs represent an essential one. From the perspective of the Styrian economy they play, despite all their difficulties, a major role in its renewal. Keeping up the pace of renewal and activation will, to a large extent, depend on the development of NTBFs.

References

Binks, M. (1996), 'The Relationship Between UK Banks and their Small Business Customers', in R. Cressy, B. Gandemo, C. Olofsson (eds), *Financing SMEs - a Comparative Perspective*, NUTEK, Stockholm.

Brooks, H. (1994), 'The Relationship between Science and Technology', *Research Policy*, 5 (23), pp. 477-486.

Cressy, R., Gandemo, B. and Olofsson, C. (eds) (1996), *Financing SMEs - a Comparative Perspective*, NUTEK, Stockholm.

During, W. and Oakey, R. (eds) (1998), *New Technology-Based Firms in the 1990s, Vol IV*, Paul Chapman, London.

European Venture Capital Association (EVCA) (1995), *Yearbook*, Zaventem.

Geenhuizen, M. van, and Ratti, R. (1998), 'Managing Openness in Transport and Regional Development: An Active Space Approach', in K. Button, P. Nijkamp, H. Priemus (eds),*Transport Networks in Europe. Concepts, Analysis and Policies*, Edward Elgar, Cheltenham, pp. 84-102.

Geldner, N. (1989), 'Das Altern von Wirtschaftsregionen', in *WIFO Monatsberichte*, 5 (89), Wien, pp. 364-356.

Gruber, M., Hartmann, Ch., Schrittwieser, W. (1996), 'Glück im Winkel oder Zentrum Südost', *Raum*, 21 (96), pp. 8-10.

Gruber, M., Mörtlbauer, U., Habsburg-Lothringen, C. and Pieber, E. (1998), *Wirtschaftsbericht Steiermark*, Joanneum Research, Graz.

Gundrum, U. and Walter G. (1995), 'Die Rolle von kleinen und jungen Technologieunternehmen in der industriellen Innovation', in M. Steiner (ed), *Regionale Innovation*, Leykam, Graz, pp. 137-143.

Heertje, A. (ed) (1989), *Technische und Finanzinnovation*, Frankfurt am Main, pp. 64-91.

Jud, Th. and Piber, H. (1996), 'Der steirische KFZ-Cluster - Chancen und Risiken für Graz', *Raum*, 21 (96), pp. 11-13.

Kay, J. (1991), 'Economics and Business', *Economic Journal*, 101, pp. 57-63.

Koschatzky, K., Breiner S., Gundrum, U. and Reger, G. (1993), *Standortvorausetzungen und Fördermaßnahmen für High-Tech-Unternehmen in der Region Rhein Main*, Fraunhofer-Institut für Systemtechnik und Innovationsforschung (ISI), Karlsruhe.

Koschatzky, K. and Kulicke, M. (1993), Policies towards Technology-Based Companies in a Regional Context, paper presented at the International NITSEP Workshop on Regional S&T Policy Research "Regionalisation of S&T Resources in the Context of Globalization", 14th.-15th. June 1993, Iwate.

Kulicke, M. (1990), Entstehungsmuster junger Technologieunternehmen; Projektbegleitung zu den Modellversuchen "Beteiligungskapital für junge Technologieunternehmen (BJTU)" und "Förderung technologieorientierter Unternehmensgründungen (TOU)", Fraunhofer-Institut für Systemtechnik und Innovationsforschung (ISI), Karlsruhe

Kulicke, M. (1993), *Chancen und Risiken junger Technologieunternehmer - Ergebnisse der Modellversuchs "Förderung technologieorientierter Unternehmensgründungen"*, Heidelberg.

Mowery, D. and Rosenberg N. (1979), 'The Influence of Market Demand upon Innovation: A Critical Review of some Recent Empirical Studies', in *Research Policy*, 8.

Murray, G. C. and Lott, J. (1995), 'Have UK Venture Capitalists a Bias against Investment in New Technology-based Firms?', *Research Policy*, 24 (95), pp. 283-299.

Oakey, R. (ed) (1994), *New Technology-Based Firms in the 1990s, Vol. I*, Paul Chapman, London.

Oakey, R. and During, W. (eds) (1998), *New Technology-Based Firms in the 1990s, Vol. V*, Paul Chapman, London.

OECD (1993), Small and Medium-sized Enterprises: Technology and Competitiveness, OECD, Paris.

Palme, G. (1989), 'Entwicklungsstand der Industrieregionen Österreichs', *WIFO Monatsberichte*, 89 (5), Wien, pp. 331-345.

Philpott, T. (1994), 'Banking and New Technology-Based Small Firms: A Study of Information Exchanges in the Financing Relationship', in R. Oakey (ed), New Technology-Based Firms in the 1990s, Paul Chapman, London.

Prakke, F. (1989), 'Die Finanzierung technischer Innovationen', in A. Heertje (ed), *Technische und Finanzinnovation*, Frankfurt am Main, pp. 64-91.

Ratti, R. and Reichman, S. (eds) (1993), *Theory and Strategy of Border Areas Development*, Helbig & Lichtenhahn, Basel.

Ratti, R. (1997), 'L'espace regionale actif: une response paradigmatique des regionalistes au debat local-glocal'. *Revue d'Economie Regionale et Urbaine* (4).

Steiner, M. (1984), *Alte Industriegebiete - theoretische Ansätze und wirtschaftspolitische Folgerungen* (PhD Dissertation), Uni Graz.

Steiner, M. and A. Belschan (1991), 'Technology Life Cycles and Regional Types: An Evolutionary Interpretation and Some Stylized Facts', in *Technovation*, 11 (8), pp. 483-498.

Steiner, M., Sturn, W. and Wendner R. (1993), *Wirtschaftspark Obersteiermark*, Joanneum Research, Graz.

Steiner, M., Jud, T., Pöschl, A. and Sturn, D. (1996), *Technologiepolitisches Konzept Steiermark*, Leykam, Graz.

Sweeney, G., (ed) (1985), *Innovation Policies - An International Perspective*, Frances Pinter, London.

Sweeney, G. P. (1987), *Innovation, Entrepreneurs and Regional Development*, St. Martin's Press, New York.

Tichy, G. (1982), *Regionalstudie Obersteiermark*, Österreichisches Institut für Raumplanung (ÖIR) / Österreichisches Wirtschaftsforschungsinstitut (WIFO), Bericht der Gutachter, Wien.

Young, M. (1996), 'Banks and Smaller Firms: the British Experience from the Banks' Viewpoint', in R. Cressy, B. Gandemo and C. Olofsson (eds), *Financing SMEs - A Comparative Perspective*, NUTEK, Stockholm.

17 The Search for the Best Location of a Free Zone at the Eastern Border of Germany

ARNDT SIEPMANN

In this chapter an attempt is made to evaluate future locations for a free zone in Saxony, Germany. The idea is to create a market place for international business where import-export-savings can be realised as a result of the beneficial framework conditions given by a special custom status. Three potential locations for the free zone are discussed. First, Leipzig which is often characterised as the gateway to the East; secondly, the Euregions, because they are seen as the best locations for cross-border interaction; and thirdly, Dresden which has never been seen as a logistic node in this respect, although various data suggest that it has the potential to be the best location within Saxony. The free zone discussed in this chapter is a strictly transport related regional development tool. Transport related indicators are analysed as a case study for the "active space approach" to identify the location with the highest level of openness and sustainability possible.

Introduction

With the political changes in Europe and Germany in 1989 and 1990, the Saxonian economy faced an enormous downward trend. Companies began to experience strong competition because of the opening and integration of the market with former West Germany and the European Union. At the same time, the companies lost their old markets in Eastern Europe due to the change of currency brought about by monetary union in Germany. As a result, there was a breakdown of the industrial base to approximately 20 per cent of the former production level. Although the export-quota of the gross domestic product stayed more or less the same, 20 per cent (1989) and 26.5 per cent (1995), the trade balance of the New Länder decreased tremendously from a balanced situation in 1989 to a deficit of DM 220,5 billion in 1995, tendency which is still rising. As a consequence, 30 to 40 per cent of the workforce in Saxony is unemployed. Given that former East Germany had full employment as well as a much higher activity rate, one can imagine the economic and social destruction caused by the break down of the economy (e.g. Lange and Pugh, 1998).

Given the long tradition in manufacturing and the broad industry base established in the area since the 1950s, the economic development authorities of the re-established state of Saxony, which covers the largest territory of the industrial South of the former German Democratic Republic (GDR), strongly prioritised the support of the secondary sector. Alongside marketing efforts, the strategy has been to help local companies to find suitable partners in low-income countries like the Czech Republic and Poland. This would enable companies in Saxony to outsource labour-intensive parts of their production and to use this as a price advantage. The region in Saxony with a relevant number of interactions between local business and companies in the adjacent states in the form of joint ventures etc., or a region that already functions as a node for East-West- and West-East-flow of goods, would be seen to be the "candidate" most able to provide an innovative environment to realise the benefits given by a free zone.

Globalisation of the Markets and the Logistic Sector of the Future

Based on the new political situation and changes in supply and demand, trade volumes between national economies world-wide have increased more rapidly than the growth rate of production over the last 30 years. The volume of global trade of goods and services was measured at $60 billion in 1950, at $300 billion in 1970 and exceeds $5 trillion today (Mervosh, 1996). This development is seen as one of the two main indicators for the internationalisation of economic activity (Dicken, 1992). The scale and the complexity of international investment are the second key indicator for this development. Both indicators reflect decision patterns which are essential to the question of this chapter: what is the role of a free zone for a regional economy in the light of the globalisation of economic activity and how does this tie into the best location of a free zone?

The rising trade volume includes to a large extent intra-company trade by trans-national companies (TNCs)[1]. These companies are the most influential actors in the accelerated development of a global economic system. Dicken sees an indication for this importance in the growth rate of TNCs which surpassed the growth rates of the world gross national product and the world exports in the 1960s. Companies built new strategic alliances and restructured their production process horizontally and vertically to accommodate the globalisation process. The requirements for suppliers have also changed. Hence, over time, the production chain has become organised differently in space.

The "space-shrinking" technologies, transport and communication, facilitate the globalisation of economic activity by enabling operation over vast geographic distances (Dicken, 1992: 103, 192; Mervosh, 1996: 18). The development forces cause fundamental changes in the demand of transport services. Changing requirements of storage, volume, time, distances and responsibility of logistic procedures cause a structural shift in the sector. Although this means more transportation services overall, the market is also becoming more competitive. Accordingly, cost reduction potential plays a major role in the competition among shippers, carriers, modes and locations. The "space shrinking" technologies can also be seen as the driving, not the enabling, factor for the globalisation, as companies have to react to increased competition from companies in other regional markets. Again, TNCs play a major role in this globalisation process for all industries as they manifest the internationalisation of capital (Dicken, 1992).

If locations become the subject of evaluation, TNCs are a very lucrative target for regions competing for their share of Foreign Direct Investment (FDI). This despite the fact that the economic effect brought about by FDI for the selected region is subject to controversial discussions in the literature (see also Krätke in this volume). Authorities in charge for the development of each region have to answer the question of what role they want their region to play in a global system. With this vision of regional development the efforts and the investment put into infrastructural assets (like a free zone) can be streamlined to attract a share of the nearly unrestricted flow of

---

1 Dicken uses the term intra-*firm* trade in distinction to international trade (Dicken, 1992: 48).

capital. By doing so, specific industries will be provided for better than others with a solution for segments of their production chain, the latter in its widest sense, at a specific location.

The question arises: what role will the low income regions adjacent to Saxony play? On the one side the relatively low income locations in Poland and the Czech Republic, although both with different strengths and developments of their industries in the past, on the other side other parts of the former GDR, since 1990 reunified with the German economy with its general characteristics of a strong currency and a high income level. Over the last years, the development of substantial investment in fairly labour intensive manufacturing operations in Poland and the Czech Republic and the potential match of added value services like design, marketing, quality control, distribution, etc., and high value / low volume manufacturing goods in Saxony, has given hope for high quality linkages in the future[2].

To accommodate the logistic demands of industries, regional authorities invest in their infrastructure to position themselves as a capable partner for the logistic industry. So-called downstream services, which include transport and distribution aspects in the production chain, have been acknowledged as a source of competitive strength and added value for complex operating companies. It will be shown in more detail in Section 3, how the free zone can play an instrumental role for companies with globally integrated competitive strategies. From a sectoral perspective, TNCs with leading activities in mass-production of technology products or technologically more sophisticated products show a strong affinity with free zone features.

The decision by the economic development agency in Saxony to evaluate the establishment of a free zone was made in the light of the understanding, that the geographic distribution of economic activity is partly the outcome of the interaction between the strategies of TNCs and the settings of the national economies ruled by governments. A free zone can be seen as a tool which offers strategic opportunities for business reorganisation. The characteristic openness of such a free zone, which is under the control of the local custom authorities, creates opportunities for easier access to the region from abroad or distant markets and savings when handling specific logistic procedures. At the same time the free zone allows interaction with local business. In this matter, a free zone (in Saxony) seems

---

2   Dicken talks about these linkages as the result of a complex division of labour (Dicken, 1992: 87).

to follow the trend seen by Dicken (1992: 220) towards a mixture of arrangements to nearby and further located businesses. Consequently, a free zone can be viewed as an active factor functioning as a logistic platform between different regional economies: the selected location in Saxony on the one side and the adjacent Eastern European countries on the other side (see also Cappelin and Batey, 1993). This flourishing interaction would not only be beneficial from an economic standpoint. By creating an organisational structure in the production chain which makes use of the specific geographic system of an adjacent low labour cost country and a high labour cost country, transport distances as well as negative environmental impacts may shrink to a minimum.

The question of social sustainability does not receive a final answer in this chapter, mainly because the research concentrates on economic figures only; however, based on data on the development of unemployment and the development of the population, one might get a first idea on the perspective for the people in the regions under discussion.

Forms, Function and Use of Free Zones World-Wide and Within the German Context

Various forms of free zones function world-wide as an instrument of regional policy to boost the regional economy (Grubel, 1982). Globally there are approximately 400-500 free zones, of different character and size[3]. A standard classification of these zones has never been provided because of the variety of type; however, a general distinction can be made between export-processing zones, mostly established in developing countries, and free ports / free trade zones (Dicken, 1992). The latter are to be found traditionally in seaports and in some cases at inland locations as part of logistic hubs. This pattern can be easily explained by the general functions of these two forms. Export-processing forms are set up for actual manufacturing. Typically, governments offer a wide variety of tax-reductions and deregulation of social and environmental issues for companies when investing within these specially designed, and sometimes developed, zones. The aim is to attract mostly multinational manufacturing

---

3 This figure is the result of calculations based on several international statistics and a variety of sources like country studies etc. and should be seen as the best estimate possible.

operations for the consumer sector serving the global market. In these cases, a high level of openness is reached very quickly and a boost in export figures is achieved while it is not based on social and economic sustainability - this not only because of a lack of promising spin-offs and mixed experiences with foreign investment in terms of competitiveness of established local industries (see different development trajectories in van Geenhuizen and Ratti, 1998, and see Hunya, 1999). The oldest and most famous special zones are Hong-Kong and Singapore (Spinanger, 1984); however, today's media focus quite extensively on the more recently developed free zones in China and Mexico. Free zones in China cover relatively large regions and have had major economic impact on large territories of the state. In contrast, the impact of free zones in Mexico is much more local. So far, the important role of free zones in other less developed countries in Asia and Africa has not received appropriate attention.

From another perspective, free ports and free trade zones in the terminology of Dicken are involved only in warehousing and light manufacturing and are therefore seen dominantly as a tool to support the integrated logistics of companies. In that they leverage the transport potential of a region. This form of a free zone in the German context[4] is the subject of evaluation in this chapter where various indicators are used within the "active space" approach. The term free zone is used as an umbrella term for both general types. Both types of free zones can be seen as a special tool for companies with developed globally integrated competitive strategies. A company can locate a business unit in a free zone as part of its organisation of its economic activity. As an integrated part of the production chain, a location in a free zone can mean an advanced solution of a company's organisational and geographical dimensions.

Generally speaking, the success of any free zone depends on a variety of criteria: location, infrastructure, regulations, co-operation between authorities in charge and the operating corporations (see also Coopers & Lybrand, 1994). Given the fact that a zone's success is based on the performance of the company units contained in the zone more than the general setting of such a zone, the key for success lies in the right concept

---

4 The German language traditionally uses the term *Freihafen* based on all the long established port-based free zones, while the actual judical term for these special zones is *Freizone*; both will be translated here as free zone.

and successful recruitment of selected industries for that zone matching the strengths and weaknesses of the region. The Republic of Ireland provides two excellent examples of good programs. In each case, competitive advantages to investors are offered: the Shannon Free Market Zone in Ireland, established in the 1960s, is a good example of how, in the long run, a free zone can function beneficially. The authorities managed to create a suitable framework which made perfect use of internal and external systems capabilities. In this case, they used the transportation potential of the cross-Atlantic stops of the carriers, as well as the inexpensive labour force and the human capital provided by the local universities. Although the Shannon Free Market Zone has had to face substantial changes in response to the external system in the last decade, as today there is no need for planes to stop over in Ireland on the North Atlantic Route, it has been possible to stabilise the development through a long term learning process. Shannon Airport has become a very attractive location for airline maintenance, i.e. Lufthansa from Germany is one of the carriers which has relocated some of its maintenance to Shannon Airport.

Further, the Dublin Free Banking Zone is an extremely successful project from the 1990s; by offering a flat capital tax of 10 per cent until 2010 for financial operations in the special zone, it has attracted nearly 100 offices of banks and financial agencies, such as accounting firms and investment companies, most of them of foreign origin, and accordingly copied the success of the London Docklands. In a similar way, parts of the run down port area in walking distance from the main street in Dublin City have received substantial investment in form of an attractive office park with adjacent modern apartment complexes, a hotel, and a few restaurants and stores. The integrated design of the new business centre has added to the city as a whole. The main benefits for Dublin lie in the internationalisation of the market place through foreign representations. Next to the benefit from the direct investment and the ongoing tax revenue, the firms bring people with well paid jobs into Dublin; these then spend part of their income in the Irish economy. The firms on their side take advantage of a young Irish graduated workforce to whom they offer a start for an international career.The fact that the development agencies focused their recruitment efforts on such an internationally oriented and flexible industry was the key to the success of this well managed endeavour[5].

5 Information based on visits by the author in 1995 and 1997.

Examining the locations of free zones, which is of major relevance to this analysis, one learns that most of them are in coastal regions linked to big ports. Another location is the famous Industrias Maquiladoras all along the Mexican-U.S. border line. The only two inland free zones in Germany were set up in the late 1980s, both along two major waterways: the Freihafen Duisburg on the Rhine and the Freihafen Deggendorf on the Danube. Interviews with entrepreneurs located here were used to draw a picture of their experience. The conclusions were that some of the companies benefited significantly from their investment; however, others used their facilities only occasionally or without the specific customs advantages of a free zone. This observation ties into the analysis of this study. The location of a free zone in Saxony is only of interest for companies with a relevant amount of trade with non-EU countries and countries of Eastern Europe which have not signed associated agreements with the EU, like the Czech Republic, Hungary or Poland, since these agreements indicate a preferred trading status in their very nature (Fratcher, 1993). The savings for a business when handling its interactions through a free zone depend on the relevant customs benefits. The key factor used to measure this is to know the local content of a product throughout its production chain. For this reason, each operation in a free zone has to be examined separately.

Location Analysis for a Free Zone in Saxony

The traditional market-place Leipzig

Due to its image as the *Gateway to the East*, Leipzig has always been considered to be the best location for a free zone. This idea was based on complex prejudice based on traditions, promotions, image and perception: in overall opinion Leipzig has been judged to be *the Boomtown of the East*. Needless to say, this is not the result of a proper analysis of the situation. The good location, i.e. the configuration of the networks, especially the physical and the social-economic networks, and the city's carrying capacity in terms of human and capital resources, seem to draw one single positive picture of the Leipzig location.

In a study done for the Economic Development Agency of Saxony (Coopers & Lybrand, 1994) which was considering promoting the establishment of a free zone, the economic position of Leipzig was analysed by assessing primary data on (i) the activity around the Leipzig Fair, (ii) the role of freight transport handled by the Leipzig Airport and (iii) the planning concept and development of the Intermodal Centre Leipzig. The promotion of the Leipzig Fair as a place of communication, with a strong emphasis on functioning as a nodal point between Eastern and Western Europe is one of the most successful marketing efforts completed in the region. The numbers of companies present and experts visiting the fair from Eastern Europe are quite reasonable, especially considering the new concept of the fair and the competitive market of the trade show business over the last few years. By contrast, air freight forwarding at Leipzig Airport and the Leipzig Intermodal Centre have not fulfilled their roles in contributing to Leipzig as an international hub for freight transport. The Leipzig Airport is often overestimated in its role as a switch between ground and air transport for freight. Firstly, an analysis proved, with regard to the small volume (about 2,000 t/month) and the narrow range of goods, e.g. documents, medical goods, fair goods, and personal belongings, that there is no industrial customer using air carriers via Leipzig as a regular mode of transport; however, when *Quelle*, Europe's largest mail-order house, begins to operate fully at its new centre in Leipzig, it will triple the freight volume at Leipzig Airport within a few weeks. Secondly, it is apparent, that Leipzig Airport does not function as a gateway between Eastern and Western Europe. There are a few direct flights to two or three destinations in Eastern Europe per week. Freight coming into Leipzig is mainly of German origin or comes through major international hubs like Frankfurt, London and Zurich. The major role of Leipzig Airport is within the regional distribution function; destinations for freight are located all over the Land Saxony, major parts of Thuringia, and northwards including Magdeburg and up to Berlin.

Similarly, the Leipzig Intermodal Centre will probably not contribute to Leipzig's position as an international freight hub; primarily, because of its concept which is too much focused on the city logistic part and is not arranged to attract international flows of goods to go through Leipzig. The concept was based on unrealistic assumptions concerning the development of the industrial base in the Leipzig Region and the legitimate, but underlying, interest of the developer of this project to maximise revenue from the land. In a study by CONMOTO (1993) there was therefore no promotion of the establishment of an international freight hub, but of a city logistic project. Furthermore, the planning procedure was characterised by political dissent which caused the project to be delayed for years. Hence, it is at present too late to attract investment through (inter)national and regional carriers. Meanwhile, most of the investment decisions for distribution and storage facilities in the region have been made and the demand for further industrial sites for logistic use is low[6].

Table 17.1 shows the conditions in the Leipzig Region in terms of "active space" development using the framework suggested by van Geenhuizen and Ratti (1998). From a transport angle, the framework aims to identify the main conditions for economic development regarding the region's capacity to manage the consequences of changes in openness. A similar approach is followed for the two other regional cases in this chapter.

Table 17.1 Conditions in terms of active space development in the Leipzig Region

Production Space	-	Breakdown of industrial base after reunification
	-	Crossing of main European highways (A4 and A9)
	-	Largest German railway station
	-	Airport without gateway function between East and West, despite highest air freight volume in Saxony and in the region
	-	Modernised transport and telecommunication infrastructure, but intermodal centre concept has mainly city logistic character
	-	Little interest shown by local industry for cross-border co-operation
Market Space	-	Trade place with long tradition
	-	Leipzig Fair well known world-wide
	-	Excellent location to reach German markets
	-	Development towards the second largest banking place in Germany, but fall back a few years later – speculative boomtown
Support Space	-	Public support for huge projects, but major dissent in planning process for intermodal centre
	-	State-of-the-art marketing within the New Länder, but little and superficial follow-up
	-	Dense high education and cultural network, but under-representation of high paying jobs

6 The best example for an early investment decision is the new main European distribution centre being set up by the above mentioned company *Quelle* near Leipzig.

The Saxonian border regions: participating in the cross-border networks of the Euregions

The border regions were seen as a promising alternative for a location of a free zone, mainly because of the expected cross-border benefits based on the interaction between regions belonging to countries, which vary significantly in standard of living and economic performance. Beside the difference in wealth, it is quite interesting that the German side of the border is one of the least developed regions in the national context, while the east side of the border, in Poland and the Czech Republic, is always seen as one of the most prosperous regions in their national context (*Figure 17.1*). The hypothesis was quickly drawn that the border regions on the *eastern* side of the border could benefit significantly from proximity to and interaction with the adjacent wealthier region while the *western* side was lacking that capability.

The main problem of the border regions on the western side is their periphery status in relation to the market system they belong to. Additionally, in the case of the eastern border of Saxony, as for all other regions along the outer European border, the regions are located at the edge of the relatively homogenous market of the European Union. The problems of this severe peripherality can only be overcome by an integrative approach of substantial public support. In the majority of the literature[7] one can find a strategy with the following four components to overcome the obstacles for economic development in the border regions:

- enlarge social-cultural interactions and understanding of the administrative systems and legislative structures,
- start cross-border spatial and infrastructure planning,
- support cross-border economic and regional policy, and
- reduce negative environmental impact.

Hence, the institutional approach was to follow the concept of Euregions. The first application of this instrument was put into practice at inner European Union borders about forty years ago and was extensively used as a part of the upcoming regional policy by the European

---

7 See for example Bundesministerium für Raumordnung, Bauwesen und Städtebau (Bundesrepublik Deutschland) and Ministerium für Raumwirtschaft und Bauwesen (Polen) (1993).

Commission. Through a special program established by the European Commission, non-European Union areas of the established Euregions were also legally qualified for program funds. In just a few years, Euregions were established in Saxony's border regions and the neighbouring regions of Poland and the Czech Republic. All cities and counties along the Saxony border participate in one of the four Euregions, one with Polish counterparts, three with Czech counterparts.

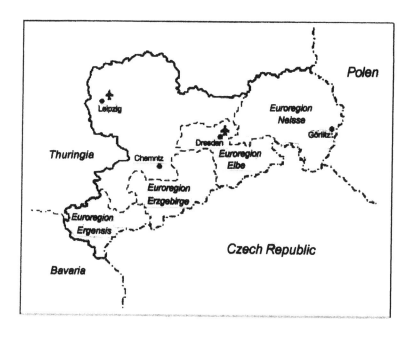

Figure 17.1   Location of Euregions in Saxony

In the past there was quite some success in the administrative co-operation of early established Euregions, such as along the Dutch-German border. The previously mentioned Mexican-U.S. border is also a promising case of economic revitalisation through cross-border activity for both sides of the border. As widely discussed in business magazines and in daily news, American corporations have identified these *Maquiladoras* on the Mexican side of the border as prime locations for labour intensive units of *their* production chains. The shift of operations from many industries in the U.S. into this border region has created ten thousands of jobs.

Three of the four Euregions in Saxony, however, are unsuccessful, despite the fact that locations in border regions in general have proved to be a promising alternative in a united Europe. An illustration of the latter statement is the decision by the Micro Compact Car AG, a joint venture between Daimler-Benz AG and the Swiss watch maker Nicolas Hayek (Swatch watches), to locate the manufacturing facility for the new SMART-car in the French-German border region of Lorraine. As a fairly rural region, positioned between two old industrial regions, the area provided great site conditions for this venture: a low cost location with strong local support, accessible industrial expertise through workforce and the existing nearby industry. The central location in Europe provides a good base to set up Just-in-Time supply structures; also the fact, that the plant is located in a French-German cultural environment helps to create a more European car for the European market. At the same time, the company can easily create an identity for the product for both the French and the German public. These soft factors should be instrumental to market the product in the two major national markets.

A strength-weakness analysis for each of the four Euregions in Saxony made clear[8], that only the Euregion Elbe-Labe will have the endogenous economic potential to benefit from the cross-border activity within the Euregion. In all other cases, the German parts of the Euregions cannot take advantage of the economic development of the neighbouring regions. The eastern side of the border regions, especially in Poland but also in the Czech Republic, do benefit from the close location to Germany. As the western side of the border is endowed with a much stronger currency, investment and consumer money flow noticeable to the East.

The positive judgement of the Euregion Elbe-Labe was based on the promising development factors of the city of Dresden and the neighbouring communities: large enough in population, with high-level central functions, strong educational institutes, solid companies and a rich cultural base, altogether excellent conditions for sustainable growth. This is mirrored in recent investment decisions by companies like AMD and Siemens in the semiconductor industries with most demanding site selection criteria

---

8 Four unpublished studies on each of the Euregions prepared for the Ministry of Economic Affairs of the State Saxony, Dresden.

including high availability of a skilled workforce, exceptional environmental conditions, research capabilities and quality of life.

Table 17.2 Conditions in terms of active space development in the German part of Euregions (Elbe-Labe excluded)

Production Space	-	Narrow economic base
	-	Poor infrastructure
	-	Little potential for cross-border benefits (joint ventures / outsourcing)
Market Space	-	Very little international business
	-	Location at the edge of the German and EU market
Support Space	-	Lack of marketing efforts
	-	Much emphasis on building administrative structures (Euregions)
	-	First priority by financial assistance (recruitment / expansion / tourism)
	-	National and supra-national (EU) help

The progressive economic performance by the Euregion Elbe-Labe as the only Saxonian Euregion with a population of roughly half a million citizens, gives strong support to the hypothesis, that economic development in the western part of border region is strongly dependent on the centrality of locations (Christaller). In other words, the institutional framework of the Euregion is not sufficient to stimulate growth, mainly because they are lacking capacity in respect to creative learning. Secondly one can expect major positive impacts of the international axis going through the Euregion Elbe-Labe connecting Hamburg and Berlin with Prague in the Czech Republic. The next section will give an in-depth analysis of the region of Dresden, including further evidence supporting the selection of this region for the location of a free zone.

Dresden: a Solid Alternative Location

In the light of the previous results, Dresden needs to be considered as an alternative to Leipzig as well as to any location along the Saxonian border within the Euregions Neisse, Erzgebirge or Egrensis. To find some more evidence, the analysis now proceeds with the size of freight flow into or out of either (a) Dresden Region, (b) Leipzig Region, (c) Nürnberg Region or (d) Görlitz Region. The regions are the *Verkehrsbezirke* (*Table 17.3*).

The freight considered in the analysis originates from the Czech Republic, Poland and from Eastern Europe in total, and covers the transport modes truck, rail and water (basically only the river Elbe). The first data are for the years 1992 and 1993 as new statistics have become available for the territory of the former GDR. The Nürnberg Region in Bavaria, which has not experienced a dynamic transformation to the free market system like the three regions in Saxony, was included for reasons of comparison. To prove the high potential of the Dresden location as a logistic hub for incoming goods, i.e. products with a structural relevance for any free zone activity, the analysis focused on two groups of products, namely (a) semi-finished and finished products and (b) machines and iron, sheet and metal ware, for 1992 and 1993 for the Czech Republic and Poland.

Table 17.3 Freight originating from the Czech Republic and Poland with destinations in four traffic regions in Germany*

	Leipzig		Görlitz		Dresden		Nürnberg	
	1992	Change 1992-93	1992	Change 1992-93	1992	Change 1992-93	1992	Change 1992-93
I	3,981	+ 33.7%	3,825	- 2.5%	7,708	+ 17.5%	6,207	+ 20.8%
II	3,987	- 0.9%	2,792	+ 32.9%	4,270	- 10.9%	14,962	+ 10.4%
III	1,870	- 9.4%	2,816	- 37.9%	3,494	+ 25.1%	2,487	+ 47.0%
IV	2,066	- 45.6%	844	+ 24.5%	9,745	- 67.3%	6,905	+ 5.2%
V	190,723	+ 59.2%	133,551	+ 89.2%	155,137	+ 57.9%	113,166	- 25.7%
VI	80,141	- 12.6%	148,075	- 26.4%	863,619	- 10.6%	205,482	- 25.4%

* Freight volume in tonnes for 1992 and percentage change in 1993. **I** = Semi and finished products originating from Poland; **II** = Semi and finished products originating from Czech Republic; **III** = Machines and iron, sheet and metal ware originating from Poland; **IV** = Machines and iron, sheet and metal ware originating from Czech Republic; **V** = All products originating from Poland; **VI** = All products originating from Czech Republic.

Source: Deutche Bahn

The analysis of freight originating from Eastern Europe proved that the Dresden Region had higher freight volumes than the other two Saxonian locations. The infrastructure in the Görlitz Region, which is almost equivalent to the Euregion Neisse, is apparently too weak to attract a higher freight volume. The figures for the Leipzig Region prove that its image as the logistic node for most of the trade flow with Eastern Europe is incorrect. The Nürnberg Region that attracts most of the monitored freight from the Czech Republic, but not Poland, can be defined as the major freight node in our analysis. The network approach can explain the good performance of the Dresden and Nürnberg regions due to the physical environment and the useable infrastructure. It is worthwhile mentioning the relatively high percentage of freight transport by rail: 20-30 per cent.

Dresden's high potential as a location for the transport sector is best explained from the perspective of openness, sustainability, and creative learning. It can be attributed to the excellent location, as well as the political support, human capital resources, and a number of successful local businesses. Success depends much on the step-by-step development within the region. *Table 17.4* contains a summary of the main indicators for a high degree of "active space development" of the Dresden Region.

Dresden's Development Trajectory in 1990-1995

In this section we will take a closer look at the actual socio-economic development path of the Dresden Region in the recent past, to find some more evidence for the selection of this region. The development of the Dresden Region needs to be compared with the Leipzig Region as well as the state's average figures for Saxony in order to draw reasonable conclusions. It is worth mentioning that the use of statistical data for the New Länder can only produce compromises and broad trend descriptions. This is due to various limitations of the statistical data which include: the changing shape of territories of counties and communities and other statistical units over time, the different approaches for data gathering, incoherent terminology and inconsistent data groupings. Given this situation, the most meaningful existing data are concerned with newly established and businesses shut downs over time. With the end of the planned economy, a major wave of business start-ups occurred, followed by an increased number of business shut downs. In addition, unemployment and population growth will be considered.

**Table 17.4   Conditions in terms of active space development in the
Dresden Region (including Euregion Elbe-Labe)**

Production Space	-	Located along the traditional E-W corridor into Poland
	-	Located along the international N-S corridor, Dresden - Prague
	-	Decision to build the A13 between Dresden and Prague
	-	Plans for a decentralised intermodal centre which also integrates the waterway Elbe
	-	The most frequent border crossing for freight transport on track along the German East border, bringing the freight flow straight into Dresden; Bad Schandau as border point
	-	Road-rail service for lorries from Dresden to Liberec (Czech Republic) and vice versa supports transport crossing to overcome the bottleneck of the road system
	-	Excellent location for high-tech companies, recently: Siemens and AMD, both in semiconductors
Market Space	-	Companies of the Chamber of Commerce Dresden show the strongest interest within Saxony for business co-operation in the Czech Republic and Poland
	-	Solid development of a balanced regional vision after changes of political and economic system
	-	Success of companies stand for healthy business climate
Support Space	-	Capital of the Land Saxony
	-	One of the top cities for culture and tourism in Germany
	-	An active modern and diverse university environment, a technical university with a transport department and many other research institutions
	-	An attractive surrounding landscape (high quality of life)

With regard to the spatial units of analysis, it was decided that the data on the actual cities would not mirror the development for the region correctly. The city boundaries are reasonably tight and new industries are mostly found outside the actual city district, due to reasons such as traffic congestion in cities, environmental concerns, etc. In the case of the New Länder, the uncertainty of ownership of land and in many cases pollution of the soil, are also forcing (re)location outside the towns. Hence, the best statistical territory for the investigation is the Regierungsbezirke. Saxony is divided into three of these administrative regions, for the present purpose two of them will be called the Greater Dresden and the Greater Leipzig

Region. The analysis of business establishments and shut downs (*Figure 17.2*) does not take into account the actual number of active businesses in these regions, but it compares the yearly difference between newly opened and closed businesses for each of the regions.

Over the first five years of available data, the Greater Dresden Region performed significantly better than the Greater Leipzig Region. This trend was so strong that it was not overshadowed by the fact that the Greater Dresden Region includes the declining area along the Polish border and the fact that the Greater Leipzig Region had a larger industrial base to start from. Dresden's development is similar to Saxony's overall performance. Saxony showed a plus of 2,500 companies for each of the years 1991 to 1993, and a firm increase of 2,914 in 1994 and nearly 3,200 in 1995. The estimates of the actual capital investment follow very much the same patterns.

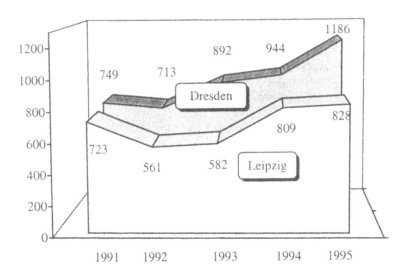

Figure 17.2 Surplus of business establishments and shut downs in the Greater Dresden and Leipzig Region

Source: Chamber of Commerce Dresden and Leipzig

The regional unemployment rates lend further support to the outlined trend. Data are available for labour market regions which include the neighbouring counties to each of the cities. These regions are slightly larger than what is called the metropolitan area in the American literature. Therefore, they give a representative picture of the economic impact of the city as a driving force for the adjacent regions. As a result of the planned economic system, in 1989 unemployment was by definition 0 per cent. In 1990, the year of reunification, the collapse of companies was imminent, but lay-offs occurred slowly. Unemployment rates for 1990 are not available for several statistical reasons, but official unemployment rates are available for 1991 to 1995 (*Figure 17.3*).

The unemployment rates appear to be fairly high, despite the fact that the labour market is highly subsidised by the Federal Agency for Labour.

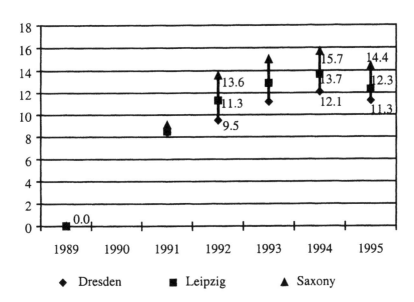

Figure 17.3 Unemployment rates for the Dresden District, Leipzig District and Saxony

Source: Sächsisches Staatsministerium für Wirtschaft und Arbeit

Using early retirement and training programs the unemployment rates are kept artificially low. In addition to the rate of official unemployment as shown in *Figure 17.3*, one would have to add at least twice as many people without jobs to get to the actual number of people who were working before. Hence, at the peak of the development approximately one third or more of the workforce were actually without a job, unemployed. Nevertheless, one finds proof that the Dresden labour market constantly performed slightly better than the Leipzig labour market.

Finally, the most basic indicator to describe the development of a region is the population trend as a reflection of people's perception of their future in a specific region. Both cities follow the general pattern of a heavily declining population in the New Länder (*Figure 17.4*). Although the population trend in the adjacent counties is stable with slight growth due to the mass housing projects on former farmland around the cities, the trend of population decline has not yet stopped.

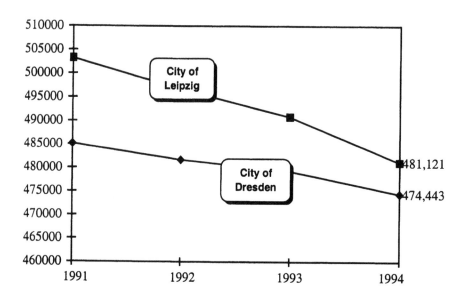

Figure 17.4 Population of the city of Dresden and the city of Leipzig

Source: Sächsisches Staatsministerium für Wirtschaft und Arbeit

The population decline is more detrimental to Leipzig. The causes for the continuous decline are well documented: the exodus of the population, the drop in birth rates, and the increase in death rates due to the ageing of the population. These factors also cause a shift in population structure, which makes matters worse for development.

It has become clear for all the selected indicators that the Dresden Region has performed significantly better for the last five to six years than the Leipzig Region, its most serious competitor in terms of our analysis. Although the transformation process for Dresden has not ended, the region is managing the rapidly unfolding openness in an active way. This is due to the fact that Dresden's capabilities to create a supportive market-place through knowledge and institutional structures have been a steady factor for the last few years. Sufficient support for sustainable projects in transportation, education, and environment have caused not only economic growth but have led to stabilised social structures.

Saxony as all other New Länder is still dependent on substantial financial transactions from the national and supranational institutions and agencies. As long as the costs of transformation are not reduced and the financial burden is not covered by tax revenue from the New Länder themselves, the costs of openness in terms of market and regional competition, and the costs of sustainability in terms of regional development and social stability, the territory will remain dependent on external sources of finance. The Dresden Region is nevertheless showing the potential to create enough momentum for sustainable regional development.

Conclusion

The analysis presented in this chapter indicates that the Dresden Region is the most amenable region to establish a free zone in Saxony. First, the flow of international freight was analysed. Secondly, indicators within the "Production Space, Market Space, and Support Space" approach were used. At this point, it could be clearly seen that the Dresden Region has established the highest capacity for using the given regional potential. Looking at the logistic sector, neither the Leipzig Region nor the Euregions were in the position to offer the necessary infrastructure, the transport potential, the economic growth, or the social stability for a free zone; however, it seems, that the political will to promote a free zone for Saxony does not exist at the present time.

Reviewing the application of the "active space" approach, the integrative concept appears to be provocative. The strength of the approach certainly lies in the fact that the classification of regional development gives a systematic picture of the possible development trajectories of regions. Evidence on the use of the broader transport potential has confirmed a favourable development path for the Dresden Region; however, to operationalise the regional development path based strictly on transport was seen as a difficulty. Here lies an interesting field for further research, i.e. the integration of transport and various other data to monitor the capacity of a region to grow in an "active way". A true regional impact analysis of free zones can only be achieved by focusing on the performance of the companies in such zones. One cannot describe the success of a free zone without knowing *the free zone specific business* of companies. As global corporations increasingly make use of the benefits of free zones for their international business, there is a need for a comprehensive study of a free zone's impact on the regional economy based on company data on their free zone specific activities.

References

Bundesforschungsanstalt für Landeskunde und Raumordung (1992), *Räumliche Folgen des politischen und gesellschaftlichen Strukturwandels in Osteuropa. Interne und Externe Auswirkungen*, Bonn.

Bundesministerium für Raumordnung, Bauwesen und Städtebau (Bundesrepublik Deutschland); Ministerium für Raumwirtschaft und Bauwesen (Polen) (1993), *Raumordnungskonzept für den deutsch-polnischen Grenzraum*. 1. Zwischenbericht, Essen-Warschau.

Cappelin, R. and Batey, P.W.J. (eds) (1993), *Regional Networks, Border Regions and European Integration*, Pion, London.

CONMOTO (1993), GVZ Leipzig, *Logistische Bedarfe und Anforderungen seitens potentieller Betriebe*, München (unpublished paper).

Coopers & Lybrand (1994), *Gutachten über die zoll- und außenwirtschaftsrechtlichen Rahmenbedingungen und die wirtschaftliche Relevanz einer Freizone im Freistaat Sachsen*, Hamburg.

Dicken, P. (1992), *Global Shift. The Internationalisation of Economic Activity*, New York.

Fratzscher, O. (1993), *The political economy of the Central European free-trade-agreement - a catalyst towards pan-European integration?*, Cambridge.

Gabbe, J. and Malchus, V. von (1993), *Grenzüberschreitende Zusammenarbeit an der deutsch-polnischen Grenze. Entwicklungsstand, Perspektiven und Anforderungen*, Gronau.

Geenhuizen, M. van, and Ratti, R. (1998), 'Managing Openness in Transport and Regional Development: An Active Space Approach', in K. Button, P. Nijkamp and H. Priemus (eds), *Transport Networks in Europe. Concepts, Analysis and Policies*, Edward Elgar, Cheltenham, pp. 84-102.

Grubel, H. G. (1982), *Towards a Theory of Free Economic Zones*, Weltwirtschaftsliches Archiv Bd. 118, S. 39-61, Tübingen.

Hunya, G. (ed) (1999), *Integration Through Foreign Direct Investment*, Edward Elgar, Cheltenham.

Kommission der Europäischen Gemeinschaften (1994), *Wettbewerbsfähigkeit und Kohäsion: Tendenzen in den Regionen. Fünfter Periodischer Bericht über die sozioökonomische Lage und Entwicklung der Regionen der Gemeinschaft*, Brüssel-Luxemburg.

Lange, T. and Pugh, G. (1998), *The Economics of German Unification*, Edward Elgar, Cheltenham.

Lichtenberger, E. (ed) (1991), *Die Zukunft von Osteuropa. Vom Plan zum Markt*, Wien.

Mervosh, E. M.(1996), Globalization, *Global Sites and Logistics*, New York, vol.1, 4, pp. 14-21.

Organizacion International Del Trabajo (1987), *La Industria Maquiladora en Mexico*, (Documenta de trabajo 49), Genf.

Ratti, R. and Reichman, S. (eds) (1993), *Theory and Practice of Transborder Cooperation*, Helbing & Lichtenhahn, Basel.

Sächsisches Staatsministerium für Umwelt und Landesentwicklung (1994), *Landesentwicklungsbericht 1994*, Dresden.

Scott, James W. (1993), *Transboundry Regionalism on the Polish-German Border: Problems and Perspectives*, Berlin.

Spinanger, D. (1984), 'Objectives and Impact of Economic Activity Zones - Some Evidence form Asia', *Weltwirtschaftliches Archiv*, Bd. 120 , S. 64-89, Tübingen.

Statistisches Landesamt des Freistaats Sachsen, *Statistische Jahrbücher 1992, 1993, 1994, 1995*, Kamenz.

18 European Policies for European Border Regions: An Active Space Approach?

FABIENNE CORVERS

This chapter contains an analysis of two policy programmes initiated at Community level and intended to foster the regional development potential of European regions and border regions. The first programme is called INTERREG and is directed towards European border regions. The second programme is called RITS/RIS and deals with regional innovation policy. In order to assess whether the "active space" approach could be of use to European border regions and their economic development potential, this chapter takes a closer look at both programmes. Both INTERREG and RITTS/RIS correspond with a bottom-up approach to regional development in policy formulation and implementation. Based on the experiences from INTERREG and RITTS/RIS, this chapter outlines some tips for border regions on how to reinforce their development potential and, in conclusion, how academic research could help them by answering the right questions.

Introduction

The European Union's membership has been expanding in the 1990s, with Austria, Finland and Sweden joining in 1995, and it can be anticipated to expand further in the near future. As a result of the Amsterdam Summit of June 1997, it was agreed that accession discussions would start with Poland, Hungary, Czech Republic, Slovenia, Estonia and Cyprus, as they stand the best chances to be able to fulfil the economic and political conditions required for membership. In March 1998 the first European Conference was held for all ten applicant countries from Central and Eastern Europe, setting an historic date for Europe.

It is beyond doubt that Europe is going through an unprecedented change, not only politically but also in a regional economic respect. The removal of political borders is bound to be one of the most important dynamics of regional change in Europe in the 1990s. The Single European Act and consequent elimination of borders since January first, 1993, have paved the way for free movement of people, capital, goods and services. Simultaneously, Eastern Europe has opened its borders and looks towards integration with a new Europe; however, vanishing borders do not automatically imply openness to the benefit of all regions involved. Vanishing borders also means the opening of regional and urban economies to many new social, economic, technological and political influences.

Regions, therefore, are no longer merely administrative or geo-political areas, but tend to become spatial mappings of socio-economic force fields (see also van Geenhuizen and Ratti, 1998). The traditional distinction into economic and geopolitical regions has become more relevant in recent years, in particular in Europe where the completion of the internal market has led to a declining importance of national borders. As a consequence, new modes of co-operation are emerging in which nation states play a less prominent role, but cross-border regions are increasingly gaining importance (see also Nijkamp, 1993; Ohmae, 1990, 1995). Examples can be found in new modes of cross-border co-operation (e.g. Euregions, the Alps-Adria area, the Rhone-Alps area). As a result, spatial competitiveness depends increasingly on the organisation of all human forces in a given space-economy (see also Martinotti, 1996).

The changes underlying spatial competitiveness have also encouraged a search for new regional development strategies, on the basis of a blend of self-reliance and network alliance (cf. Naastepad and Storm, 1996). Regions are conceived as a planning framework aimed at a better use of socio-economic potentials, by means of improved activity networks and improved competitiveness of regions, leading to new opportunities, particularly related to technologically innovative economic activity. The "active space" approach is a new approach to regional development which connects changes in openness of regions (related with transport) with the capability of regions to manage the consequences of these changes in openness. The key force in "active space" development is creative learning, based on a variety of local assets and the capability of the region to use these assets strategically and to their fullest potential.

In order to assess whether the active space approach could be of use to European border regions and their economic development potential, this chapter will take a closer look at the INTERREG initiative, short for "Inter + Regions" in this chapter. This is a Community initiative for European border regions intended to develop cross-border co-operation, contribute to the solution of problems of border regions and complete energy networks and link them to larger European networks. The budget for INTERREG II (1994-1999) equals 3,518 million ECU. The RITTS/RIS scheme is a Community action for regional innovation policy. The acronym RITTS stands for "Regional Innovation and Technology Transfer Strategies and Infrastructures" and RIS is short for "Regional Innovation Strategies". These two programmes are jointly managed by DG XIII (telecommunications, technology transfer and innovation policy) and DG XVI (regional policy). Their aim is to help regional authorities and regional innovation support agencies to develop an overall innovation strategy for the region taking into account the real (innovation) needs of the business sector.

Both INTERREG and RITTS/RIS correspond to a particular approach to policy formulation and implementation. It is this bottom-up approach that fits within the "active space" approach to regional development. Based on the experiences from INTERREG and RITTS/RIS, an attempt will be made in this chapter to give some tips on how border regions might reinforce their creative learning by distinguishing some rules of thumb and, in conclusion, how academic research could help in finding answers to burning questions (see also Ratti, 1997; van Geenhuizen and Ratti, 1998).

European Border Regions and the Emergence of INTERREG

After the ratification of the Single European Act in 1987, the completion of the Internal Market became a reality. The European Commission realised that the disappearance of the European internal borders on January 1, 1993, also implied the disappearance of the European border regions. Border-related problems however, would not disappear automatically, but could very well prevent the European integration from becoming a success.

Historically, border regions in Western Europe have almost never played an important role in the industrial development of a country. The few times that they have been integrated in the process of industrialisation occurred because of their natural resources, for example coal. Examples of industrialised border regions based on coal-fields can be found on the borders of Belgium and northern France, of The Netherlands and Germany, of Saarland and Lorraine and, prior to 1939, in the Upper Silesia industrial region on the Prussian German-Polish border (Mikus, 1986). Other examples of border regions whose economic position has been very important in their respective national spatial-economic structure are Basel and Antwerp as they are "economic core or intermediately located" border cities.

In most cases, border regions in Western Europe remained economically underdeveloped because of the danger of a military conflict, the agglomeration tendency of industry elsewhere and the impossibility of market expansion (Mikus, 1986). To this can be added their peripheral location (from the viewpoint of the respective capitals), the lopsided production structure and the shortage of (cross-border) infrastructure (Corvers, 1992)[1].

Although Western European borders have not been insurmountable barriers in the past, the Single European Market initiative has brought border regions to our attention. The completion of the internal market in 1993 gave a new and important impulse to enterprises, promoting a more European orientation, with important advantages of scale and scope. The completion of the internal market also stimulated a renewed interest in the economic potential and dynamics of border regions among policy-makers as much as among researchers. Firms located in border regions were no longer confronted with an incomplete hinterland and were given the unprecedented opportunity to take advantage of bigger sales potential, a larger supply of labour, cheaper purchasing possibilities and a greater number of knowledge suppliers such as universities, research institutes, laboratories, and other firms. This potential of hitherto unused sources of knowledge across the border could change the future economic development of the region as a whole dramatically.

As far as the European Commission is concerned, it was convinced that the border is the ultimate place where the success of the European integration will be proven. Large differences in production structure and production environment between areas on either side of the border were considered to hinder that success. The European Community has therefore

---

1 It should be noted, however, that border regions in economically weak countries bordering economically strong countries can benefit from their location as this location can stimulate a catching-up process. Examples of this advantageous position can currently be found in Eastern European countries along the borders of Germany and Austria.

put a lot of (financial) effort into promoting cross-border co-operation between these regions in order to solve border-related problems. The INTERREG initiative that was launched in 1990 stimulates cross-border co-operation in seven areas, namely: networking, information exchange and communication (1), traffic, transport and infrastructure (2), recreation and tourism (3), education and labour market (4), environment (5), technology transfer and innovation (6), research and project management (7). The initiative takes the form of projects and finances up to 50 per cent of the project costs. For the first INTERREG period (1991-1993) 800 million ECU was made available for all European border regions for a period of three years. The total budget for the second INTERREG period (1994-1999) will be around 3,5 billion ECU for a period of six years. INTERREG II also offers the possibility to finance projects related to drought, flooding and spatial planning issues.

The launching of this initiative led to the emergence of a new phenomenon, the "Euroregion" or according to other authors "Euregion" or "Euregio". Euregion is an abbreviation of European region and indicates some form of cross-border co-operation, mainly between (semi-)public organisations (see also Perkmann, 1999). The European Commission was willing to finance cross-border co-operation on condition that border regions had some kind of organisation, ranging from gentlemen's agreements between public authorities to organisations set up in accordance with civil or public law. An Euregion therefore can be viewed as an organised border region. Although the European Community refers to Euregions as a more or less homogeneous category, they can be quite diverse in terms of their geographical size, population density, economic characteristics and degree of development. Moreover, there are also significant differences in the bodies that have been set up to initiate, plan or implement cross-border co-operation and the degree of formal co-operation that has been established (Martinos and Caspari, 1990). Differences in the type of public authorities involved (local or regional), in administrative structure (de facto or according to civil or public law), in goals to be achieved as well as in the methods of financing cross-border co-operation, all add to this diversity (Kessen, 1992).

As far as European border regions are concerned, there exists a clear north-south divide between the more developed border regions in the North of the Community and the less developed border areas on the southern and western periphery of the EU (CEC, 1991). These border regions, in Ireland and the South of the Community along the border between Spain and

Portugal, Spain and France, France and Italy and along the external borders of Greece, are largely mountainous areas with under-developed economies. Population density in these regions varies from just over 40 to just under 100 people per square kilometre, compared to the EU average of 145 (CEC, 1991). In contrast, most border regions in the North of the Community, such as those along the Dutch-German, French-German and Dutch-Belgian frontiers, are densely populated in the range of 240 to 780 people per square kilometre (CEC, 1991; Corvers, 1992). These border regions are not divided by any dominant physical features and their economies are developed, yet divided by historical events. The border regions in the new Member States, Sweden and Finland, which are both peripherally located in the EU and under-populated with less than 8 habitants per square kilometre, are a special situation.

The situation for border regions along the eastern (external) borders of the EU is very different. What the single market opportunities for firms located in the southern and particularly the eastern border regions - along the external borders of the EU - will be is more difficult to predict than for firms along the internal borders of the EU. Firstly, because the external borders of the EU are likely to be replaced by a whole set of new borders in a few years time. Secondly, the lack of economic integration between regions on either side of the European eastern border, due to the former Iron Curtain, is likely to hinder development efforts. Thirdly, as a result of this, the administrative, legal and economic systems on either side of this border are much more divergent than those between EU countries. The ability to share public services (health, education, police) and public utilities (electricity, gas, telecommunications) to achieve economies of scale and efficient delivery will therefore continue to be severely limited because of these differences. Fourthly, the conditions for free movement of goods, services, labour and capital across the southern and western borders of the EU area rather different from those within the EU, where you have intra-European trade and cross-border infrastructure. Fifthly, border regions in the East will face (and have already faced) significant immigration flows from the East-European countries, including flows from the former Soviet-Union. All these events do not induce an economically and politically stable climate for firms. On the other hand, there are possibilities for firms in East-European border regions to produce for western firms, located on the other side of the border that want to hive off production in order to save costs, for example, the Czech-Bavarian border-regions.

Whether these opportunities generated by the European integration process are seized by the border region depends on three factors. The first factor is the level of economic development on either side of the border in terms of GDP per capita, unemployment rate, annual growth of employment and average annual household income. Big differences in economic development on either side of the border do not facilitate regional economic integration, although this of course also depends on how integration is understood (see also van Geenhuizen and Nijkamp, 1998). A second important factor, besides level of economic development, is concerned with the economic characteristics of the border region - in terms of production structure and production environment. Opportunities occur when such a region shares complementary sectors, e.g. steel industry on the one side of the border and mechanical engineering on the other side. A third factor is the role of the Euregion and particularly what local and regional actors make of it, the learning capability of the region. Cross-border co-operation between public/semi-public organisations in various policy areas can create favourable conditions for regional economic development. For the first time in the existence of border regions, cross-border planning, production and provision of public goods and services will become possible and create unprecedented opportunities for the region as a whole.

The Historic Development of INTERREG and its Change in Focus[2]

As already indicated in the previous section, the original impetus for the European Commission to get involved in the late eighties in border regions - a symbol of national sovereignty and in many countries subject to the Ministry of Foreign Affairs - was the on-going economic and political integration process. Not surprisingly, the Commission's policy was cautious: true cross-border co-operation was to be encouraged, but more along the internal borders than along the external borders. Above all, attention should be given to the development of the border area itself. Eligibility was strictly limited to areas directly adjacent to the border, and the border had to be a land border with some minor exceptions.

---

2 Based on a presentation by Robert Shotton, European Commission, DG XVI, at the RSA Summer Institute, 15-20 June 1997, Åre-Meråker.

In recent years, although the formal objectives of the INTERREG initiative have remained unchanged, the policy emphasis has shifted towards promoting true cross-border co-operation both along the internal and external borders of the EU. The intention is to diminish the sense of frontier by building wide-ranging networks of cross-border co-operation between local and regional actors, both public and private. Borders are no longer to be thought of as "at the end of the road", but on the contrary, as integrated into a regional economic area. An integration which might lead to one single economic and administrative space spanning both sides of the national border, a true "Euregion".

As far as the external borders of the EU are concerned, the aim is to contribute to the stabilisation, democratisation and prosperity of neighbouring regions, again by building wide-ranging networks of cross-border co-operation between local and regional actors. Where bordering countries are candidates for accession, there is the additional task of encouraging adaptation to Community policies by means of close association with partner regions in the Union. Another change introduced to the original INTERREG programme is the possibility to organise cross-border - even trans-national - co-operation between regions that share similar problems, namely floods or drought (INTERREG II B) or that would like jointly to develop spatial planning issues (INTERREG II C).

The accession of three new Member States in 1995, in particular Sweden and Finland, brought new challenges for the INTERREG programme due to the 1,300 km Union border with the Russian Federation and the new significance of relations with the Baltic States. In the Baltic and Barents Areas, INTERREG programmes work together with other EU funding instruments in neighbouring countries (PHARE - Poland, Hungary: Aid for the Reconstruction of the Economy - and TACIS - Technical Assistance for the Community of Independent States and Georgia -), in larger frameworks for co-operation. They participate in particular in giving substance to the follow-up to the Visby Summit on Baltic Sea Co-operation in Spring 1996. Specific budget allocations for cross-border co-operation have been established within both the PHARE and TACIS programmes.

The decision of Norway not to join EU membership in 1995, has also placed a new accent on the Nordic Co-operation; INTERREG programmes with neighbouring regions in Norway have been understood to make an important contribution to the future organisation of that Co-operation. Further, the restriction to land frontiers has been softened in the Baltic Sea Area to cover INTERREG programmes for co-operation with Poland and the Baltic States across the Baltic Sea involving Denmark, Finland, Sweden and northern Germany.

Along the eastern land border of the Union, from the borders with Poland, the Russian Federation, Slovenia down to the Balkans and the neighbours of Italy and Greece, mirror programmes have been set up by PHARE, called multi-annual indicative programmes, to organise cross-border co-operation in partnership with INTERREG funding. For the Russian Federation, this degree of ambition is not yet possible, but the groundwork is being laid for a possible future move in the same direction. Already there exists a close co-ordination of the annual project selection cycle under TACIS cross-border action with INTERREG management committee decisions.

Fostering Innovative European Regions: the Emergence of RITTS/RIS

The INTERREG programme - particularly since its recent shift in emphasis - contains elements of an "active space" approach in the sense that it is intended to bring together relevant actors to achieve synergies that otherwise would not be achieved. Overall, the European Single Market initiative, facilitating the free movement of goods, services, labour and capital, promoted greater concern with the competitiveness and productivity of industries and firms. Against this background, Community regional policy - of which INTERREG forms part - has focused increasingly on assisting the restructuring of regional production systems by means of research and technological development and innovation support to make regions more competitive. In order to succeed in this "technology-based, innovation-led restructuring", the White Paper on Growth, Competitiveness and Employment from 1994 identified the need to define a global strategy bringing together the public authorities, research bodies and the various sectors of society concerned. Whereas the Green Paper on Innovation from 1995 stressed the importance of the regional level in the formulation and implementation of such a strategy.

The RITTS/RIS programme (RITTS: "Regional Innovation and Technology Transfer Strategies and Infrastructures" and RIS: "Regional Innovation Strategies") is another scheme of the European Commission that is concerned with regional competitiveness in which border regions can participate. The RITTS/RIS scheme has introduced an approach to bring all relevant regional actors together to define a regional innovation strategy, thereby testing their capability to co-operate, deal with conflicts and anticipate changes within and outside their region. Similar to INTERREG, elements of an "active space" approach can be detected. Since its start in 1993/1994, some hundred regions in the European Union and the European Economic Area have been participating in these schemes. The main idea behind RITTS and RIS is:

- to improve the capability of regional actors to formulate regional economic policy which takes into account the real needs of the business sector, particularly small and medium-sized firms, and the strengths and capabilities of the regional research and technological development and innovation community, and
- to provide a framework within which both the EU and the regions can optimise policy decisions regarding future investments in research and technological development, innovation and technology transfer initiatives at regional level.

Innovation is here defined as '…the necessary steps, managerial, commercial, technical and financial, to introduce a new product or process into the market place'. (RIS Guidebook, 1996: 5). The policy approach of RITTS and RIS is to support the development of a regional innovation strategy which identifies the strengths and weaknesses in the innovative capability of the region, including management, training and organisational issues as well as purely technological ones. The development of a regional innovation strategy should be the outcome of a process that involves all the regional actors related to research and technological development, innovation and associated business support activities, such as local and regional governments, local and regional economic development organisations, regional representatives of national agencies in charge of innovation, technology, science, economic and/or regional policy, central government ministries in those areas, research organisations, higher education institutes, technology transfer organisations, innovation support organisations, large businesses, R and D laboratories, business associations, trade unions.

A regional innovation strategy developed in the framework of RITTS and RIS should reflect the following approaches:

- *A bottom-up approach*: it should be demand-driven, based on strengthened dialogue between firms, particularly SMEs, regionally-based research and technology transfer organisations and the public sector.

- *A regional approach*: there should be a specific territorial dimension which takes full account of the national and international context. Perhaps more importantly, RIS should build a consensus at the regional level on the priorities for action between the principal actors involved.
- *A strategic approach* should be applied to regional development in the fields of technological progress and innovation. They should plan for short and medium term actions that fit with the long-term objectives and priorities defined by the region.
- *An integrated approach*: the efforts of the public sector (local, regional, national and European) and the private sector should be linked towards the common goal of increasing regional productivity and competitiveness. They should try to maximise the economic impact of regional, national and European programmes.
- *An international approach*: a RIS should adopt an international perspective in terms of the analysis of global economic trends as well as on the need to co-operate nationally and internationally to be more effective in the field of research and technological development and innovation.

In its design, RITTS and RIS experiment with a new approach to regional development placing innovation at the core of it with an emphasis on the necessity to involve all regional actors, that have an assignment in strengthening the region's innovation capability in the formulation phase, the decision-making process, the implementation of practical actions, and the financing of regional innovation policies. Its approach is also new in connecting the region to external changes: what global economic trends will affect the region and how should these be tackled? Finally, its approach is new in underlining the learning aspect of a RITTS/RIS exercise. It is due to this built-in dependence in the programme that actors are forced to interact with one another to achieve the goals of a RITTS/RIS project. This situation where actors depend on the co-operation (or at least non-opposition) of others provides ample opportunity for conflicts to occur and serves as a testing-ground for the problem-solving capability of actors. Experience with the implementation of the RITTS/RIS programme in almost hundred regions in the European Union and the European Economic Area have shown that the success of a RITTS/RIS project is often less related to the official tasks and assignments of the organisation put in charge than to the motives for participation and the scope of manoeuvre of the project promoter as well as his or her personal qualities. The scope of manoeuvre relates to the resources an actor is able to mobilise to influence the process, such as money, responsibilities, authority, personnel, infrastructure, contacts, access to networks, knowledge, information, negotiation skills, persuasion techniques, expertise, political support, the capability to solve conflicts, to achieve compromises, and to motivate others etc. Resources can therefore be interpreted widely which gives hope to those regional organisations, besides regional authorities, that have an assignment in innovation, technology transfer and technology-based regional development policies to get successfully involved in Community programmes.

Regions have to change their way of thinking from regarding themselves as a sub-national, merely administrative tier of government to thinking of themselves as a strategic entity that is linked to the rest of the world. New organisation and management structures can turn the region from an abstract spatial construct into an instrument of endogenous development. The "active space" approach which emphases networks could provide a better framework to succeed in this; INTERREG and RITTS/RIS have already experimented with such a bottom-up policy approach.

INTERREG and RITTS/RIS: Characteristics of a Policy Network

Policy networks share four characteristics: multiplicity, pluriformity, interdependence and interaction. Firstly, a policy network is made up of a *multitude* of actors who are formally or informally interrelated. Both INTERREG and RITTS/RIS only function when all relevant regional actors are involved in the design, implementation and financing of policy measures. Secondly, a policy network is characterised by its *multiformity*. The actors differ from one another in terms of assignments, budget, personnel, power, legal status etc. The network can be formed around certain policy problems or clusters of resources (Klijn, Koppenjan and Termeer, 1995). In a network organised around one policy field, e.g. regional policy, the actors can *ceteris paribus* also differ from one another in terms of policy outcomes. The European regions do not only differ widely on economic performance indicators, such as GDP per capita, unemployment rate, annual growth of employment and average household income per region, but they also make up a wide variety of different types of depressed areas. The large economic disparities among regions in the EU greatly exceed those inside the United States (Corvers, 1994). The policy practices developed in the EU to tackle these regional inequalities vary strongly per region and per member state.

Policy networks are also characterised by the *interdependence* of the actors. Interdependence means that the actors depend upon each other to acquire the means they need to achieve the policy objectives they have set out. Actors will therefore interact with one another as they assume that by means of interacting they will acquire the resources they need. An actor can be considered powerful in a policy network not just because of the resources he/she might be able to mobilise, but also because of the actor's strategic abilities to put these resources to use. Policy is the result of the interaction between the actors. Interdependence does not only refer to the difference between the resources an actor has and the resources an actor needs to achieve objectives. The fact that actors can undertake activities that facilitate or obstruct the achievement of objectives of other actors also points at interdependence.

Finally, *interaction* in a policy network is displayed in various forms, namely co-operation, collusion, competition and coercion (Bish, 1978). Interaction involves some form of negotiation to exchange resources. Actors can interact if they think that by means of co-operation they will both be better off. The mutual benefit can come from jointly seeking a common objective, for example reducing regional economic disparities will make the EU more competitive as a whole, or from entering into an exchange agreement, for example cross-national co-operation between the police forces of two neighbouring regions. Collusion is almost similar to co-operation, but imposes costs on third parties that do not take part in the interaction, for example infant industry protection. The term collusion is not optimal to describe this type of co-operation, as it implies secrecy or conspiracy. Many co-operation agreements, especially in the public sector, may generate unintended rather than planned negative consequences for others (which cannot serve as an excuse of course). The police forces of two adjacent regions may agree to stay out of each others territory even if this is legally allowed which is at the expense of the citizens' options for calling upon either of two police forces instead of only one. Competition - as a form of interaction - is related to the two above mentioned forms. Competition concerns rivalry between two (or more) actors to co-operate with a third actor. Competition is related to co-operation as entering into an agreement with another actor means that this actor's offer was superior to competing alternatives. Collusion is related to competition as collusion is undertaken by potential competitors to eliminate the competition for the favours of a third party. Research funds that require co-operation among researchers in a consortium trigger off competition between researchers for the favours of the main research contracting party to join the consortium. Coercion, finally, distinguishes between a coerced party and a coercing party. In this form of interaction the coerced party has no option but to meet the requirements of the coercing party or bear some sanction imposed by that party if the requirements are not met. Taxes and laws are two typical examples of coercion: sanctions such as fines or jail are used to encourage citizens to pay their taxes and obey the laws; however, there is no guarantee given whether these taxes and laws are used for the benefit of these citizens. Cross-border co-operation between neighbouring regions, e.g. to solve specific border-related problems such as the pollution of a trans-national river, can serve as another example of coercion. Traditionally, national governments have considered border affairs a matter of national security which fell under the responsibility of either the Ministry of Defence or Foreign Affairs; regional authorities were not allowed to act on their own.

In their attempts to make Europe more competitive, both INTERREG and RITTS/RIS reflect a shift in policy approach which stresses the capability of regional actors to combine forces and create synergies. The success of such policy network types of setting can be tested by verifying whether the policy objectives set out are actually achieved, although this is not the only criterion to measure the success of policy networks. Equally important is whether the policy network has succeeded in promoting co-operation between the actors, in strengthening their problem-solving capability, in organising mechanisms for more structural information exchange, in developing a more democratic, more transparent policy-making process, in formulating policies that are more responsive to the needs of citizens, in devising policy actions that anticipate and incorporate external changes, etc. The indicators used in ex-post policy evaluation to date, however, are still strongly focused on measuring input and output effects, instead of such behavioural effects.

Some Tips for Policy Makers Implementing an Active Space Approach based on INTERREG and RITTS/RIS Experiences

Different patterns of interaction between actors have different effects on the course of policy processes and produce therefore different policy outcomes.

Under what conditions will policy networks favour successful policy outcomes? Given the fact that policy networks do not have one central decision and control centre, co-operation problems are likely to occur. Policy networks can succeed in achieving co-operation between autonomous yet interdependent actors and other behavioural effects mentioned above using the following rules of thumb (Klijn et al., 1995):

- *Project champion*: the policy network needs a "project champion", an actor who organises the structure in which a policy network can develop. Besides organising the network meetings, sending the invitations, preparing the agenda, diffusing information and the like, a project champion has to be able to open doors, to get political support, to keep an overview of the process(es), to motivate people, to get people to follow their promises, etc. The organisation which takes the lead is therefore less important than the qualities and capacities of the project champion. Policy networks are dynamic processes; the functioning of policy networks involves social engineering.
- *Achieving win-win situations*: the policy network should facilitate bringing about a situation which represents an improvement on the starting position for all actors concerned. This does not mean that all those involved will achieve their objectives to the same extent. What is important is to foster a situation that makes participation in the network more interesting than non-participation.
- *Activating actors and resources*: interaction involves some form of negotiation to exchange resources. The underlying assumption is that actors will be willing to invest their resources in this particular negotiation process. As the success of a policy network - or any network - depends heavily on this willingness, the initial interest and enthusiasm of actors should be stimulated. Specifying the benefits of participation versus the costs of non-participation is an effective way to achieve this.
- *Limiting interaction costs*: as interaction and negotiation processes also involve costs, it is necessary to prevent actors from pulling out in disillusionment after an enthusiastic start. Interaction costs should be proportionate to the stakes, but win-lose or even lose-lose situations should be avoided, restructured or ended in time. Conflicts can perform an important role in policy networks as they increase the transparency of the issues at stake and the true interests of the quarrelling parties, but they should be prevented from becoming dysfunctional and destructive.

- *Procuring commitment*: the policy network should bring about a situation where actors will make a (serious) commitment to the joint undertaking. Actors will always be aware of the danger that the effect of the joint actions may benefit others or that actors may pull out at crucial moments and leave others (= themselves) with the risks. Commitment can be procured via informal or more formal arrangements, such as convenants, contracts or establishing autonomous legal persons. Another way to procure commitment is to involve actors more in the actual policy process by giving them (shared) responsibilities, including a budget, and/or by setting up a representative steering committee.

These rules of thumb cannot guarantee better interaction and better policy outcomes, but they do increase the chances of these things occurring. One could argue, however, that border regions are different - as we have seen in the previous sections - and that the introduction of a network approach to policy making may have different success rates. Indeed, the role of national governments remains a crucial factor in to what extent border regions can determine their own development course. If border affairs are considered to be a matter of national security, regional authorities will not be allowed to act upon their own. Regions are reminded of their position as a sub-national, merely administrative tier of government. If the regional tradition - devolution of power - developed more strongly, then there is room for new organisation and management structures that can turn the region into an instrument of endogenous development.

The democratic tradition in a region, the level of entrepreneurship both in private and public sector, the "milieu innovateur", i.e. the extent to which the region offers conditions that favour innovations, are all factors that influence the success of a network approach to policy making. Despite all these conditions, it is, in the end, the region that has to change its way of thinking from regarding itself as a sub-national, merely administrative tier of government to a strategic entity that is linked with the rest of the world. This requires a whole new way of thinking, a culture change, in public sector organisations at all levels as these actors are used in thinking in terms of hierarchical relations, but if they manage to do this, then the abstract concept of "region" can be turned into a strategic instrument promoting regional endogenous growth.

Further Academic Research

INTERREG is a Community initiative for European border regions, and the RITTS/RIS scheme is a Community action for regional innovation policy. Both programmes contain elements of an "active space" approach which could be of use to European border regions and their economic development potential. Both schemes are geared to the specific European situation of border regions, but other regions could also take advantage of the experiences gathered over the past years. Both INTERREG and RITTS/RIS apply a systems approach considering regions as territorial organisations. Regional development in accordance with an "active space" approach connects changes in openness of regions with the capability of regions to manage the consequences of these changes in openness. Openness within an "active space" approach is defined as potential interaction between various components of the system (internal openness) and between the system and external systems (Ratti, 1997; van Geenhuizen and Ratti, 1998). This interaction can be shaped as competition, co-operation or complementarity. In order for a region to manage the consequences of internal and external changes, it needs to possess a learning capability which rests on a variety of local assets such as human capital (skills, experiences, etc.), trans-national relationships between manufacturers, local suppliers and customers, informal contact networks, and synergy effects between various local actors such as universities, higher education institutes, firms and Chambers of Commerce. Experiences from INTERREG and RITTS/RIS show that the creative learning of regions - world-wide - could be reinforced by applying some rules of thumb, placed within the specific regional context, when formulating and implementing policies.

Further academic research could help find answers to burning questions, such as: what distinguishes an innovative border region from a backward border region in terms of policy-making processes? What critical success factors can be distinguished? How can good practices be transferred to less favoured (border) regions? Is it possible to design a model of a successful policy network for border regions (archetype) which takes into account the differences and similarities of border regions and which could be applied in a non-European context (a universal model)?

Finally, due to the lack of satisfactory new indicators, the success of policy networks is still measured in terms of classical input-output indicators, such as goal(s) achieved, cost efficiency, etc. New, more behaviourally oriented indicators should be added to measure other desirable policy outcomes, such as the development of a strategic framework for regional development and regional innovation, the creation of networks and the promotion of inter- and intra-regional co-operation, and the start of trans-national relationships between manufacturers, local suppliers and previously non-existent customers. Other potential indicators are the identification and preparation of innovation projects in firms, the strengthening of regional research and technological development and innovation centres located in a region, optimisation of a design for new public-private programmes for the promotion of innovation, strengthening of the region's problem-solving capability in general, and developing a more democratic, more transparent policy-making process that anticipates and incorporates external changes, etc. (Corvers, 1995).

References

Bish, R. (1978), 'Intergovernmental Relations in the United States', in K. Hanf and F. Scharpf (eds), *Interorganizational Policy Making. Limits to Coordination and Central Control*, Sage Publications, London/Beverly Hills.

CEC (1991), Commission of the European Communities, *Europe 2000: Outlook for the Development of the Community's Territory*, Office for Official Publications of the European Communities, Directorate-General for Regional Policies, Luxembourg.

Corvers, F. (1992), *Grensregionale Samenwerking als Institutioneel Arrangement. De Nederlandse grensregio Euregio Maas-Rijn als voorbeeld*, Master Thesis, Leiden University, Department of Public Administration, Leiden.

Corvers, F. (1994), *Economic Integration the European Way: Stronger Firms, Stronger Region*, MERIT Research Paper nr.94-040, Maastricht.

Corvers, F. (1995), *The Linkage between Innovation and Regional Policy. Experiences made with the Regional Technology Plan*, paper prepared for the RETI Conference on Economic Development by Innovation, October 12-13, Magdeburg.

Geenhuizen, M. van, and Nijkamp, P. (1998), 'Potentials for East-West Integration: The Case of Foreign Direct Investment', *Environment and Planning C*, 16, pp. 105-120.

Geenhuizen, M. van, and Ratti, R. (1998), 'Managing Openness in Transport and Regional Development', in K. Button, P. Nijkamp and H. Priemus (eds), *Transport Networks in Europe. Concepts, Analysis and Policies*, Edward Elgar, London, pp. 84-102.

Kessen, A.A.L.G.M. (1992), *Bestuurlijke Vernieuwing in Grensgebieden: Intergemeentelijke Grensoverschrijdende Samenwerking*, PhD Thesis, Nijmegen University, Faculty of Policy Sciences, Nijmegen.

Klijn, E-H., Koppenjan, J. and Termeer, K. (1995), 'Managing Networks in the Public Sector: a Theoretical Study of Management Strategies in Policy Networks', *Public Administration*, vol. 73, 3, pp. 437-454.

Martinos, H. and Caspari, A. (1990), *Cooperation between Border Regions for Local and Regional Development*, Final Report prepared by the Innovation Development Planning Group for the Commission of the European Communities, DG XVI, Brussels.

Martinotti, G. (1996) 'Four Populations: Human Settlements and Social Morphology in the Contemporary Metropolis', *European Review*, vol. 4, 1, pp. 3-23.

Mikus, W. (1986) 'Industrial Systems and Change in the Economies of Border Regions: Cross Cultural Comparisons', in I. Hamilton (ed), *Industrialization in Developing and Peripheral Regions*, Croom Helm, London.

Naastepad, C. and Storm, S. (eds) (1996), *The State and the Economic Press*, Edward Elgar, London.

Nijkamp, P. (1993), 'Towards a Network of Regions: The United States of Europe', *European Planning Studies*, vol. 1, 2, pp. 149-169.

Ohmae, K. (1990), *The Borderless World*, Sage, New York.

Ohmae, K. (1995), *The End of the Nation State: the Rise of Regional Economies*, Sage, New York.

Perkmann, M. (1999), 'Building Governance Institutions Across European Borders', *Regional Studies*, vol. 33.7, pp. 657-667.

Ratti, R. (1997), 'L'espace regionale actif: une response paradigmatique des regionalistes au debat local-glocal', *Revue d'Economie Regionale et Urbaine*, no.4, pp. 525-544.

RIS Guidebook (1996), *Regional Innovation Strategies Guidebook*, DGXVI/DGXIII, Brussels/Luxembourg.

Shotton, R. (1997), 'Cross-border co-operation and the INTERREG programmes: present and future perspectives', in K.I. Westeren (ed), *Cross Border Cooperation and Strategies for Development in Peripheral Regions*, Proceedings from the European Regional Science Association's Summer Institute 1997, Åre-Meråker, NTF-fagserie 1998:1, Steinkjer.

19 Reflections on Active Space Development: Emerging Issues and New Research Paths

MARINA VAN GEENHUIZEN AND REMIGIO RATTI

In this concluding chapter we first review the main issues of gaining advantage from open borders and explore the implications of these issues for regional development and regional policy. With regard to openness, particular attention is given to the role of ICTs and the emerging e-economy in regional development. Secondly, we reflect on the "active space" paradigm as a generic framework to analyse regional development and policy. We conclude the chapter with a reflection on new research paths identified in this volume.

Gaining Advantage from Open Borders

The past decades have shown the disappearance of many political borders and the removal of other obstacles in a free movement of persons, goods, capital, and information. It is not easy to evaluate whether increased interaction due to this new openness is advantageous or disadvantageous. The gains from open borders depend, for example, on the time horizon used and on the segments of society taken into account. From the viewpoint of the economy we may say that open borders enable greater economic efficiency in the medium to long term, but this works partially along the lines of a greater competition from the outside for certain economic sectors, leading to negative developments on the short term. We may also say that for one segment in society the new interaction is a solution to problems whereas this is not the case in other segments. For example, the influx of large groups of low skilled migrants contributes to solving shortages of labour force in particular industries but it may also cause problems of social cohesion between those communities of migrants and the rest of society. From an "active space" approach we may say that for society at large interactions are advantageous when these lead to a more efficient use of resources and at the same time are accompanied by a sufficient concern for sustainability, in an economic, social and environmental sense.

To what extent advantages can be gained from greater openness is dependent upon three factors which are strongly coloured by historic and regional specificity. First, the availability of opportunities for interaction, such as complementarity in economic activity or advantageous wage and tax differentials; secondly, the strength of remaining barriers for interaction that have often grown in association with political borders; and thirdly, the way regional actors cope with these circumstances in terms of their management skills and - more broadly - their learning capability.

In this volume various regions could be identified with limited opportunities for interaction due to the specific economic structure in the area and low population density, i.e. Karelia, along the Russian Finnish border (Chapter 10) and, on a smaller spatial scale, particular regions along the Dutch-German border (Chapter 6). In addition, different barriers could be identified in gaining advantage from open borders, ranging from mental and image barriers (or attitudes) (Chapter 7) to deficient service levels in transport and communication (Chapters 5 and 6), and institutional deficiency in foreign capital investment (Chapter 9). In a few studies an attempt has been made to measure barrier impacts from borders. For frequencies in traffic services between West European cities a reduction was found to a level between 70 per cent and 24 per cent in cross-border railway connections compared with railway connections within the country (Chapter 5). For foreign direct investment by western firms in Central and Eastern Europe a reduction was found to levels ranging between 60 per cent and almost zero investment; and this was attributed to differences in institutional barriers between the countries (Chapter 9). The levels of reduction are substantial and call for more systematic comparative research and further causal analysis.

In this volume, a number of regions have been studied that face or have faced the need for a radical policy change in responding to new openness and for an improved use of creative learning. Two cases are connected with transport and one with labour migration. For example, the Chiasso area in Switzerland was unable to react in a timely manner to logistic changes and competition from nearby Milano in the 1980s (Chapter 2). The region of Rotterdam in The Netherlands (Chapter 14) is nowadays struggling in policy making for its future economic role. A shift is taking place from a policy that confirms traditional goals set in the past (old logic of mass transport) to a policy that incorporates creative learning both in the aims set and the methods used to design policy. In a similar vein, in the Canton Ticino (Switzerland) there is a need for a learning policy to "upgrade" particular segments of the labour market instead of using those segments as a reservoir in times of shortage of low-skilled workers (Chapter 8).

Implications for Regional Development and Policy

Divergence

The above findings have clear implications for regional development. Opportunities for interaction, impacts from barriers, and learning capability differ widely between border regions in the European Union. This means that border regions where these circumstances are positive may develop favourably, whereas other regions may seriously lag behind. In other words, there may be an increasing *divergence* between border regions within the EU. Such a situation seems true for inner and outer border areas of the EU, and will become all the more true after some Central and Eastern European countries have joined the EU. Such divergence is undesirable, particularly if it manifests itself within one and the same country. The concomitant socio-economic gaps may cause social and political unrest, and may lead to negative sentiments and discrimination. The policy answer, however, should not simply be creating jobs and new infrastructures in lagging regions. There is a need for a far more comprehensive policy, with a major focus on enhancing the learning capability in the region. The EU regional innovation policy by the RITTS/RIS schemes points to such new directions in regional policy making (Chapter 18). On a higher spatial scale we find the question of different ways of integration of transition countries with EU countries (see e.g. Hunya, 1999).

Multilevel and multisector setting

It has also become clear that border region development needs a *multilevel* and *multisector* approach of governance for an appropriate analysis and policy design, the latter particularly with regard to the establishment of cross-border co-operation (Chapter 12 and Chapter 13). Today the national state is losing power and this affects border regions more strongly than other regions. At the same time, informal actors and networks (lobbies) are emergent factors in governance, and most of their interaction is based on loose mechanisms (Saez et al., 1997; Ratti, 2000) (Chapter 4). It is not clear to what extent these forms of networks and interaction need to be institutionalised, but it seems that informal power relationships are difficult to correct if their influence produces undesirable impacts. In this context, the national government might stimulate the capacity of regional (local) society to develop bottom-up approaches that matches the historical and spatial specificity of the region, and increases support and commitment from actors involved. Policy-making within networks, using the specific dynamics between actors and the creativity of them, seems particularly promising (e.g. Morgan, 1997; de Bruijn and ten Heuvelhof, 2000). Such an approach would also mean that large projects need to be avoided. Starting with small pilot projects is preferred to reduce the risks of institutional complexity and the involvement of too many actors.

Proximity

An emerging (or re-emerging) concept in regional development studies is *proximity*, understood as physical proximity. It is often taken for granted that physical proximity between actors is a *conditio sine qua non* for creative learning, particularly if tacit knowledge is involved. Similarly, it is often taken for granted that with the disappearance of borders, actors in close proximity in adjacent countries would use new opportunities for interaction. However if analysis focuses too strongly on physical proximity, it tends to overlook the influence of proximity in terms of culture and mental maps. The latter types of proximity might be more influential and might end up being crucial in the question whether a new interaction is established or not (Chapter 7). In this framework it is important to emphasise that another type of proximity is looming and that is *virtual proximity* or proximity in electronic space. Virtual space might, like physical space, be divided by borders with different levels of permeability and access, and concomitant filtering-, contact- and conditioning effects. Disadvantages of border region locations in physical space might be compensated by free access in virtual space or, in contrast, be reproduced and reinforced in virtual space by established economic power relations (van Geenhuizen, 2000). This ties into the use of ICTs and the rise of the *e-economy*, and the way in which firms may respond to the attendant challenges and threats.

ICTs and the Emerging E-Economy

Information and communication technologies play a vital role in economic processes today. It seems that modern ICTs increase regional openness without any limits. Electronic communication takes place where computers are and where these are connected with larger networks. There are however clear indications that large cities, in the present time of market-driven infrastructure investment, are achieving the best connections in terms of fast communication and global linkages, and as a consequence may reinforce their economic position (Graham, 1999). This does not exclude however that smaller towns, such as university towns and border "hubs", can become well-connected in global networks as well. For border regions, this means that the development of specialised ICT-based clusters in market niches is most realistic, provided that these are embedded in a particular economic specialisation of the region. We can mention as an example the town of Lugano in the Canton Ticino (Switzerland), which has developed an expertise in the use of ICTs for financial services. A major prerequisite, however, remains a sufficient learning capability in the region.

There is not only an increased application of new ICTs which move information increasingly faster around the globe. The wide use of Internet appears to have far-reaching repercussions on the organisation of production, at least in a number of product and service chains (e.g. Westland and Clark, 1999). These are connected with e-commerce and fit in with the emergence of the *e-economy* as follows (van Geenhuizen and Nijkamp, 2001):

- An increased control of customers over chains and concomitantly a growing demand for product differentiation and tailor-made products and services. There are various implications for logistic organisation and the location of logistic nodes.
- A strong need for firms to develop customer management, particularly customer services. E-commerce has led to growing competition and the supply of customer services is a strategy to improve (re-establish) customer relations, often using face-to-face contact. It seems likely that

such services cannot be supplied in small places, perhaps rendering medium-sized towns more important.

- Desintermediation, i.e. the disappearance of particular segments of chains. Chain segments at the end (customer-side) may be gradually eliminated, such as bank offices, travel agencies, etc. This may also become true for some other chains segments, due to electronic auctions. There are no signs of full substitution but these developments imply that central place functions of towns are subject to erosion for routine parts.
- Re-intermediation, i.e. the creation of new, virtual, segments of chains. There are a number of new roles for firms here, particularly connected to abundant information supply on Internet, namely to select information, provide access to information, provide connections between customers and suppliers, and even negotiate between them using intelligent agents. Thus, existing firms have to consider which roles they want to play and how, alone or in networks with other firms.
- Entry of new e-firms. Due to the disappearance of various entry barriers a number of new firms appear on screen. Because these virtual firms can avoid the costs of physical outlets (shops, offices, etc.) they can offer cheaper prices, meaning more competition for established firms.

The above changes have implications for the location of different segments of chain activity and for new roles for firms in virtual space. Border regions will be affected by these changes like any other region. Accordingly, there is a new pressing reason to enhance the learning capability of regions. It seems true that without a "fertile breeding ground", e.g. skills of the population, capability for networking, etc., no advantages can be gained from opportunities given by ICTs and the e-economy, and firms will lose competitive strength.

Performance of the Active Space Approach

What makes a space or a region "active" is the behaviour of its actors and networks based on synergy and creation of territoriality (Ratti, 1996). It is the capacity to respond internally and externally to changes in openness, driven by sets of principles, rules and strategic behaviours designed to support any level and any sector of society through dynamic cohesion; and it manifests itself in different degrees of "active space". We may conclude that the approach performs satisfactorily in this volume for the following reasons:

- It is supported by a "family" of research approaches. The "active space" approach touches many new perspectives and backgrounds in an *integrative* way. The book illustrates this by means of contributions from institutional economics (Steiner, Bramanti), public finance, transport economics, micro-behaviour of firms, etc. Such an integrative and multidisciplinary approach is needed just because of the very nature of space (Benko, 1995; Benko and Lipietz, 2000).
- It allows for a high level of generalisation due to the concept of "degree of active space". Thus, peripheral regions functioning merely as a transport corridor may equally well be approached from an "active space" perspective as centrally located high technology regions.
- It produces insights into the outcomes and process characteristics of policy making. We need to stress that policy making according to the "active space" paradigm contains a number of *normative* elements, such as the recognition of uncertainty (complexity) and responses to this in terms of using creativity, bottom-up (network) approaches and flexibility in policy making. At the same time, it is recognised that the extent to which these elements can be incorporated in policy making is dependent upon the specific (historic) governance tradition and the available level of self-organisation in the regions at hand.

Future Research

The above observations lead to research questions in three broad fields, i.e. (1) empirical study of border region development and regional development in general, (2) more theoretically oriented analysis concerning key concepts in regional development, and (3) policy analysis for border regions and for regions in general.

As concerns *empirical* studies, an important research path would focus on the rise of divergence between border regions in the EU, and the implications for social cohesion in the countries with lagging regions. Such cross-comparative research requires a further international standardisation of databases on the regional level, among others concerning traffic flow. On a higher spatial scale, there is the question of the pace and level of integration of transition economies with the EU, and the difference in trajectories between these countries. A further path is the multilevel power situation of border region development, more precisely, the dynamics in the relationships between national and regional governments in the context of increasing power of the latter. It is important to analyse empirically the way in which actors deal with power relations in terms of co-ordination and integration, and the type of strategy and practical forms that are most successful. With regard to innovative behaviour of firms, there is a need for more insight into the circumstances under which governance structure makes the region perform better. This relates to the best role for governments in such governance structure and the degree of institutionalisation of loose mechanisms of interaction. There is a another reason to analyse the relationships between national and regional (border) governments, and that is in the context of national state formation and formation of national identity, for those countries established after 1989.

In addition, there are a number of empirical questions concerning practice of cross-border interaction, particularly the influence of remaining border-barriers. For example, the barrier of attitudes and images in not well investigated in terms of the way these are created in the minds of people and influence social and economic behaviour. A relatively new question is concerned with the role of ICTs as a factor that eases information dissemination and information gathering, and may overcome particular barriers, but may also add new ones. Finally, we would like to draw attention to the fact that most border region studies deal with land-border regions. However, sea-border regions might be an equally interesting subject of analysis, particularly because these seem to be in a less advantageous position compared with land-border regions in the past decades, due to far more progress in transport over land compared with transport over sea.

Regarding the second field, i.e. *theoretically oriented studies*, we propose a further conceptualisation and operationalisation of the key concepts of the "active space" paradigm, i.e. openness, creative learning and sustainability, the latter in a broad sense. In terms of measuring, there is a need for a refining and testing of indicators. At the same time we realise that various other concepts are at the base of the paradigm, proximity being one of them. This concept equally needs a further elaboration, possibly in connection with the related concepts of distance and borders. In a (new) conceptualisation the different dimensions of proximity need to be taken into consideration, i.e. physical proximity, cultural proximity and mental proximity. An interesting question is the way in which these different types of proximity may influence interaction. A theoretical reflection on proximity, distance and borders and their influence on interaction ties into an equally relevant issue that needs more clarification, i.e. the spatial boundaries of a border region; in other words, the nature of border impacts and the specific character of a border region.

A relatively new field of conceptualisation is virtual space, including virtual proximity, virtual distance and virtual barriers, and the power relations through which these are created. Furthermore, the relation between physical and virtual space is an interesting issue. Are power positions of actors and regions in physical space reproduced in virtual space, or are there new chances independent of power positions in physical space?

In the field of *policy analysis* there is a need to achieve insights into the results of relatively new ways of policy making (process design) based on network approaches. This holds particularly for those policies aimed at cross-border co-operation and for those aimed at improving the use of learning capability. Important issues are bottom-up (participatory) methods, coping with historical and regional specificity, ways to link the local and the global, and coping with uncertainty and actors' complexity in multilevel and multisector situations. There is a need to identify best practice and best strategy, given particular historical and spatial conditions. It is however important to note that evaluation studies by themselves need to be improved in the sense that the focus is preferably on behavioural (or process) criteria and not merely on (economic) output. This requires a further refining of criteria.

References

Benko, G. (1995), 'Les theories du développement local', Sciences Humaines, *Numéro spécial Régions et mondialisation*, Paris, Février/Mars.

Benko, G. and Lipietz, A. (eds) (2000), 'La Richesse des Régions', *La Nouvelle Géographie Socio-économique*, Presses Universitaires de France, Paris.

Bruijn, J.A. de, and Heuvelhof, E.F. ten (2000), *Networks and Decision Making*, Lemma, Utrecht.

Cocossis, H., and Nijkamp, P. (eds) (1995), *Overcoming Isolation. Information and Communication Networks in Development Strategies for Peripheral Areas*. Springer, Berlin.

Geenhuizen, M. van (2001), 'ICT and Regional Policy: Experiences in The Netherlands', in M. Heitor (ed), *Innovation and Regional Development*, Edward Elgar, Cheltenham (forthcoming).

Geenhuizen, M. van and Nijkamp, P. (2001), 'Electronic Banking and the City System in the Netherlands', in S. Brunn and T. Leimbach (eds) *The Wired Worlds of Electronic Commerce*, John Wiley, London, pp. 181-201.

Graham, S. (1999), 'Global grids of Glass; On Global Cities, Telecommunications and

Planetary Urban Networks, *Urban Studies,* 36 (5-6): pp. 929-949.

Hunya, G. (ed) (1999), *Integration through Foreign Direct Investment,* Edward Elgar, Cheltenham.

Morgan, K. (1997), 'The Learning Region: Institutions, Innovation and Regional Renewal', *Regional Studies,* 31.5, pp. 491-503.

Ratti, R. (1996), 'Global versus local: Lessons from the Swiss Experience', *Swiss Journal of Economics and Statistics,* vol.132 (3), pp.241-256.

Ratti, R. (2000), 'Die Globalisierung und die politische Kleinraumigkeit' in G. Neugebauer (ed) *Föderalismus in Bewegung – Wohin steuert Helvetia?,* Franz Ebner, Zurich, pp. 19-27.

Saez, G., Leresche, J-P. and Bassand, M. (eds) (1997) *Gouvernance Métropolitaine et Transfrontaliere. Action publique territoriale,* L'Harmattan, Paris.

Shapiro, C. and Varian, H.R. (1999), *Information Rules. A Strategic Guide to the Network Economy,* The MIT Press, Cambridge.

Westland, J.C. and Clark, T.H.K. (1999), *Global Electronic Commerce. Theory and Case Studies,* The MIT Press, Cambridge.